Contemporary
Communication
Systems
Using MATLAB®

John G. Proakis
Masoud Salehi
Northeastern University

Brooks/Cole
Thomson Learning™

Australia • Canada • Mexico • Singapore • Spain • United Kingdom • United States

Publisher: *Bill Stenquist*
Sponsoring Editor: *Heather Woods*
Project Development Editor: *Suzanne Jeans*
Marketing Team: *Nathan Wilbur, Christina
De Veto, Samantha Cabaluna*
Editorial Assistant: *Shelley Gesicki*
Production Editor: *Laurel Jackson*

Manuscript Editor: *Carol Reitz*
Cover Design: *Denise Davidson*
Art Editor: *Lisa Torri*
Print Buyer: *Tracy Brown*
Typesetting: *WestWords, Inc.*
Printing and Binding: *Webcom Limited*

MATLAB and Simulink are registered trademarks of The MathWorks, Inc. Further information about MATLAB, Simulink, and related publications may be obtained from The MathWorks, Inc., 3 Apple Hill Drive, Natick, MA 01760. Phone: (508) 647-7001; e-mail: www.info@mathworks.com, http://www.mathworks.com

BookWare Companion Series is a trademark of Brooks/Cole Publishing Company.

Printed in Canada

10   9   8   7   6   5   4   3

Library of Congress Cataloging-in-Publication Data

Proakis, John G.
  Contemporary communication systems using MATLAB / John G. Proakis,
Masoud Salehi. — Updated print.
    p. cm.
  Includes bibliographical references and index.
  ISBN 0–534–37173–6 (alk. paper)
   1. Data transmission systems — Computer simulation. 2. Telecommunication
systems — Computer simulation. 3. MATLAB. I. Salehi, Masoud. II. Title.

TK5105 .P744 2000
621.382′16 — dc21
                                                                99-048385

About the Series

"The purpose of computing is insight, not numbers."
—R. W. Hamming, *Numerical Methods for*
Engineers and Scientists, McGraw-Hill, Inc.

It is with this spirit in mind that we present the BookWare Companion Series.™

Increasingly, the latest technologies and modern methods are crammed into courses already dense with important theory. As a result, many instructors now ask, "Are we simply teaching students the latest technology, or are we teaching them to reason?" We believe that these two alternatives need not be mutually exclusive. In fact, this series was founded on the belief that computer solutions and theory can be mutually reinforcing. Properly applied, computing can illuminate theory and help students to think, analyze, and reason in meaningful ways. It can also help them to understand the relationships and connections between new information and existing knowledge and to cultivate problem-solving skills, intuition, and critical thinking. The BookWare Companion Series was developed in response to this mission.

Specifically, the series is designed for educators who want to integrate computer-based learning tools into their courses and for students who want to go further than their textbook alone allows. The former will find in the series the means by which to use powerful software tools to support their course activities without having to customize the applications themselves. The latter will find relevant problems and examples quickly and easily available and will have electronic access to them. Important for both educators and students is the premise on which the series is based: that students learn best when they are actively involved in their own learning. The BookWare Companion Series will engage them, provide a taste of real-life issues, demonstrate clear techniques for solving real problems, and challenge them to understand and apply these techniques on their own.

To serve your needs better, we are continually looking for ways to improve the series. Toward that end, please join us at our BookWare Companion Resource Center Web site: **http://www.brookscole.com/engineering/ee/bookware.html**

You can recommend ways to make the series even better, share your ideas about using technology in the classroom with your colleagues, suggest a specific problem or example for the next edition, or just let us know what's on your mind. We look forward to hearing from you, and we thank you for your continuing support.

Bill Stenquist	Publisher	*bill.stenquist@brookscole.com*
Heather Woods	Sponsoring Editor	*heather.woods@brookscole.com*
Nathan Wilbur	Marketing Manager	*nathan.wilbur@brookscole.com*
Christina DeVeto	Marketing Assistant	*christina.deveto@brookscole.com*

Contents

Preface

Many textbooks on the market today treat the basic topics in analog and digital communication systems, including coding and decoding algorithms and modulation and demodulation techniques. By necessity, the focus of most of these textbooks is on the theory that underlies the design and performance analysis of the various building blocks (e.g., coders, decoders, modulators, and demodulators) that constitute the basic elements of a communication system. However, relatively few of the textbooks, especially those written for undergraduate students, include a variety of applications designed to motivate the students.

Scope

The objective of this book is to serve as a companion or supplement to any of the comprehensive textbooks in communication systems. The book provides a variety of exercises that may be solved on the computer (generally, a personal computer is sufficient) using the popular student edition of MATLAB®. The book is intended primarily for senior-level undergraduate students and graduate students in electrical engineering, computer engineering, and computer science. We assume that the student (or user) is familiar with the fundamentals of MATLAB. Those topics are not covered because several tutorial books and manuals on MATLAB are available.

By design, the treatment of the various topics is brief. We provide the motivation and a short introduction to each topic, establish the necessary notation, and then illustrate the basic concepts by means of an example. The primary text and the instructor are expected to provide the required depth of the topics treated. For example, we introduce the matched filter and the correlator and assert that these devices result in the optimum demodulation of signals corrupted by additive white Gaussian noise (AWGN), but we do not provide a proof of this assertion. Such a proof is generally given in most textbooks on communication systems.

Organization

This book consists of nine chapters. The first two chapters, on signals and linear systems and on random processes, present the basic background that is generally required in the study of communication systems. One chapter covers analog communication techniques, and the remaining six chapters are focused on digital communications.

Chapter 1: Signals and Linear Systems

This chapter provides a review of the basic tools and techniques from linear systems analysis, including both time-domain and frequency-domain characterizations. Frequency-domain analysis techniques are emphasized because these techniques are most frequently used in the treatment of communication systems.

Chapter 2: Random Processes

In this chapter, we illustrate methods for generating random variables and samples of random processes. The topics include the generation of random variables with a specified probability distribution function, the generation of samples of Gaussian and Gauss-Markov processes, and the characterization of stationary random processes in the time domain and the frequency domain.

Chapter 3: Analog Modulation

The performances of analog modulation and demodulation techniques in the presence and absence of additive noise are treated in this chapter. Systems studied include amplitude modulation (AM), such as double-sideband AM, single-sideband AM, and conventional AM; and angle-modulation schemes, such as frequency modulation (FM) and phase modulation (PM).

Chapter 4: Analog-to-Digital Conversion

In this chapter, we examine various methods used to convert analog source signals into digital sequences in an efficient way. Conversion allows us to transmit or store the signals digitally. We consider both lossy data compression schemes, such as pulse-code modulation (PCM), and lossless data compression, such as Huffman coding.

Chapter 5: Baseband Digital Transmission

In this chapter, we introduce baseband digital modulation and demodulation techniques for transmitting digital information through an AWGN channel. Both binary and nonbinary modulation techniques are considered. The optimum demodulation of these signals is described, and the performance of the demodulator is evaluated.

Chapter 6: Digital Transmission Through Bandlimited Channels

In this chapter, we consider the characterization of bandlimited channels and the problem of designing signal waveforms for such channels. We show that channel distortion results in intersymbol interference (ISI), which causes errors in signal demodulation. Then we treat the design of channel equalizers that compensate for channel distortion.

Chapter 7: Digital Transmission via Carrier Modulation

We discuss four types of carrier-modulated signals that are suitable for transmission through bandpass channels: amplitude-modulated signals, quadrature-amplitude-modulated signals, phase-shift keying, and frequency-shift keying.

Chapter 8: Channel Capacity and Coding

In this chapter, we consider appropriate mathematical models for communication channels and introduce a fundamental quantity, called the channel capacity, that gives the limit on the amount of information that can be transmitted through the channel. In particular, we consider two channel models: the binary symmetric channel (BSC) and the additive white Gaussian noise (AWGN) channel. These channel models are used in treating block and convolutional codes to achieve reliable communication through such channels.

Chapter 9: Spread Spectrum Communication Systems

The basic elements of a spread spectrum digital communication system are treated in this chapter. In particular, direct-sequence (DS) spread spectrum and frequency-hopped (FH) spread spectrum systems are considered in conjunction with phase-shift keying (PSK) and frequency-shift keying (FSK) modulation, respectively. The generation of pseudonoise (PN) sequences for use in spread spectrum systems is also treated.

About the Software

MATLAB files for this book are available at the BookWare Companion Resource Center, online at **http://www.brookscole.com/engineering/ee/bookware.htm**. The files include all the MATLAB files used in the text. In most instances, we have added numerous comments to the MATLAB files to make them easier to understand. It should be noted, however, that in developing the files, our main objective was the clarity of the MATLAB code rather than its efficiency. In cases where the most efficient code would have made the files difficult to follow, we have chosen to use a less efficient but more readable code.

The BookWare Companion Series Resource Center

New to this updated printing is the BookWare Companion Series Resource Center, a central online Web site that supports the entire series. There you will find downloadable MATLAB files for this book. We intend to keep these files current, thus making the most of the advantages of Web delivery. At the Resource Center, you will also find other resources, such as additional information about our series, links to other helpful MATLAB sites, and ideas on teaching technology in the classroom from BookWare authors and other engineering educators.

Over time, we plan to expand this site into a clearinghouse for the exchange of reliable teaching ideas and for ongoing commentary. Do you have an idea for a unique problem or example that you would like us to consider for the next edition of this book? If so, visit our site and click on the Open Manuscript icon to join an ongoing discussion with the authors, peers, students, and the publisher. The Resource Center is located at **http://www.brookscole.com/engineering/ee/bookware.htm**.

John G. Proakis
Masoud Salehi

Chapter 1

Signals and Linear Systems

1.1 Preview

In this chapter, we review the basic tools and techniques from linear system analysis used in the analysis of communication systems. Linear systems and their characteristics in the time and frequency domains, together with probability and analysis of random signals, are the two fundamental topics that must be understood in the study of communication systems. Most communication channels and many subblocks of transmitters and receivers can be well modeled as linear time-invariant (LTI) systems, and so the well-known tools and techniques from linear system analysis can be employed in their analysis. We emphasize frequency-domain analysis tools, since these are the most frequently used techniques. We start with the Fourier series and transforms; then we cover power and energy concepts, the sampling theorem, and lowpass representation of bandpass signals.

1.2 Fourier Series

The input-output relation of a linear time-invariant system is given by the convolution integral defined by

$$y(t) = x(t) \star h(t) \tag{1.2.1}$$

$$= \int_{-\infty}^{\infty} h(\tau)x(t - \tau)\, d\tau$$

where $h(t)$ denotes the impulse response of the system, $x(t)$ is the input signal, and $y(t)$ is the output signal. If the input $x(t)$ is a complex exponential given by

$$x(t) = Ae^{j2\pi f_0 t} \tag{1.2.2}$$

1

then the output is given by

$$y(t) = \int_{-\infty}^{\infty} A e^{j2\pi f_0(t-\tau)} h(\tau) \, d\tau$$

$$= A \left[\int_{-\infty}^{\infty} h(\tau) e^{-j2\pi f_0 \tau} \, d\tau \right] e^{j2\pi f_0 t} \tag{1.2.3}$$

In other words, the output is *a complex exponential with the same frequency as the input.* The (complex) amplitude of the output, however, is the (complex) amplitude of the input amplified by

$$\int_{-\infty}^{\infty} h(\tau) e^{-j2\pi f_0 \tau} \, d\tau$$

Note that the above quantity is a function of the impulse response of the LTI system, $h(t)$, and the frequency of the input signal, f_0. Therefore, computing the response of LTI systems to exponential inputs is particularly easy. Consequently, it is natural in linear system analysis to look for methods of expanding signals as the sum of complex exponentials. *Fourier series and Fourier transforms are techniques for expanding signals in terms of complex exponentials.*

A Fourier series is the orthogonal expansion of periodic signals with period T_0 when the signal set $\left\{ e^{j2\pi nt/T_0} \right\}_{n=-\infty}^{\infty}$ is employed as the basis for the expansion. With this basis, any periodic signal[1] $x(t)$ with period T_0 can be expressed as

$$x(t) = \sum_{n=-\infty}^{\infty} x_n e^{j2\pi nt/T_0} \tag{1.2.4}$$

where the x_n's are called the *Fourier series coefficients* of the signal $x(t)$ and are given by

$$x_n = \frac{1}{T_0} \int_{\alpha}^{\alpha+T_0} x(t) e^{-j2\pi nt/T_0} \, dt \tag{1.2.5}$$

Here α is an arbitrary constant chosen in such a way that the computation of the integral is simplified. The frequency $f_0 = 1/T_0$ is called the *fundamental frequency* of the periodic signal, and the frequency $f_n = nf_0$ is called the *n*th *harmonic*. In most cases either $\alpha = 0$ or $\alpha = -T_0/2$ is a good choice.

This type of Fourier series is known as the *exponential Fourier series* and can be applied to both real-valued and complex-valued signals $x(t)$ as long as they are periodic. In general, the Fourier series coefficients $\{x_n\}$ are complex numbers even when $x(t)$ is a real-valued signal.

---

[1] A sufficient condition for the existence of the Fourier series is that $x(t)$ satisfy the Dirichlet conditions. For details, see [1].

When $x(t)$ is a *real-valued* periodic signal, we have

$$
\begin{aligned}
x_{-n} &= \frac{1}{T_0} \int_{\alpha}^{\alpha+T_0} x(t) e^{j2\pi nt/T_0} \, dt \\
&= \frac{1}{T_0} \left[\int_{\alpha}^{\alpha+T_0} x(t) e^{-j2\pi nt/T_0} \, dt \right]^* \\
&= x_n^*
\end{aligned}
\tag{1.2.6}
$$

From this it is obvious that

$$
\begin{cases}
|x_n| = |x_{-n}| \\
\angle x_n = -\angle x_{-n}
\end{cases}
\tag{1.2.7}
$$

Thus, the Fourier series coefficients of a real-valued signal have *Hermitian symmetry;* that is, their real part is even and their imaginary part is odd (or, equivalently, their magnitude is even and their phase is odd).

Another form of Fourier series, known as *trigonometric Fourier series,* can be applied only to real, periodic signals and is obtained by defining

$$
x_n = \frac{a_n - jb_n}{2}
\tag{1.2.8}
$$

$$
x_{-n} = \frac{a_n + jb_n}{2}
\tag{1.2.9}
$$

which, after using Euler's relation

$$
e^{-j2\pi nt/T_0} = \cos\left(2\pi t \frac{n}{T_0}\right) - j\sin\left(2\pi t \frac{n}{T_0}\right)
\tag{1.2.10}
$$

results in

$$
a_n = \frac{2}{T_0} \int_{\alpha}^{\alpha+T_0} x(t) \cos\left(2\pi t \frac{n}{T_0}\right) dt
$$

$$
b_n = \frac{2}{T_0} \int_{\alpha}^{\alpha+T_0} x(t) \sin\left(2\pi t \frac{n}{T_0}\right) dt
\tag{1.2.11}
$$

and, therefore,

$$
x(t) = \frac{a_0}{2} + \sum_{n=1}^{\infty} a_n \cos\left(2\pi t \frac{n}{T_0}\right) + b_n \sin\left(2\pi t \frac{n}{T_0}\right)
\tag{1.2.12}
$$

Note that for $n = 0$, we always have $b_0 = 0$, so $a_0 = 2x_0$.

By defining

$$
\begin{cases}
c_n = \sqrt{a_n^2 + b_n^2} \\
\theta_n = -\arctan \frac{b_n}{a_n}
\end{cases}
\tag{1.2.13}
$$

and using the relation

$$a \cos \phi + b \sin \phi = \sqrt{a^2 + b^2} \cos\left(\phi - \arctan\frac{b}{a}\right) \tag{1.2.14}$$

we can write Equation (1.2.12) in the form

$$x(t) = \frac{a_0}{2} + \sum_{n=1}^{\infty} c_n \cos\left(2\pi t \frac{n}{T_0} + \theta_n\right) \tag{1.2.15}$$

which is the third form of the Fourier series expansion for real and periodic signals. In general, the Fourier series coefficients $\{x_n\}$ for real-valued signals are related to a_n, b_n, c_n, and θ_n through

$$\begin{cases} a_n = 2\operatorname{Re}[x_n] \\ b_n = -2\operatorname{Im}[x_n] \\ c_n = |x_n| \\ \theta_n = \angle x_n \end{cases} \tag{1.2.16}$$

Plots of $|x_n|$ and $\angle x_n$ versus n or nf_0 are called the *discrete spectra* of $x(t)$. The plot of $|x_n|$ is usually called the *magnitude spectrum*, and the plot of $\angle x_n$ is referred to as the *phase spectrum*.

If $x(t)$ is real and even—that is, if $x(-t) = x(t)$—then taking $\alpha = -T_0/2$, we have

$$b_n = \frac{2}{T_0} \int_{-T_0/2}^{T_0/2} x(t) \sin\left(2\pi t \frac{n}{T_0}\right) dt \tag{1.2.17}$$

which is zero because the integrand is an odd function of t. Therefore, for a real and even signal $x(t)$, all x_n's are real. In this case, the trigonometric Fourier series consists of all cosine functions. Similarly, if $x(t)$ is real and odd—that is, if $x(-t) = -x(t)$—then

$$a_n = \frac{2}{T_0} \int_{\alpha}^{\alpha+T_0} x(t) \cos\left(2\pi t \frac{n}{T_0}\right) dt \tag{1.2.18}$$

is zero and all x_n's are imaginary. In this case, the trigonometric Fourier series consists of all sine functions.

──── ILLUSTRATIVE PROBLEM ────

Illustrative Problem 1.1 [Fourier series of a rectangular signal train] Let the periodic signal $x(t)$, with period T_0, be defined by

$$x(t) = A\Pi\left(\frac{t}{2t_0}\right) = \begin{cases} A, & |t| < t_0 \\ \dfrac{A}{2}, & t = \pm t_0 \\ 0, & \text{otherwise} \end{cases} \tag{1.2.19}$$

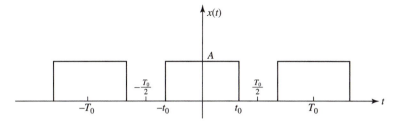

Figure 1.1   The signal $x(t)$ in Illustrative Problem 1.1

for $|t| \leq T_0/2$, where $t_0 < T_0/2$. The rectangular signal $\Pi(t)$ is, as usual, defined by

$$\Pi(t) = \begin{cases} 1, & |t| < \frac{1}{2} \\ \frac{1}{2}, & t = \pm\frac{1}{2} \\ 0, & \text{otherwise} \end{cases} \tag{1.2.20}$$

A plot of $x(t)$ is shown in Figure 1.1. Assuming $A = 1$, $T_0 = 4$, and $t_0 = 1$:

1. Determine the Fourier series coefficients of $x(t)$ in exponential and trigonometric form.

2. Plot the discrete spectrum of $x(t)$.

SOLUTION

1. To derive the Fourier series coefficients in the expansion of $x(t)$, we have

$$x_n = \frac{1}{4} \int_{-1}^{1} e^{-j2\pi nt/4} \, dt$$

$$= \frac{1}{-2j\pi n} \left[e^{-j2\pi n/4} - e^{j2\pi n/4} \right] \tag{1.2.21}$$

$$= \frac{1}{2} \text{sinc}\left(\frac{n}{2} \right) \tag{1.2.22}$$

where $\text{sinc}(x)$ is defined as

$$\text{sinc}(x) = \frac{\sin(\pi x)}{\pi x} \tag{1.2.23}$$

A plot of the sinc function is shown in Figure 1.2. Obviously, all the x_n's are real (since $x(t)$ is real and even), so

$$\begin{cases} a_n = \text{sinc}\left(\frac{n}{2} \right) \\ b_n = 0 \\ c_n = \left| \text{sinc}\left(\frac{n}{2} \right) \right| \\ \theta_n = 0, \pi \end{cases} \tag{1.2.24}$$

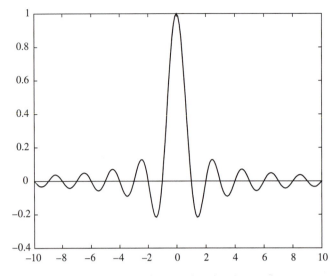

Figure 1.2   The sinc signal

Note that for even n's, $x_n = 0$ (with the exception of $n = 0$, where $a_0 = c_0 = 1$ and $x_0 = \frac{1}{2}$). Using these coefficients, we have

$$x(t) = \sum_{n=-\infty}^{\infty} \frac{1}{2}\mathrm{sinc}\left(\frac{n}{2}\right) e^{j2\pi nt/4}$$

$$= \frac{1}{2} + \sum_{n=1}^{\infty} \mathrm{sinc}\left(\frac{n}{2}\right) \cos\left(2\pi t \frac{n}{4}\right) \tag{1.2.25}$$

A plot of the Fourier series approximations to this signal over one period for $n = 0, 1, 3, 5, 7, 9$ is shown in Figure 1.3. Note that as n increases, the approximation becomes closer to the original signal $x(t)$.

2. Note that x_n is always real. Therefore, depending on its sign, the phase is either zero or π. The magnitude of the x_n's is $\frac{1}{2}\left|\mathrm{sinc}\left(\frac{n}{2}\right)\right|$. The discrete spectrum is shown in Figure 1.4.

The MATLAB script for plotting the discrete spectrum of the signal is given next.

── **M-FILE** ──────────────────────────────

```
% MATLAB script for Illustrative Problem 1.1.
n=[-20:1:20];
x=abs(sinc(n/2));
stem(n,x);
```

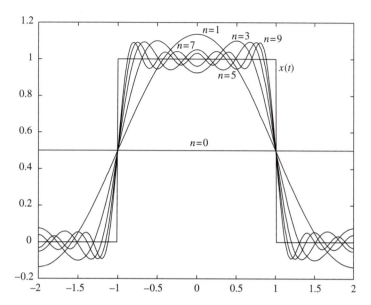

Figure 1.3   Various Fourier series approximations for the rectangular pulse in
Illustrative Problem 1.1

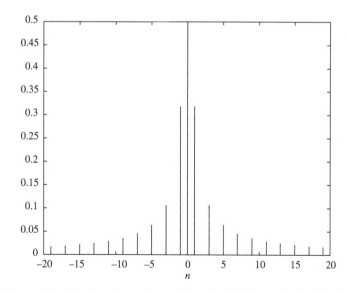

Figure 1.4   The discrete spectrum of the signal in Illustrative Problem 1.1

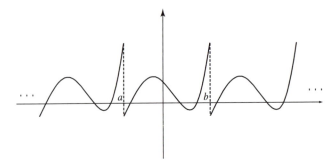

Figure 1.5   A periodic signal

When the signal $x(t)$ is described on one period between a and b, as shown in Figure 1.5, and the signal in the interval $[a, b]$ is given in an m-file, the Fourier series coefficients can be obtained using the m-file fseries.m given next.

◖ **M-FILE** ◗ ────────────────────────────────

```
function xx=fseries(funfcn,a,b,n,tol,p1,p2,p3)
%FSERIES      Returns the Fourier series coefficients.
%             XX=FSERIES(FUNFCN,A,B,N,TOL,P1,P2,P3)
%             funfcn=The given function, in an m-file.
%             It can depend on up to three parameters
%             p1, p2, and p3. The function is given
%             over one period extending from 'a' to 'b'
%             xx=vector of length n+1 of Fourier series
%             coefficients, xx0,xx1,...,xxn.
%             p1,p2,p3=parameters of funfcn.
%             tol=the error level.

j=sqrt(-1);
args0=[ ];
for nn=1:nargin-5
  args0=[args0,',p',int2str(nn)];
end
args=[args0,')'];
t=b-a;
xx(1)=eval(['1/(',num2str(t),').*quad(funfcn,a,b,tol,[]',args]) ;

for i=1:n
  new_fun = 'exp_fnct' ;
  args=[',', num2str(i), ',', num2str(t), args0, ')' ] ;
  xx(i+1)=eval(['1/(',num2str(t),').*quad(new_fun,a,b,tol,[],funfcn',args]);
end
```

---(ILLUSTRATIVE PROBLEM)--

Illustrative Problem 1.2 [**The magnitude and the phase spectra**] Determine and plot the discrete magnitude and phase spectra of the periodic signal $x(t)$ with a period equal to 8 and defined as $x(t) = \Lambda(t)$ for $|t| \leq 4$.

---(SOLUTION)--

Since the signal is given by an m-file lambda.m, we can choose the interval $[a, b] = [-4, 4]$ and determine the coefficients. Note that the m-file fseries.m determines the Fourier series coefficients for nonnegative values of n, but since here $x(t)$ is real-valued, we have $x_{-n} = x_n^*$. In Figure 1.6 the magnitude and the phase spectra of this signal are plotted for a choice of $n = 24$.

The MATLAB script for determining and plotting the magnitude and the phase spectra is given next.

---(M-FILE)--

```
% MATLAB script for Illustrative Problem 1.2.
echo on
fnct='lambda';
a=-4;
b=4;
n=24;
tol=0.1;
xx=fseries(fnct,a,b,n,tol);
xx1=xx(n+1:-1:2);
xx1=[conj(xx1),xx];
absxx1=abs(xx1);
pause % Press any key to see a plot of the magnitude spectrum
n1=[-n:n];
stem(n1,absxx1)
title('The Discrete Magnitude Spectrum')
phasexx1=angle(xx1);
pause % Press any key to see a plot of the phase spectrum
stem(n1,phasexx1)
title('The Discrete Phase Spectrum')
```

---(ILLUSTRATIVE PROBLEM)--

Illustrative Problem 1.3 [**The magnitude and the phase spectra**] Determine and plot the magnitude and the phase spectra of a periodic signal with a period equal to 12 that is given by

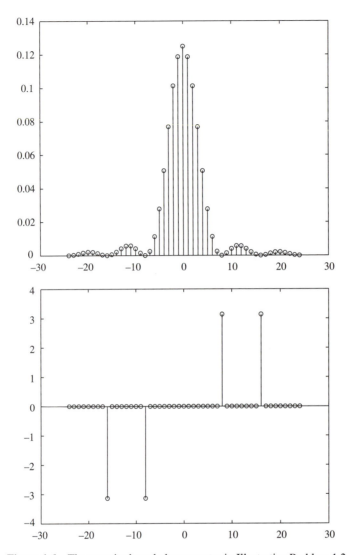

Figure 1.6   The magnitude and phase spectra in Illustrative Problem 1.2

$$x(t) = \frac{1}{\sqrt{2\pi}} e^{-t^2/2}$$

in the interval $[-6, 6]$. A plot of this signal is shown in Figure 1.7.

SOLUTION

The signal is equal to the density function of a zero-mean unit-variance Gaussian (normal) random variable given in the m-file normal.m. This file requires two parameters, *m* and

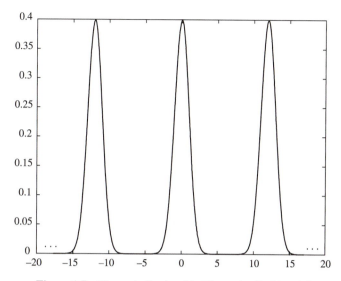

Figure 1.7   The periodic signal in Illustrative Problem 1.3

s, the mean and the standard deviation of the random variable, which in the problem are 0 and 1, respectively. Therefore, we can use the following MATLAB script to obtain the magnitude and the phase plots shown in Figure 1.8.

M-FILE

```
% MATLAB script for Illustrative Problem 1.3.
echo on
fnct='normal';
a=−6;
b=6;
n=24;
tol=0.1;
xx=fseries(fnct,a,b,n,tol,0,1);
xx1=xx(n+1:−1:2);
xx1=[conj(xx1),xx];
absxx1=abs(xx1);
pause % Press any key to see a plot of the magnitude
n1=[−n:n];
stem(n1,absxx1)
title('The Discrete Magnitude Spectrum')
phasexx1=angle(xx1);
pause % Press any key to see a plot of the phase
stem(n1,phasexx1)
title('The Discrete Phase Spectrum')
```

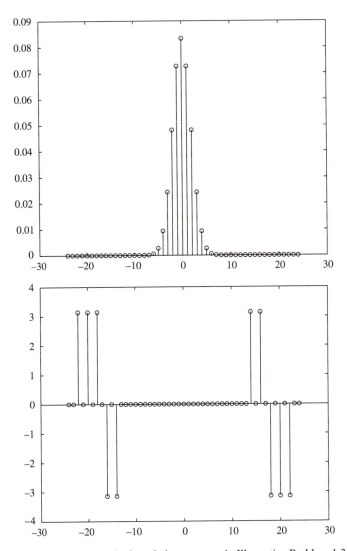

Figure 1.8   The magnitude and phase spectra in Illustrative Problem 1.3

1.2.1   Periodic Signals and LTI Systems

When a periodic signal $x(t)$ is passed through a linear time-invariant (LTI) system, as shown in Figure 1.9, the output signal $y(t)$ is also periodic, usually with the same period as the input signal[2] (why?), and therefore it has a Fourier series expansion.

---

[2] We say *usually* with the same period as the input signal. Can you give an example where the period of the output is different from the period of the input?

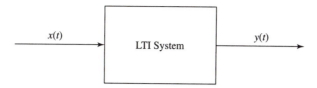

Figure 1.9 Periodic signals through LTI systems

If $x(t)$ and $y(t)$ are expanded as

$$x(t) = \sum_{n=-\infty}^{\infty} x_n e^{j2\pi nt/T_0} \qquad (1.2.26)$$

$$y(t) = \sum_{n=-\infty}^{\infty} y_n e^{j2\pi nt/T_0} \qquad (1.2.27)$$

then the relation between the Fourier series coefficients of $x(t)$ and $y(t)$ can be obtained by employing the convolution integral

$$
\begin{aligned}
y(t) &= \int_{-\infty}^{\infty} x(t-\tau)h(\tau)\,d\tau \\
&= \int_{-\infty}^{\infty} \sum_{n=-\infty}^{\infty} x_n e^{j2\pi n(t-\tau)/T_0} h(\tau)\,d\tau \\
&= \sum_{n=-\infty}^{\infty} x_n \left(\int_{-\infty}^{\infty} h(\tau)e^{-j2\pi n\tau/T_0}\,d\tau \right) e^{j2\pi nt/T_0} \\
&= \sum_{n=-\infty}^{\infty} y_n e^{j2\pi nt/T_0} \qquad (1.2.28)
\end{aligned}
$$

From the preceding relation, we have

$$y_n = x_n H\left(\frac{n}{T_0}\right) \qquad (1.2.29)$$

where $H(f)$ denotes the transfer function[3] of the LTI system given as the Fourier transform of its impulse response $h(t)$:

$$H(f) = \int_{-\infty}^{\infty} h(t)e^{-j2\pi ft}\,dt \qquad (1.2.30)$$

---

[3] Also known as the *frequency response* of the system.

ILLUSTRATIVE PROBLEM

Illustrative Problem 1.4 [Filtering of periodic signals] A triangular pulse train $x(t)$ with period $T_0 = 2$ is defined in one period as

$$\Lambda(t) = \begin{cases} t+1, & -1 \le t \le 0 \\ -t+1, & 0 \le t \le 1 \\ 0, & \text{otherwise} \end{cases} \tag{1.2.31}$$

1. Determine the Fourier series coefficients of $x(t)$.

2. Plot the discrete spectrum of $x(t)$.

3. Assuming that this signal passes through an LTI system whose impulse response is given by

$$h(t) = \begin{cases} t, & 0 \le t < 1 \\ 0, & \text{otherwise} \end{cases} \tag{1.2.32}$$

   plot the discrete spectrum and the output $y(t)$. Plots of $x(t)$ and $h(t)$ are given in Figure 1.10.

SOLUTION

1. We have

$$x_n = \frac{1}{T_0} \int_{-T_0/2}^{T_0/2} x(t) e^{-j2\pi nt/T_0}\, dt \tag{1.2.33}$$

$$= \frac{1}{2} \int_{-1}^{1} \Lambda(t) e^{-j\pi nt}\, dt \tag{1.2.34}$$

$$= \frac{1}{2} \int_{-\infty}^{\infty} \Lambda(t) e^{-j\pi nt}\, dt \tag{1.2.35}$$

$$= \frac{1}{2} \mathcal{F}[\Lambda(t)]_{f=n/2} \tag{1.2.36}$$

$$= \frac{1}{2} \operatorname{sinc}^2\left(\frac{n}{2}\right) \tag{1.2.37}$$

   where we have used the facts that $\Lambda(t)$ vanishes outside the $[-1, 1]$ interval and that the Fourier transform of $\Lambda(t)$ is $\operatorname{sinc}^2(f)$. This result can also be obtained by using the expression for $\Lambda(t)$ and integrating by parts. Obviously, we have $x_n = 0$ for all even values of n except for $n = 0$.

2. A plot of the discrete spectrum of $x(t)$ is shown in Figure 1.11.

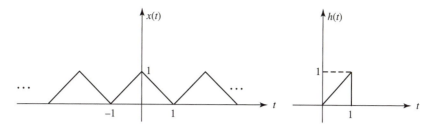

Figure 1.10   The input signal and the system impulse response

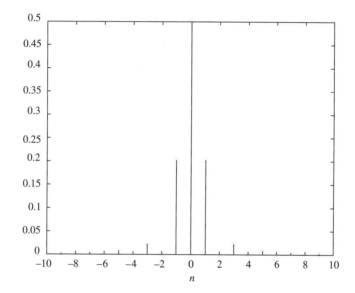

Figure 1.11   The discrete spectrum of the signal

3. First we have to derive $H(f)$, the transfer function of the system. Although this can be done analytically, we will adopt a numerical approach. The resulting magnitude of the transfer function and also the magnitude of $H(n/T_0) = H(n/2)$ are shown in Figure 1.12. To derive the discrete spectrum of the output, we employ the relation

$$y_n = x_n H\left(\frac{n}{T_0}\right) \tag{1.2.38}$$

$$= \frac{1}{2}\operatorname{sinc}^2\left(\frac{n}{2}\right) H\left(\frac{n}{2}\right) \tag{1.2.39}$$

The resulting discrete spectrum of the output is shown in Figure 1.13.

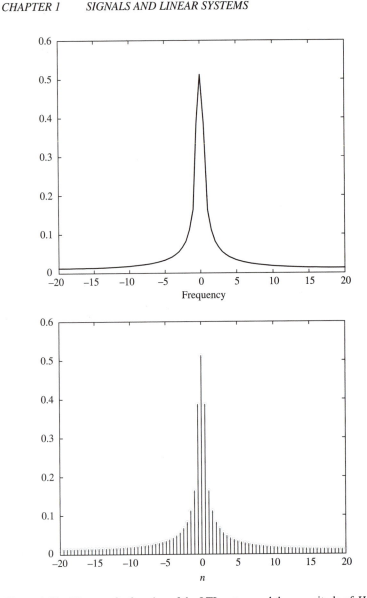

Figure 1.12   The transfer function of the LTI system and the magnitude of $H(n/2)$

The MATLAB script for this problem follows.

M-FILE

```
% MATLAB script for Illustrative Problem 1.4.
echo on
n=[−20:1:20];
% Fourier series coefficients of x(t) vector
```

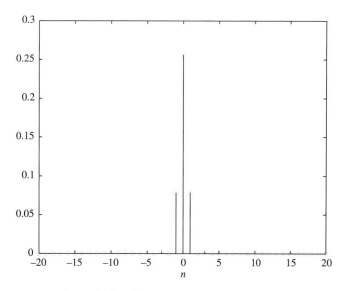

Figure 1.13   The discrete spectrum of the output

```
x=.5*(sinc(n/2)).^2;
% sampling interval
ts=1/40;
% Time vector
t=[−.5:ts:1.5];
% impulse response
fs=1/ts;
h=[zeros(1,20),t(21:61),zeros(1,20)];
% transfer function
H=fft(h)/fs;
% frequency resolution
df=fs/80;
f=[0:df:fs]−fs/2;
% rearrange H
H1=fftshift(H);
y=x.*H1(21:61);
% Plotting commands follow
```

1.3   Fourier Transforms

The Fourier transform is the extension of the Fourier series to nonperiodic signals. The Fourier transform of a signal $x(t)$ that satisfies certain conditions, known as Dirichlet's conditions [1], is denoted by $X(f)$ or, equivalently, $\mathcal{F}[x(t)]$ and is defined by

$$\mathcal{F}[x(t)] = X(f) = \int_{-\infty}^{\infty} x(t)e^{-j2\pi ft}\, dt \tag{1.3.1}$$

The inverse Fourier transform of $X(f)$ is $x(t)$, given by

$$\mathcal{F}^{-1}[X(f)] = x(t) = \int_{-\infty}^{\infty} X(f)e^{j2\pi ft}\, df \tag{1.3.2}$$

If $x(t)$ is a real signal, then $X(f)$ satisfies the Hermitian symmetry; that is,

$$X(-f) = X^*(f) \tag{1.3.3}$$

There are certain properties that the Fourier transform satisfies. The most important properties of the Fourier transform are summarized as follows.

1. **Linearity:** The Fourier transform of a linear combination of two or more signals is the linear combination of the corresponding Fourier transforms:

$$\mathcal{F}[\alpha x_1(t) + \beta x_2(t)] = \alpha \mathcal{F}[x_1(t)] + \beta \mathcal{F}[x_2(t)] \tag{1.3.4}$$

2. **Duality:** If $X(f) = \mathcal{F}[x(t)]$, then

$$\mathcal{F}[X(t)] = x(-f) \tag{1.3.5}$$

3. **Time shift:** A shift in the time domain results in a phase shift in the frequency domain. If $X(f) = \mathcal{F}[x(t)]$, then

$$\mathcal{F}[x(t - t_0)] = e^{-j2\pi f t_0} X(f) \tag{1.3.6}$$

4. **Scaling:** An expansion in the time domain results in a contraction in the frequency domain, and vice versa. If $X(f) = \mathcal{F}[x(t)]$, then

$$\mathcal{F}[x(at)] = \frac{1}{|a|} X\left(\frac{f}{a}\right), \qquad a \neq 0 \tag{1.3.7}$$

5. **Modulation:** Multiplication by an exponential in the time domain corresponds to a frequency shift in the frequency domain. If $X(f) = \mathcal{F}[x(t)]$, then

$$\begin{cases} \mathcal{F}[e^{j2\pi f_0 t} x(t)] = X(f - f_0) \\ \mathcal{F}[x(t)\cos(2\pi f_0 t)] = \dfrac{1}{2}[X(f - f_0) + X(f + f_0)] \end{cases} \tag{1.3.8}$$

6. **Differentiation:** Differentiation in the time domain corresponds to multiplication by $j2\pi f$ in the frequency domain. If $X(f) = \mathcal{F}[x(t)]$, then

$$\mathcal{F}[x'(t)] = j2\pi f X(f) \tag{1.3.9}$$

$$\mathcal{F}\left[\frac{d^n}{dt^n} x(t)\right] = (j2\pi f)^n X(f) \tag{1.3.10}$$

7. **Convolution:** Convolution in the time domain is equivalent to multiplication in the frequency domain, and vice versa. If $X(f) = \mathcal{F}[x(t)]$ and $Y(f) = \mathcal{F}[y(t)]$, then

$$\mathcal{F}[x(t) \star y(t)] = X(f)Y(f) \tag{1.3.11}$$

$$\mathcal{F}[x(t)y(t)] = X(f) \star Y(f) \tag{1.3.12}$$

8. **Parseval's relation:** If $X(f) = \mathcal{F}[x(t)]$ and $Y(f) = \mathcal{F}[y(t)]$, then

$$\int_{-\infty}^{\infty} x(t)y^*(t)\,dt = \int_{-\infty}^{\infty} X(f)Y^*(f)\,df \tag{1.3.13}$$

$$\int_{-\infty}^{\infty} |x(t)|^2\,dt = \int_{-\infty}^{\infty} |X(f)|^2\,df \tag{1.3.14}$$

The second relation is also referred to as *Rayleigh's relation.*

Table 1.1 gives the most useful Fourier transform pairs. In this table, $u_{-1}(t)$ denotes the unit step function; $\delta(t)$ is the impulse signal; sgn(t) is the *signum function,* defined as

$$\text{sgn}(t) = \begin{cases} 1, & t > 0 \\ 0, & t = 0 \\ -1, & t < 0 \end{cases} \tag{1.3.15}$$

and $\delta^{(n)}(t)$ denotes the nth derivative of the impulse signal.

For a periodic signal $x(t)$, with period T_0, whose Fourier series coefficients are given by x_n; that is,

$$x(t) = \sum_{n=-\infty}^{\infty} x_n e^{j2\pi nt/T_0}$$

the Fourier transform is obtained by

$$X(f) = \mathcal{F}[x(t)]$$

$$= \mathcal{F}\left[\sum_{n=-\infty}^{\infty} x_n e^{j2\pi nt/T_0}\right]$$

$$= \sum_{n=-\infty}^{\infty} x_n \mathcal{F}\left[e^{j2\pi nt/T_0}\right]$$

$$= \sum_{n=-\infty}^{\infty} x_n \delta\left(f - \frac{n}{T_0}\right) \tag{1.3.16}$$

In other words, the Fourier transform of a periodic signal consists of impulses at multiples of the fundamental frequency (harmonics) of the original signal.

It is also possible to express the Fourier series coefficients in terms of the Fourier transform of the *truncated signal* by

$$x_n = \frac{1}{T_0} X_{T_0}\left(\frac{n}{T_0}\right) \tag{1.3.17}$$

Table 1.1   Table of Fourier transform pairs

$x(t)$	$X(f)$
$\delta(t)$	1
1	$\delta(f)$
$\delta(t - t_0)$	$e^{-j2\pi f t_0}$
$e^{j2\pi f t_0}$	$\delta(f - f_0)$
$\cos(2\pi f_0 t)$	$\frac{1}{2}\delta(f - f_0) + \frac{1}{2}\delta(f + f_0)$
$\sin(2\pi f_0 t)$	$\frac{1}{2j}\delta(f - f_0) - \frac{1}{2j}\delta(f + f_0)$
$\Pi(t)$	$\mathrm{sinc}(f)$
$\mathrm{sinc}(t)$	$\Pi(f)$
$\Lambda(t)$	$\mathrm{sinc}^2(f)$
$\mathrm{sinc}^2(t)$	$\Lambda(f)$
$e^{-\alpha t}u_{-1}(t), \quad \alpha > 0$	$\dfrac{1}{\alpha + j2\pi f}$
$te^{-\alpha t}u_{-1}(t), \quad \alpha > 0$	$\dfrac{1}{(\alpha + j2\pi f)^2}$
$e^{-\alpha|t|}, \quad \alpha > 0$	$\dfrac{2\alpha}{\alpha^2 + (2\pi f)^2}$
$e^{-\pi t^2}$	$e^{-\pi f^2}$
$\mathrm{sgn}(t)$	$\dfrac{1}{j\pi f}$
$u_{-1}(t)$	$\dfrac{1}{2}\delta(f) + \dfrac{1}{j2\pi f}$
$\delta'(t)$	$j2\pi f$
$\delta^{(n)}(t)$	$(j2\pi f)^n$
$\displaystyle\sum_{n=-\infty}^{\infty}\delta(t - nT_0)$	$\dfrac{1}{T_0}\displaystyle\sum_{n=-\infty}^{\infty}\delta\left(f - \dfrac{n}{T_0}\right)$

where, by definition, $X_{T_0}(f)$ is the Fourier transform of $x_{T_0}(t)$, the truncated signal, defined by

$$x_{T_0}(t) = \begin{cases} x(t), & -\frac{T_0}{2} < t \leq \frac{T_0}{2} \\ 0, & \text{otherwise} \end{cases} \tag{1.3.18}$$

The Fourier transform of a signal is called the *spectrum* of the signal. The spectrum of a signal in general is a complex function $X(f)$; therefore, to plot the spectrum, usually two plots are provided: the magnitude spectrum $|X(f)|$ and the phase spectrum $\angle X(f)$.

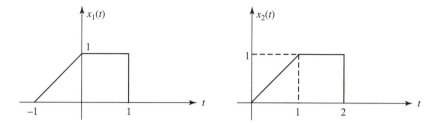

Figure 1.14   Signals $x_1(t)$ and $x_2(t)$

ILLUSTRATIVE PROBLEM

Illustrative Problem 1.5 [Fourier transforms] Plot the magnitude and the phase spectra of signals $x_1(t)$ and $x_2(t)$ shown in Figure 1.14.

SOLUTION

Since the signals are similar except for a time shift, we would expect them to have the same magnitude spectra. The common magnitude spectrum and the two phase spectra plotted on the same axes are shown in Figures 1.15 and 1.16, respectively.

The MATLAB script for this problem follows. In Section 1.3.1, we show how to obtain the Fourier transform of a signal using MATLAB.

M-FILE

```
% MATLAB script for Illustrative Problem 1.5
df=0.01;
fs=10;
ts=1/fs;
t=[−5:ts:5];
x1=zeros(size(t));
x1(41:51)=t(41:51)+1;
x1(52:61)=ones(size(x1(52:61)));
x2=zeros(size(t));
x2(51:71)=x1(41:61);
[X1,x11,df1]=fftseq(x1,ts,df);
[X2,x21,df2]=fftseq(x2,ts,df);
X11=X1/fs;
X21=X2/fs;
f=[0:df1:df1*(length(x11)−1)]−fs/2;
plot(f,fftshift(abs(X11)))
figure
plot(f(500:525),fftshift(angle(X11(500:525))),f(500:525),fftshift(angle(X21(500:525))),'−−')
```

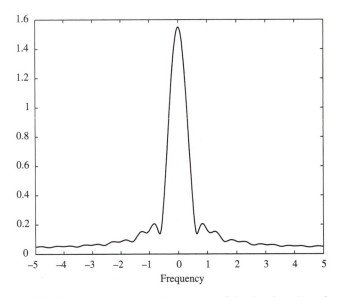

Figure 1.15   The common magnitude spectrum of the signals $x_1(t)$ and $x_2(t)$

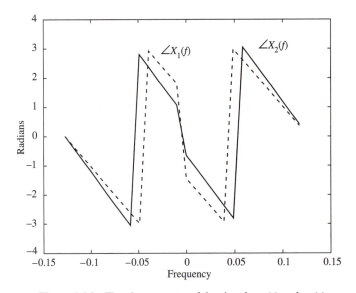

Figure 1.16   The phase spectra of the signals $x_1(t)$ and $x_2(t)$

1.3.1   Sampling Theorem

The sampling theorem is one of the most important results in signal and system analysis; it forms the basis for the relation between continuous-time signals and discrete-time signals. The sampling theorem says that a bandlimited signal—that is, a signal whose Fourier transform vanishes for $|f| > W$ for some W—can be completely described in terms of its sample values taken at intervals T_s as long as $T_s \leq 1/2W$. If the sampling is done at intervals $T_s = 1/2W$, known as the *Nyquist interval* (or *Nyquist rate*), the signal $x(t)$ can be reconstructed from the sample values $\{x[n] = x(nT_s)\}_{n=-\infty}^{\infty}$ as

$$x(t) = \sum_{n=-\infty}^{\infty} x(nT_s)\text{sinc}\,(2W(t - nT_s)) \qquad (1.3.19)$$

This result is based on the fact that the sampled waveform $x_\delta(t)$ defined as

$$x_\delta(t) = \sum_{n=-\infty}^{\infty} x(nT_s)\delta(t - nT_s) \qquad (1.3.20)$$

has a Fourier transform given by

$$X_\delta(f) = \frac{1}{T_s} \sum_{n=-\infty}^{\infty} X\left(f - \frac{n}{T_s}\right) \qquad \text{for all } f$$

$$= \frac{1}{T_s}X(f) \qquad \text{for } |f| < W \qquad (1.3.21)$$

so passing it through a lowpass filter with a bandwidth of W and a gain of T_s in the passband will reproduce the original signal.

Figure 1.17 is a representation of Equation (1.3.19) for $T_s = 1$ and $\{x[n]\}_{n=-3}^{3} = \{1, 1, -1, 2, -2, 1, 2\}$. In other words,

$$x(t) = \text{sinc}(t + 3) + \text{sinc}(t + 2) - \text{sinc}(t + 1) + 2\,\text{sinc}(t)$$
$$- 2\,\text{sinc}(t - 1) + \text{sinc}(t - 2) + 2\,\text{sinc}(t - 2)$$

The *discrete Fourier transform (DFT)* of the discrete-time sequence $x[n]$ is expressed as

$$X_d(f) = \sum_{n=-\infty}^{\infty} x[n]e^{-j2\pi f n T_s} \qquad (1.3.22)$$

Comparing Equations (1.3.22) and (1.3.21), we conclude that

$$X(f) = T_s X_d(f) \qquad \text{for } |f| < W \qquad (1.3.23)$$

which gives the relation between the Fourier transform of an analog signal and the discrete Fourier transform of its corresponding sampled signal.

Numerical computation of the discrete Fourier transform is done via the well-known *fast Fourier transform (FFT)* algorithm. In this algorithm a sequence of length N of samples of the signals $x(t)$ taken at intervals of T_s is used as the representation of the signal. The result is a sequence of length N of samples of $X_d(f)$ in the frequency interval $[0, f_s]$, where $f_s = 1/T_s = 2W$ is the Nyquist frequency. When the samples are $\Delta f = f_s/N$ apart, the value of Δf gives the frequency resolution of the resulting Fourier transform.

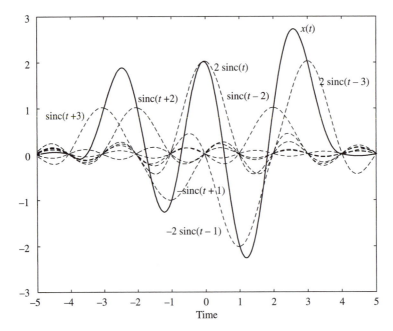

Figure 1.17   Representation of the sampling theorem

The FFT algorithm is computationally efficient if the length of the input sequence, N, is a power of 2. In many cases, if this length is not a power of 2, it is made to be a power of 2 by techniques such as zero-padding. Note that since the FFT algorithm essentially gives the DFT of the sampled signal, in order to get the Fourier transform of the analog signal, we have to employ Equation (1.3.23). This means that after computing the FFT, we have to multiply it by T_s or, equivalently, divide it by f_s in order to obtain the Fourier transform of the original analog signal.

The MATLAB function fftseq.m, given next, takes as its input a time sequence m, the sampling interval t_s, and the required frequency resolution df and returns a sequence whose length is a power of 2, the FFT of this sequence M, and the resulting frequency resolution.

M-FILE

```
function [M,m,df]=fftseq(m,ts,df)
%              [M,m,df]=fftseq(m,ts,df)
%              [M,m,df]=fftseq(m,ts)
%FFTSEQ        Generates M, the FFT of the sequence m.
%              The sequence is zero-padded to meet the required frequency resolution df.
%              ts is the sampling interval. The output df is the final frequency resolution.
%              Output m is the zero-padded version of input m. M is the FFT.
fs=1/ts;
```

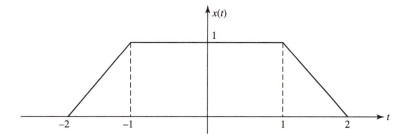

Figure 1.18 The signal $x(t)$

```
if nargin == 2
  n1=0;
else
  n1=fs/df;
end
n2=length(m);
n=2^(max(nextpow2(n1),nextpow2(n2)));
M=fft(m,n);
m=[m,zeros(1,n−n2)];
df=fs/n;
```

ILLUSTRATIVE PROBLEM

Illustrative Problem 1.6 [Analytical and numerical derivation of the Fourier transform] The signal $x(t)$ is described by

$$x(t) = \begin{cases} t+2, & -2 \leq t \leq -1 \\ 1, & -1 < t \leq 1 \\ -t+2, & 1 < t \leq 2 \\ 0, & \text{otherwise} \end{cases} \tag{1.3.24}$$

and is shown in Figure 1.18.

1. Determine the Fourier transform of $x(t)$ analytically and plot the spectrum of $x(t)$.

2. Using MATLAB, determine the Fourier transform numerically and plot the result.

SOLUTION

1. The signal $x(t)$ can be written as

$$x(t) = 2\Lambda\left(\frac{t}{2}\right) - \Lambda(t) \tag{1.3.25}$$

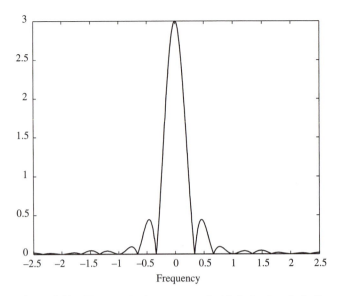

Figure 1.19   The magnitude spectrum of $x(t)$ derived analytically

and therefore

$$X(f) = 4 \operatorname{sinc}^2(2f) - \operatorname{sinc}^2(f) \tag{1.3.26}$$

where we have used linearity, scaling, and the fact that the Fourier transform of $\Lambda(t)$ is $\operatorname{sinc}^2(f)$. Obviously the Fourier transform is real. The magnitude spectrum is shown in Figure 1.19.

2. To determine the Fourier transform using MATLAB, we first give a rough estimate of the bandwidth of the signal. Since the signal is relatively smooth, its bandwidth is proportional to the inverse time duration of the signal. The time duration of the signal is 4. To be on the safe side we take the bandwidth as 10 times the inverse time duration, or

$$BW = 10 \times \frac{1}{4} = 2.5 \tag{1.3.27}$$

and therefore the Nyquist frequency is twice the bandwidth and is equal to 5. Hence, the sampling interval is $T_s = 1/f_s = 0.2$. We consider the signal on the interval $[-4, 4]$ and sample it at T_s intervals. With this choice, using a simple MATLAB script employing the fftseq.m function, we can derive the FFT numerically. We have chosen the *required* frequency resolution to be 0.01 Hz, so the *resulting* frequency resolution returned by fftseq.m is 0.0098 Hz, which meets the requirements of the problem. The signal vector x, which has length 41, is zero-padded to a length of 256 to meet the frequency-resolution requirement and also to make it a power of 2 for computational efficiency. A plot of the magnitude spectrum of the Fourier transform is given in Figure 1.20.

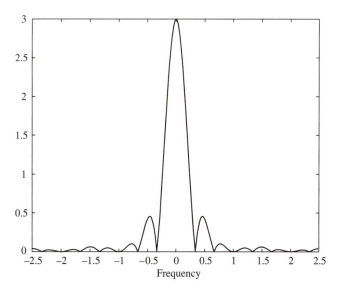

Figure 1.20   The magnitude spectrum of $x(t)$ derived numerically

The MATLAB script for this problem is given next.

M-FILE

```
% MATLAB script for Illustrative Problem 1.6
echo on
ts=0.2;                                  % set parameters
fs=1/ts;
df=0.01;
x=[zeros(1,10),[0:0.2:1],ones(1,9),[1:−0.2:0],zeros(1,10)];
[X,x,df1]=fftseq(x,ts,df);               % derive the FFT
X1=X/fs;                                 % scaling
f=[0:df1:df1*(length(x)−1)]−fs/2;        % frequency vector for FFT
f1=[−2.5:0.001:2.5];                     % frequency vector for analytic approach
y=4*(sinc(2*f1)).^2−(sinc(f1)).^2;       % Exact Fourier Transform
pause % Press a key to see the plot of the Fourier transform derived analytically
clf
subplot(2,1,1)
plot(f1,abs(y));
xlabel('Frequency')
title('Magnitude Spectrum of x(t) derived analytically')
pause % Press a key to see the plot of the Fourier transform derived numerically
subplot(2,1,2)
plot(f,fftshift(abs(X1)));
xlabel('Frequency')
title('Magnitude Spectrum of x(t) derived numerically')
```

1.3.2   Frequency-Domain Analysis of LTI Systems

The output of an LTI system with impulse response $h(t)$ when the input signal is $x(t)$ is given by the convolution integral

$$y(t) = x(t) \star h(t) \tag{1.3.28}$$

Applying the convolution theorem, we obtain

$$Y(f) = X(f)H(f) \tag{1.3.29}$$

where

$$H(f) = \mathcal{F}[h(t)] = \int_{-\infty}^{\infty} h(t)e^{-j2\pi ft}\,dt \tag{1.3.30}$$

is the transfer function of the system. Equation (1.3.29) can be written in the form

$$\begin{cases} |Y(f)| = |X(f)|\,|H(f)| \\ \angle Y(f) = \angle X(f) + \angle H(f) \end{cases} \tag{1.3.31}$$

which shows the relation between the magnitude and phase spectra of the input and the output.

ILLUSTRATIVE PROBLEM

Illustrative Problem 1.7 [LTI system analysis in the frequency domain] The signal $x(t)$ whose plot is given in Figure 1.21 consists of some line segments and a sinusoidal segment.

1. Determine the FFT of this signal and plot it.

2. If the signal is passed through an ideal lowpass filter with a bandwidth of 1.5 Hz, find the output of the filter and plot it.

3. If the signal is passed through a filter whose impulse response is given by

$$h(t) = \begin{cases} t, & 0 \le t < 1 \\ 1, & 1 \le t \le 2 \\ 0, & \text{otherwise} \end{cases} \tag{1.3.32}$$

plot the filter output.

SOLUTION

First we derive an expression for the sinusoidal part of the signal. This is a sinusoidal whose

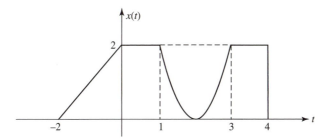

Figure 1.21   The signal $x(t)$

half-period is 2; therefore, it has a frequency of $f_0 = \frac{1}{4} = 0.25$ Hz. The signal has an amplitude of 2 and is raised by 2, so the general expression for it is $2\cos(2\pi \times 0.25t + \theta) + 2 = 2\cos(0.5\pi t + \theta) + 2$. The value of the phase θ is derived by employing the boundary conditions

$$2 + 2\cos(0.5\pi t + \theta)|_{t=2} = 0 \tag{1.3.33}$$

or $\theta = 0$. Therefore, the signal can be written as

$$x(t) = \begin{cases} t + 2, & -2 \leq t \leq 0 \\ 1, & 0 < t \leq 1 \\ 2 + 2\cos(0.5\pi t), & 1 < t \leq 3 \\ 1, & 3 < t \leq 4 \\ 0, & \text{otherwise} \end{cases} \tag{1.3.34}$$

Having a complete description of the signal, we can proceed with the solution.

1. The bandwidth of the signal has been chosen to be 5 Hz. The required frequency resolution is 0.01 Hz. The plot of the magnitude spectrum of the signal is given in Figure 1.22.

2. Here $f_s = 5$ Hz. Since the bandwidth of the lowpass filter is 1.5 Hz, its transfer function is given by

$$H(f) = \begin{cases} 1, & 0 \leq f \leq 1.5 \\ 0, & 1.5 < f \leq 3.5 \\ 1, & 3.5 < f \leq 5 \end{cases} \tag{1.3.35}$$

which is multiplied by $X(f)$ to generate $Y(f)$, the Fourier transform of the output. Using this transfer function gives the output shown in Figure 1.23.

3. Here we obtain the output of the filter by a simple convolution. The result is shown in Figure 1.24.

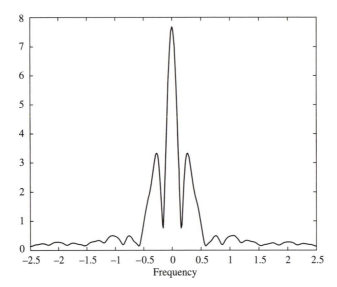

Figure 1.22   The magnitude spectrum of the signal

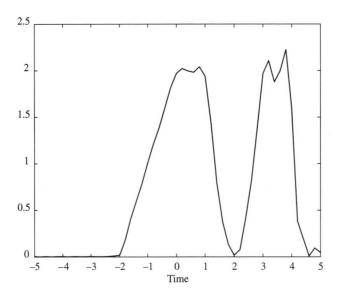

Figure 1.23   The output of the lowpass filter

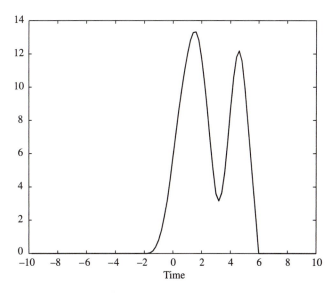

Figure 1.24   The output signal in the third part of Illustrative Problem 1.7

The MATLAB script for this problem is given next.

M-FILE

```
% MATLAB script for Illustrative Problem 1.7.
echo on
df=0.01;                                    % Freq. resolution
fs=5;                                       % Sampling frequency
ts=1/fs;                                    % Sampling interval
t=[−5:ts:5];                                % Time vector
x=zeros(1,length(t));                       % Input signal initiation
x(16:26)=t(16:26)+2;
x(27:31)=2*ones(1,5);
x(32:41)=2+2*cos(0.5*pi*t(32:41));
x(42:46)=2*ones(1,5);
% Part 1
[X,x1,df1]=fftseq(x,ts,df);                 % Spectrum of the input
f=[0:df1:df1*(length(x1)−1)]−fs/2;          % Frequency vector
X1=X/fs;                                    % Scaling
% Part 2
% Filter transfer function
H=[ones(1,ceil(1.5/df1)),zeros(1,length(X)−2*ceil(1.5/df1)),ones(1,ceil(1.5/df1))];
Y=X.*H;                                     % Output spectrum
y1=ifft(Y);                                 % Output of the filter
% Part 3
% LTI system impulse response
h=[zeros(1,ceil(5/ts)),t(ceil(5/ts)+1:ceil(6/ts)),ones(1,ceil(7/ts)−ceil(6/ts)),zeros(1,51−ceil(7/ts))];
```

```
y2=conv(h,x);                              % Output of the LTI system
pause    % Press a key to see spectrum of the input
plot(f,fftshift(abs(X1)))
pause    % Press a key to see the output of the lowpass filter
plot(t,abs(y1(1:length(t))));
pause    % Press a key to see the output of the LTI system
plot([−10:ts:10],y2);
```

1.4   Power and Energy

The energy and the power contents of a real signal $x(t)$, denoted by E_X and P_X, respectively, are defined as

$$
\begin{cases}
E_X = \displaystyle\int_{-\infty}^{\infty} x^2(t)\, dt \\[3mm]
P_X = \displaystyle\lim_{T \to \infty} \frac{1}{T} \int_{-T/2}^{T/2} x^2(t)\, dt
\end{cases}
\tag{1.4.1}
$$

A signal with finite energy is called an *energy-type signal*, and a signal with positive and finite power is a *power-type signal*.[4] For instance, $x(t) = \Pi(t)$ is an example of an energy-type signal, whereas $x(t) = \cos(t)$ is an example of a power-type signal. All periodic signals[5] are power-type signals. The *energy spectral density* of an energy-type signal gives the distribution of energy at various frequencies of the signal and is given by

$$
\mathcal{G}_X(f) = |X(f)|^2
\tag{1.4.2}
$$

Therefore,

$$
E_X = \int_{-\infty}^{\infty} \mathcal{G}_X(f)\, df
\tag{1.4.3}
$$

Using the convolution theorem, we have

$$
\mathcal{G}_X(f) = \mathcal{F}[R_X(\tau)]
\tag{1.4.4}
$$

where $R_X(\tau)$ is the *autocorrelation function* of $x(t)$ defined as

$$
R_X(\tau) = \int_{-\infty}^{\infty} x(t)x(t+\tau)\, dt
$$
$$
= x(\tau) \star x(-\tau)
\tag{1.4.5}
$$

---

[4]There exist signals that are neither energy type nor power type. One example of such signals is $x(t) = e^t u_{-1}(t)$.

[5]The only exception is those signals that are equal to zero almost everywhere.

for real-valued signals. For power-type signals, we define the *time-average autocorrelation function* as

$$R_X(\tau) = \lim_{T\to\infty} \frac{1}{T} \int_{-T/2}^{T/2} x(t)x(t+\tau)\,dt \tag{1.4.6}$$

and the *power-spectral density* is in general given by

$$S_X(f) = \mathcal{F}[R_X(\tau)] \tag{1.4.7}$$

The total power is the integral of the power-spectral density given by

$$P_X = \int_{-\infty}^{\infty} S_X(f)\,df \tag{1.4.8}$$

For the special case of a periodic signal $x(t)$ with period T_0 and Fourier series coefficients x_n, the power-spectral density is given by

$$S_X(f) = \sum_{n=-\infty}^{\infty} |x_n|^2 \delta\left(f - \frac{n}{T_0}\right) \tag{1.4.9}$$

which means all the power is concentrated at the harmonics of the fundamental frequency and the power at the nth harmonic (n/T_0) is $|x_n|^2$—that is, the magnitude square of the corresponding Fourier series coefficient.

When the signal $x(t)$ passes through a filter with transfer function $H(f)$, the output energy spectral density, or power-spectral density, is obtained via

$$\begin{cases} \mathcal{G}_Y(f) = |H(f)|^2 \mathcal{G}_X(f) \\ S_Y(f) = |H(f)|^2 S_X(f) \end{cases} \tag{1.4.10}$$

If we use the discrete-time (sampled) signal, the energy and power relations equivalents to Equation (1.4.1) in terms of the discrete-time signal become

$$\begin{cases} E_X = T_s \sum_{n=-\infty}^{\infty} x^2[n] \\ P_X = \lim_{N\to\infty} \frac{1}{2N+1} \sum_{n=-N}^{N} x^2[n] \end{cases} \tag{1.4.11}$$

and if the FFT is employed—that is, if the length of the sequence is finite and the sequence is repeated—then

$$\begin{cases} E_X = T_s \sum_{n=0}^{N-1} x^2[n] \\ P_X = \frac{1}{N} \sum_{n=0}^{N-1} x^2[n] \end{cases} \tag{1.4.12}$$

The following MATLAB function power.m gives the power content of a signal vector.

M-FILE

```
function  p=spower(x)
%                    p=spower(x)
%SPOWER         Returns  the  power  in  signal  x
p=(norm(x)^2)/length(x);
```

If $X_d(f)$ is the DFT of the sequence $x[n]$, then the energy spectral density of $x(t)$, the equivalent analog signal, is obtained by using Equation (1.3.23) and is given by

$$\mathscr{G}_X(f) = T_s^2 \, |X_d(f)|^2 \qquad (1.4.13)$$

where T_s is the sampling interval. The power-spectral density of a sequence $x[n]$ is most easily derived by using the MATLAB function spectrum.m.

ILLUSTRATIVE PROBLEM

Illustrative Problem 1.8 [Power and power spectrum] The signal $x(t)$ has a duration of 10 and is the sum of two sinusoidal signals of unit amplitude, one with frequency 47 Hz and the other with frequency 219 Hz:

$$x(t) = \begin{cases} \cos(2\pi \times 47t) + \cos(2\pi \times 219t), & 0 \le t \le 10 \\ 0, & \text{otherwise} \end{cases}$$

This signal is sampled at a sampling rate of 1000 samples per second. Use MATLAB to find the power content and the power-spectral density for this signal.

SOLUTION

With the MATLAB function spower.m the power content of the signal is found to be 1.0003 W. By using spectrum.m and specplot.m, we can plot the power-spectral density of the signal, as shown in Figure 1.25. The twin peaks in the power spectrum correspond to the two frequencies present in the signal.

The MATLAB script for this problem follows.

M-FILE

```
% MATLAB  script  for  Illustrative  Problem  1.8.
ts=0.001;
fs=1/ts;
```

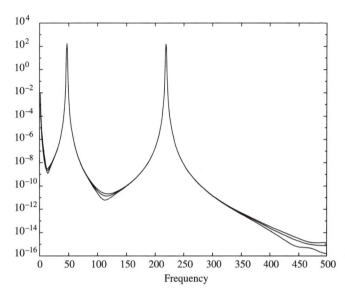

Figure 1.25   The power-spectral density of the signal consisting of two sinusoidal
signals at frequencies $f_1 = 47$ Hz and $f_2 = 219$ Hz

```
t=[0:ts:10];
x=cos(2*pi*47*t)+cos(2*pi*219*t);
p=spower(x);
psd=spectrum(x,1024);
pause    % Press a key to see the power in the signal
p
pause    % Press a key to see the power spectrum
specplot(psd,fs)
```

1.5   Lowpass Equivalent of Bandpass Signals

A bandpass signal is a signal for which all frequency components are located in the neighborhood of a *central frequency* f_0 (and, of course, $-f_0$). In other words, for a bandpass signal $X(f) \equiv 0$ for $|f \pm f_0| > W$, where $W \ll f_0$. A lowpass signal is a signal for which the frequency components are located around the zero frequency; that is, for $|f| > W$, we have $X(f) \equiv 0$.

Corresponding to a bandpass signal $x(t)$ we can define the *analytic signal* $z(t)$, whose Fourier transform is given as

$$Z(f) = 2u_{-1}(f)X(f) \tag{1.5.1}$$

where $u_{-1}(f)$ is the unit step function. In the time domain, this relation is written as

$$z(t) = x(t) + j\hat{x}(t) \tag{1.5.2}$$

where $\hat{x}(t)$ denotes the *Hilbert transform* of $x(t)$ defined as $\hat{x}(t) = x(t) \star (1/\pi t)$; in the frequency domain, it is given by

$$\hat{X}(f) = -j \operatorname{sgn}(f)X(f) \tag{1.5.3}$$

We note that the Hilbert transform function in MATLAB, denoted by hilbert.m, generates the complex sequence $z(t)$. The real part of $z(t)$ is the original sequence, and its imaginary part is the Hilbert transform of the original sequence.

The *lowpass equivalent* of the signal $x(t)$, denoted by $x_l(t)$, is expressed in terms of $z(t)$ as

$$x_l(t) = z(t)e^{-j2\pi f_0 t} \tag{1.5.4}$$

From this relation, we have

$$\begin{cases} x(t) = \operatorname{Re}[x_l(t)e^{j2\pi f_0 t}] \\ \hat{x}(t) = \operatorname{Im}[x_l(t)e^{j2\pi f_0 t}] \end{cases} \tag{1.5.5}$$

In the frequency domain, we have

$$X_l(f) = Z(f + f_0) = 2u_{-1}(f + f_0)X(f + f_0) \tag{1.5.6}$$

and

$$X_l(f) = X(f - f_0) + X^*(-f - f_0) \tag{1.5.7}$$

The lowpass equivalent of a real bandpass signal is, in general, a complex signal. Its real part, denoted by $x_c(t)$, is called the *in-phase component* of $x(t)$, and its imaginary part is called the *quadrature component* of $x(t)$ and is denoted by $x_s(t)$; that is,

$$x_l(t) = x_c(t) + jx_s(t) \tag{1.5.8}$$

In terms of the in-phase and the quadrature components, we have

$$\begin{cases} x(t) = x_c(t) \cos(2\pi f_0 t) - x_s(t) \sin(2\pi f_0 t) \\ \hat{x}(t) = x_s(t) \cos(2\pi f_0 t) + x_c(t) \sin(2\pi f_0 t) \end{cases} \tag{1.5.9}$$

If we express $x_l(t)$ in polar coordinates, we have

$$x_l(t) = V(t)e^{j\Theta(t)} \tag{1.5.10}$$

where $V(t)$ and $\Theta(t)$ are called the *envelope* and the *phase* of the signal $x(t)$. In terms of these two, we have

$$x(t) = V(t) \cos(2\pi f_0 t + \Theta(t)) \tag{1.5.11}$$

The envelope and the phase can be expressed as

$$
\begin{cases}
V(t) = \sqrt{x_c^2(t) + x_s^2(t)} \\[2mm]
\Theta(t) = \arctan \dfrac{x_s(t)}{x_c(t)}
\end{cases}
\tag{1.5.12}
$$

or, equivalently,

$$
\begin{cases}
V(t) = \sqrt{x^2(t) + \hat{x}^2(t)} \\[2mm]
\Theta(t) = \arctan \dfrac{\hat{x}(t)}{x(t)} - 2\pi f_0 t
\end{cases}
\tag{1.5.13}
$$

It is obvious from the preceding relations that the envelope is independent of the choice of f_0, whereas the phase depends on this choice.

We have written some simple MATLAB files to generate the analytic signal, the low-pass representation of a signal, the in-phase and quadrature components, and the envelope and phase. These MATLAB functions are analytic.m, loweq.m, quadcomp.m, and env_phas.m, respectively. A listing of these functions is given next.

M-FILE

```
function z=analytic(x)
%                z=analytic(x)
%ANALYTIC        Returns the analytic signal corresponding to signal x.
%
z=hilbert(x);
```

M-FILE

```
function xl=loweq(x,ts,f0)
%                xl=loweq(x,ts,f0)
%LOWEQ           Returns the lowpass equivalent of the signal x.
%                f0 is the center frequency.
%                ts is the sampling interval.
%
t=[0:ts:ts*(length(x)−1)];
z=hilbert(x);
xl=z.*exp(−j*2*pi*f0*t);
```

M-FILE

```
function [xc,xs]=quadcomp(x,ts,f0)
%                  [xc,xs]=quadcomp(x,ts,f0)
%QUADCOMP   Returns the in-phase and quadrature components of
%           the signal x. f0 is the center frequency. ts is the
%           sampling interval.
%
z=loweq(x,ts,f0);
xc=real(z);
xs=imag(z);
```

M-FILE

```
function [v,phi]=env_phas(x,ts,f0)
%                  [v,phi]=env_phas(x,ts,f0)
%                     v=env_phas(x,ts,f0)
%ENV_PHAS   Returns the envelope and the phase of the bandpass signal x.
%           f0 is the center frequency.
%           ts is the sampling interval.
%
if nargout == 2
  z=loweq(x,ts,f0);
  phi=angle(z);
end
v=abs(hilbert(x));
```

ILLUSTRATIVE PROBLEM

Illustrative Problem 1.9 [Bandpass to lowpass transformation] The signal $x(t)$ is given as

$$x(t) = \text{sinc}(100t)\cos(2\pi \times 200t) \qquad (1.5.14)$$

1. Plot this signal and its magnitude spectrum.

2. With $f_0 = 200$ Hz, find the lowpass equivalent and plot its magnitude spectrum. Plot the in-phase and the quadrature components and the envelope of this signal.

3. Repeat part 2 assuming $f_0 = 100$ Hz.

SOLUTION

Choosing the sampling interval to be $t_s = 0.001$ s, we have a sampling frequency of $f_s = 1/t_s = 1000$ Hz. Choosing a desired frequency resolution of $df = 0.5$ Hz, we have the following.

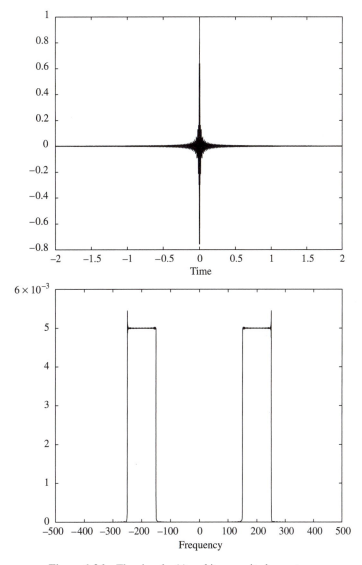

Figure 1.26   The signal $x(t)$ and its magnitude spectrum

1. Plots of the signal and its magnitude spectrum are given in Figure 1.26. Plots are generated by MATLAB.

2. Choosing $f_0 = 200$ Hz, we find the lowpass equivalent to $x(t)$ by using the loweq.m function. Then, using fftseq.m, we obtain its spectrum; we plot its magnitude spectrum in Figure 1.27. It is seen that the magnitude spectrum is an even function in this case because we can write

$$x(t) = \text{Re}[\text{sinc}(100t)e^{j2\pi \times 200t}] \tag{1.5.15}$$

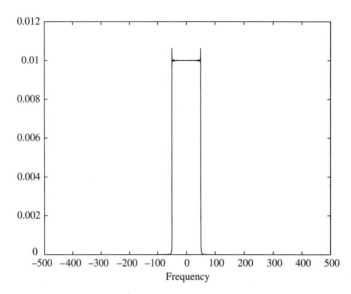

Figure 1.27 The magnitude spectrum of the lowpass equivalent to $x(t)$ in Illustrative
Problem 1.9 when $f_0 = 200$ Hz

Comparing this to

$$x(t) = \mathrm{Re}[x_l(t)e^{j2\pi \times f_0 t}] \tag{1.5.16}$$

we conclude that

$$x_l(t) = \mathrm{sinc}(100t) \tag{1.5.17}$$

which means that the lowpass equivalent signal is a real signal in this case. This, in
turn, means that $x_c(t) = x_l(t)$ and $x_s(t) = 0$. Also, we conclude that

$$\begin{cases} V(t) = |x_c(t)| \\ \Theta(t) = \begin{cases} 0, & x_c(t) \geq 0 \\ \pi, & x_c(t) < 0 \end{cases} \end{cases} \tag{1.5.18}$$

Plots of $x_c(t)$ and $V(t)$ are given in Figure 1.28. Note that choosing f_0 to be the
frequency with respect to which $X(f)$ is symmetric results in these figures.

3. If $f_0 = 100$ Hz, then the above results will not be true in general, and $x_l(t)$ will
   be a complex signal. The magnitude spectrum of the lowpass equivalent signal is
   plotted in Figure 1.29. As it is seen here, the magnitude spectrum lacks the symmetry
   present in the Fourier transform of real signals. Plots of the in-phase component of
   $x(t)$ and its envelope are given in Figure 1.30.

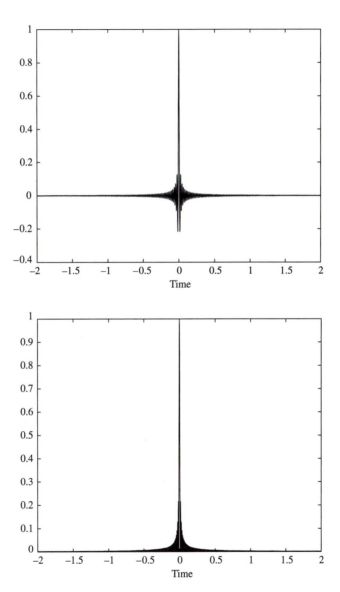

Figure 1.28   The in-phase component and the envelope of $x(t)$

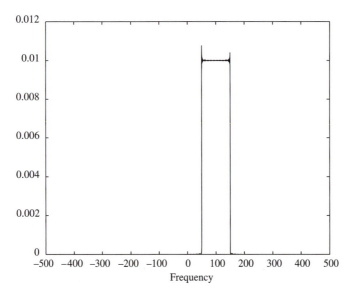

Figure 1.29   The magnitude spectrum of the lowpass equivalent to $x(t)$ in Illustrative
Problem 1.9 when $f_0 = 100$ Hz

Problems

1.1 Consider the periodic signal of Illustrative Problem 1.1 shown in Figure 1.1. Assuming $A = 1$, $T_0 = 10$, and $t_0 = 1$, determine and plot the discrete spectrum of the signal. Compare your results with those obtained in Illustrative Problem 1.1 and justify the differences.

1.2 In Illustrative Problem 1.1, assuming $A = 1$, $T_0 = 4$, and $t_0 = 2$, determine and plot the discrete spectrum of the signal. Compare your results with those obtained in Illustrative Problem 1.1 and justify the differences.

1.3 Using the m-file fseries.m, determine the Fourier series coefficients of the signal shown in Figure 1.1 with $A = 1$, $T_0 = 4$, and $t_0 = \frac{1}{2}$ for $-24 \leq n \leq 24$. Plot the magnitude spectrum of the signal. Now, using Equation (1.2.5), determine the Fourier series coefficients and plot the magnitude spectrum. Why are the results not exactly the same?

1.4 Repeat Problem 1.3 with $T_0 = 4.6$, and compare the results with those obtained by using Equation (1.2.5). Do you observe the same discrepancy between the two results here? Why?

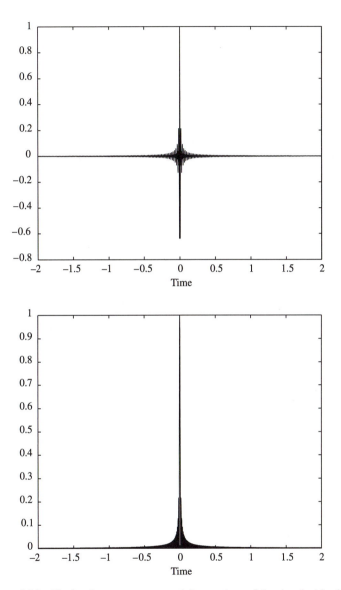

Figure 1.30 The in-phase component and the envelope of the signal $x(t)$ when $f_0 = 100$ Hz

1.5 Using the MATLAB script dis_spct.m, determine and plot the magnitude and phase spectra of the periodic signal $x(t)$ with a period of $T_0 = 4.6$ and described in the interval $[-2.3, 2.3]$ by the relation $x(t) = \Lambda(t)$. Plot the spectra for $-24 \leq n \leq 24$. Now, analytically determine the Fourier series coefficients of the signal, and show that all coefficients are nonnegative real numbers. Does the phase spectrum you plotted earlier agree with this result? If not, explain why.

1.6 In Problem 1.5, define $x(t) = \Lambda(t)$ in the interval $[-1.3, 3.3]$; the period is still $T_0 = 4.6$. Determine and plot the magnitude and phase spectra using the MATLAB script dis_spct.m. Note that the signal is the same as the signal in Problem 1.5. Compare the magnitude and the phase spectra with that obtained in Problem 1.5. Which one shows a more noticeable difference, the magnitude or the phase spectrum? Why?

1.7 Repeat Illustrative Problem 1.2 with $[a, b] = [-4, 4]$ and $x(t) = \cos(\pi t/8)$ for $|t| \leq 4$.

1.8 Repeat Illustrative Problem 1.2 with $[a, b] = [-4, 4]$ and $x(t) = \sin(\pi t/8)$ for $|t| \leq 4$ and compare your results with those of Problem 1.7.

1.9 Numerically determine and plot the magnitude and the phase spectra of a signal $x(t)$ with a period equal to 10^{-6} seconds and defined as

$$x(t) = \begin{cases} -10^6 t + 0.5, & 0 \leq t \leq 5 \times 10^{-7} \\ 0, & \text{otherwise} \end{cases}$$

in the interval $|t| \leq 5 \times 10^{-7}$.

1.10 A periodic signal $x(t)$ with period $T_0 = 6$ is defined by $x(t) = \Pi(t/3)$ for $|t| \leq 3$. This signal passes through an LTI system with an impulse response given by

$$h(t) = \begin{cases} e^{-t/2}, & 0 \leq t \leq 4 \\ 0, & \text{otherwise} \end{cases}$$

Numerically determine and plot the discrete spectrum of the output signal.

1.11 Repeat Problem 1.10 with $x(t) = e^{-2t}$ for $|t| \leq 3$ and

$$h(t) = \begin{cases} 1, & 0 \leq t \leq 4 \\ 0, & \text{otherwise} \end{cases}$$

1.12 Verify the convolution theorem of the Fourier transform for signals $x(t) = \Pi(t)$ and $y(t) = \Lambda(t)$ numerically, once by determining the convolution directly and once by using the Fourier transforms of the two signals.

1.13 Plot the magnitude and the phase spectra of a signal given by

$$x(t) = \begin{cases} 1, & -2 \le t \le -1 \\ |t|, & |t| < 1 \\ 1, & 1 \le t < 2 \\ 0, & \text{otherwise} \end{cases}$$

1.14 Determine and plot the magnitude spectrum of an even signal $x(t)$ that for positive values of t is given as

$$x(t) = \begin{cases} t+1, & 0 \le t \le 1 \\ 2, & 1 \le t \le 2 \\ -t+4, & 2 \le t \le 4 \\ 0, & \text{otherwise} \end{cases}$$

Determine your result both analytically and numerically and compare the results.

1.15 The signal described in Problem 1.14 is passed through an LTI system with an impulse response given by

$$h(t) = \begin{cases} 1, & 0 \le t \le 2 \\ 2, & 2 < t \le 3 \\ 0, & \text{otherwise} \end{cases}$$

Determine the magnitude and the phase spectra of the output signal.

1.16 The signal

$$x(t) = \begin{cases} \cos(2\pi \times 47t) + \cos(2\pi \times 219t), & 0 \le t \le 10 \\ 0, & \text{otherwise} \end{cases}$$

is considered. As in Illustrative Problem 1.8, assume this signal is sampled at a rate of 1000 samples/second. Using the MATLAB m-file butter.m, design a lowpass Butterworth filter of order 4 with a cutoff frequency of 100 Hz, and pass $x(t)$ through this filter. Determine and sketch the output power spectrum, and compare it with Figure 1.25. Now design a Butterworth filter of order 8 with the same cutoff frequency, determine the output of this filter, and plot its power spectrum. Compare your results in these two cases.

1.17 Repeat Problem 1.16, but this time design highpass Butterworth filters with the same orders and the same cutoff frequencies. Plot your results and make the comparisons.

1.18 Consider the signal

$$x(t) = \begin{cases} \cos(2\pi \times 47t) + \cos(2\pi \times 219t), & 0 \le t \le 10 \\ 0, & \text{otherwise} \end{cases}$$

a. Determine the analytic signal corresponding to this signal.

b. Determine and plot the Hilbert transform of this signal.

c. Determine and plot the envelope of this signal.

d. Once assuming $f_0 = 47$ and once assuming $f_0 = 219$, determine the lowpass equivalent and the in-phase and the quadrature components of this signal.

Chapter 2

Random Processes

2.1   Preview

In this chapter, we illustrate methods for generating random variables and samples of random processes. We begin with the description of a method for generating random variables with a specified probability distribution function. Then we consider Gaussian and Gauss-Markov processes and illustrate a method for generating samples of such processes. The third topic that we consider is the characterization of a stationary random process by its autocorrelation in the time domain and by its power spectrum in the frequency domain. Since linear filters play a very important role in communication systems, we also consider the autocorrelation function and the power spectrum of a linearly filtered random process. The final section of this chapter deals with the characteristics of lowpass and bandpass random processes.

2.2   Generation of Random Variables

Random number generators are often used in practice to simulate the effect of noiselike signals and other random phenomena that are encountered in the physical world. Such noise is present in electronic devices and systems and usually limits our ability to communicate over large distances and to detect relatively weak signals. By generating such noise on a computer, we are able to study its effects through simulation of communication systems and to assess the performance of such systems in the presence of noise.

Most computer software libraries include a uniform random number generator. Such a random number generator generates a number between 0 and 1 with equal probability. We call the output of the random number generator a *random variable*. If A denotes such a random variable, its range is the interval $0 \leq A \leq 1$.

We know that the numerical output of a digital computer has limited precision, and as a consequence, it is impossible to represent the continuum of numbers in the interval $0 \leq A \leq 1$. However, we may assume that our computer represents each output by a large number of bits in either fixed point or floating point. Consequently, for all practical

47

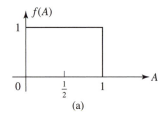 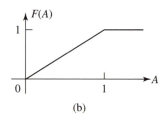

(a)   (b)

Figure 2.1   Probability density function $f(A)$ and the probability distribution function $F(A)$ of a uniformly distributed random variable A

purposes, the number of outputs in the interval $0 \le A \le 1$ is sufficiently large so that we are justified in assuming that any value in the interval is a possible output from the generator.

The uniform probability density function for the random variable A, denoted as $f(A)$, is illustrated in Figure 2.1(a). We note that the average value or mean value of A, denoted as m_A, is $m_A = \frac{1}{2}$. The integral of the probability density function, which represents the area under $f(A)$, is called the *probability distribution function* of the random variable A and is defined as

$$F(A) = \int_{-\infty}^{A} f(x)\,dx \tag{2.2.1}$$

For any random variable, this area must always be unity, which is the maximum value that can be achieved by a distribution function. Hence, for the uniform random variable A we have

$$F(1) = \int_{-\infty}^{1} f(x)\,dx = 1 \tag{2.2.2}$$

and the range of $F(A)$ is $0 \le F(A) \le 1$ for $0 \le A \le 1$. The probability distribution function is shown in Figure 2.1(b).

If we wish to generate uniformly distributed noise in an interval $(b, b + 1)$, it can be accomplished simply by using the output A of the random number generator and shifting it by an amount b. Thus, a new random variable B can be defined as

$$B = A + b \tag{2.2.3}$$

which now has a mean value $m_B = b + \frac{1}{2}$. For example, if $b = -\frac{1}{2}$, the random variable B is uniformly distributed in the interval $(-\frac{1}{2}, \frac{1}{2})$, as shown in Figure 2.2(a). Its probability distribution function $F(B)$ is shown in Figure 2.2(b).

A uniformly distributed random variable in the range $(0,1)$ can be used to generate random variables with other probability distribution functions. For example, suppose that we wish to generate a random variable C with probability distribution function $F(C)$, as illustrated in Figure 2.3. Since the range of $F(C)$ is the interval $(0,1)$, we begin by

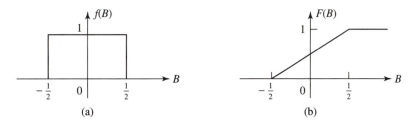

Figure 2.2   Probability density function and the probability distribution function of a zero-mean uniformly distributed random variable

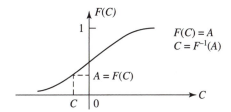

Figure 2.3   Inverse mapping from the uniformly distributed random variable A to the new random variable C

generating a uniformly distributed random variable A in the range $(0,1)$. If we set

$$F(C) = A \tag{2.2.4}$$

then

$$C = F^{-1}(A) \tag{2.2.5}$$

Thus, we solve (2.2.4) for C, and the solution in (2.2.5) provides the value of C for which $F(C) = A$. By this means, we obtain a new random variable C with probability distribution function $F(C)$. This inverse mapping from A to C is illustrated in Figure 2.3.

ILLUSTRATIVE PROBLEM

Illustrative Problem 2.1   Generate a random variable C that has the linear probability density function shown in Figure 2.4(a); that is,

$$f(C) = \begin{cases} \frac{1}{2}C, & 0 \le C \le 2 \\ 0, & \text{otherwise} \end{cases}$$

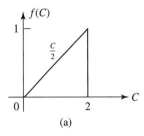 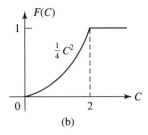

(a)　　　　　　　　　　　　　　　(b)

Figure 2.4　Linear probability density function and the corresponding probability distribution function

---

SOLUTION

This random variable has a probability distribution function

$$F(C) = \begin{cases} 0, & C < 0 \\ \frac{1}{4}C^2, & 0 \le C \le 2 \\ 1, & C > 2 \end{cases}$$

which is illustrated in Figure 2.4(b). We generate a uniformly distributed random variable A and set $F(C) = A$. Hence,

$$F(C) = \frac{1}{4}C^2 = A \qquad (2.2.6)$$

Upon solving for C, we obtain

$$C = 2\sqrt{A} \qquad (2.2.7)$$

Thus, we generate a random variable C with probability distribution function $F(C)$, as shown in Figure 2.4(b).

In Illustrative Problem 2.1, the inverse mapping $C = F^{-1}(A)$ was simple. In some cases, it is not. This problem arises in trying to generate random numbers that have a normal distribution function.

Noise encountered in physical systems is often characterized by the normal, or Gaussian, probability distribution, which is illustrated in Figure 2.5. The probability density function is given by

$$f(C) = \frac{1}{\sqrt{2\pi}\,\sigma}e^{-C^2/2\sigma^2}, \qquad -\infty < C < \infty \qquad (2.2.8)$$

where σ^2 is the variance of C, which is a measure of the spread of the probability density function $f(C)$. The probability distribution function $F(C)$ is the area under $f(C)$ over the range $(-\infty, C)$. Thus,

$$F(C) = \int_{-\infty}^{C} f(x)\,dx \qquad (2.2.9)$$

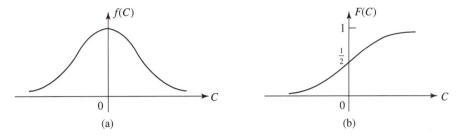

Figure 2.5   Gaussian probability density function and the corresponding probability
distribution function

Unfortunately, the integral in (2.2.9) cannot be expressed in terms of simple functions.
Consequently, the inverse mapping is difficult to achieve. A way has been found to circum-
vent this problem. From probability theory, it is known that a Rayleigh distributed random
variable R, with probability distribution function

$$F(R) = \begin{cases} 0, & R < 0 \\ 1 - e^{-R^2/2\sigma^2}, & R \geq 0 \end{cases} \tag{2.2.10}$$

is related to a pair of Gaussian random variables C and D through the transformation

$$C = R \cos \Theta \tag{2.2.11}$$
$$D = R \sin \Theta \tag{2.2.12}$$

where Θ is a uniformly distributed variable in the interval $(0, 2\pi)$. The parameter σ^2 is the
variance of C and D. Since (2.2.10) is easily inverted, we have

$$F(R) = 1 - e^{-R^2/2\sigma^2} = A \tag{2.2.13}$$

and hence,

$$R = \sqrt{2\sigma^2 \ln\left(\frac{1}{1-A}\right)} \tag{2.2.14}$$

where A is a uniformly distributed random variable in the interval $(0,1)$. Now, if we generate a second uniformly distributed random variable B and define

$$\Theta = 2\pi B \tag{2.2.15}$$

then from (2.2.11) and (2.2.12), we obtain two statistically independent Gaussian distributed random variables C and D.

The method described above is often used in practice to generate Gaussian distributed
random variables. As shown in Figure 2.5, these random variables have a mean value of
zero and a variance σ^2. If a nonzero mean Gaussian random variable is desired, then C
and D can be translated by the addition of the mean value.

The MATLAB script that implements the preceding method for generating Gaussian distributed random variables is given next.

M-FILE

```
function [gsrv1,gsrv2]=gngauss(m,sgma)
%    [gsrv1,gsrv2]=gngauss(m,sgma)
%    [gsrv1,gsrv2]=gngauss(sgma)
%    [gsrv1,gsrv2]=gngauss
%               GNGAUSS   generates two independent Gaussian random variables with mean
%               m and standard deviation sgma. If one of the input arguments is missing,
%               it takes the mean as 0.
%               If neither the mean nor the variance is given, it generates two standard
%               Gaussian random variables.
if nargin == 0,
  m=0; sgma=1;
elseif nargin == 1,
  sgma=m; m=0;
end;
u=rand;                              % a uniform random variable in (0,1)
z=sgma*(sqrt(2*log(1/(1−u))));       % a Rayleigh distributed random variable
u=rand;                              % another uniform random variable in (0,1)
gsrv1=m+z*cos(2*pi*u);
gsrv2=m+z*sin(2*pi*u);
```

2.3   Gaussian and Gauss-Markov Processes

Gaussian processes play an important role in communication systems. The fundamental reason for their importance is that thermal noise in electronic devices, which is produced by random movement of electrons due to thermal agitation, can be closely modeled by a Gaussian process. The reason for the Gaussian behavior of thermal noise is that the current introduced by movement of electrons in an electric circuit can be regarded as the sum of small currents of a very large number of sources—namely, individual electrons. It can be assumed that at least a majority of these sources behave independently, and therefore, the total current is the sum of a large number of independent and identically distributed (i.i.d.) random variables. When the central limit theorem is applied, this total current has a Gaussian distribution.

Apart from thermal noise, Gaussian processes provide rather good models for some information sources as well. Some interesting properties of the Gaussian processes, which are given below, make these processes mathematically tractable and easy to deal with. We begin with a formal definition of a Gaussian process.

Definition: A random process $X(t)$ is a *Gaussian process* if for all n and all $(t_1, t_2, \ldots, t_n)$, the random variables $\{X(t_1)\}_{i=1}^n$ have a jointly Gaussian density function, which may be expressed as

$$f(x) = \frac{1}{(2\pi)^{n/2}[\det(C)]^{1/2}} \exp\left[-\frac{1}{2}(x-m)^t C^{-1}(x-m)\right] \qquad (2.3.1)$$

where the vector $x = (x_1, x_2, \ldots, x_n)^t$ denotes the n random variables $x_i \equiv X(t_i)$, m is the mean value vector; that is, $m = E(X)$, and C is the $n \times n$ covariance matrix of the random variables $(x_1, x_2, \ldots, x_n)$ with elements

$$c_{ij} = E[(x_i - m_i)(x_j - m_j)] \qquad (2.3.2)$$

The superscript t denotes the transpose of a vector or a matrix, and C^{-1} is the inverse of the covariance matrix C.

From the definition it is seen, in particular, that at any time instant t_0 the random variable $X(t_0)$ is Gaussian, and at any two points t_1, t_2 the random variables $(X(t_1), X(t_2))$ are distributed according to a two-dimensional Gaussian random variable. Moreover, since a complete statistical description of $\{X(t_1)\}_{i=1}^n$ depends on only the mean vector m and the covariance matrix C, we have the following property.

Property 1: For Gaussian processes, knowledge of the mean m and covariance C provides a complete statistical description of the process.

Another very important property of a Gaussian process is concerned with its characteristics when passed through a linear time-invariant system. This property may be stated as follows.

Property 2: If the Gaussian process $X(t)$ is passed through a linear time-invariant (LTI) system, the output of the system is also a Gaussian process. The effect of the system on $X(t)$ is simply reflected by a change in the mean value and the covariance of $X(t)$.

───(ILLUSTRATIVE PROBLEM)──────────────────────────────

Illustrative Problem 2.2 [Generation of samples of a multivariate Gaussian process]
Generate samples of a multivariate Gaussian random process $X(t)$ having a specified mean value m_x and a covariance C_x.

SOLUTION

First, we generate a sequence of n statistically independent, zero-mean and unit-variance Gaussian random variables by using the method described in Section 2.2. Let us denote this sequence of n samples by the vector $Y = (y_1, y_2, \ldots, y_n)^t$. Second, we factor the desired $n \times n$ covariance matrix C_x as

$$C_x = C_x^{1/2}(C_x^{1/2})^t \tag{2.3.3}$$

Then, we define the linearly transformed $(n \times 1)$ vector X as

$$X = C_x^{1/2}Y + m_x \tag{2.3.4}$$

Thus, the covariance of X is

$$\begin{aligned} C_x &= E[(X - m_x)(X - m_x)^t] \\ &= E[C_x^{1/2}YY^t(C_x^{1/2})^t] \\ &= C_x^{1/2}E(YY^t)(C_x^{1/2})^t \\ &= C_x^{1/2}(C_x^{1/2})^t \end{aligned} \tag{2.3.5}$$

The most difficult step in this process is the factorization of the covariance matrix C_x. Let us demonstrate this procedure by means of an example that employs the bivariate Gaussian distribution. Suppose we begin with a pair of statistically independent Gaussian random variables y_1 and y_2, which have zero mean and unit variance. We wish to transform these into a pair of Gaussian random variables x_1 and x_2 with mean $m = 0$ and covariance matrix

$$\begin{aligned} C &= \begin{bmatrix} \sigma_1^2 & \rho\sigma_1\sigma_2 \\ \rho\sigma_1\sigma_2 & \sigma_2^2 \end{bmatrix} \\ &= \begin{bmatrix} 1 & \frac{1}{2} \\ \frac{1}{2} & 1 \end{bmatrix} \end{aligned} \tag{2.3.6}$$

where σ_1^2 and σ_2^2 are the variances of x_1 and x_2, respectively, and ρ is the normalized covariance, defined as

$$\rho = \frac{E[(X_1 - m_1)(X_2 - m_2)]}{\sigma_1\sigma_2} = \frac{c_{12}}{\sigma_1\sigma_2} \tag{2.3.7}$$

The covariance matrix C can be factored as

$$C = C^{1/2}(C^{1/2})^t$$

where

$$C^{1/2} = \frac{1}{2\sqrt{2}}\begin{bmatrix} \sqrt{3}+1 & \sqrt{3}-1 \\ \sqrt{3}-1 & \sqrt{3}+1 \end{bmatrix} \tag{2.3.8}$$

Therefore,

$$X = \begin{bmatrix} x_1 \\ x_2 \end{bmatrix}$$

$$= C^{1/2} \begin{bmatrix} y_1 \\ y_2 \end{bmatrix}$$

$$= \frac{1}{2\sqrt{2}} \begin{bmatrix} \sqrt{3}+1 & \sqrt{3}-1 \\ \sqrt{3}-1 & \sqrt{3}+1 \end{bmatrix} \begin{bmatrix} y_1 \\ y_2 \end{bmatrix}$$

$$= \frac{1}{2\sqrt{2}} \begin{bmatrix} (\sqrt{3}+1)y_1 + (\sqrt{3}-1)y_2 \\ (\sqrt{3}-1)y_1 + (\sqrt{3}+1)y_2 \end{bmatrix} \qquad (2.3.9)$$

The MATLAB scripts for this computation are given next.

M-FILE

```
% MATLAB script for Illustrative Problem 2.2.
echo on
mx=[0 0]';
Cx=[1 1/2;1/2 1];
x=multi_gp(mx,Cx);
% Computation of the pdf of (x1,x2) follows
delta=0.3;
x1=-3:delta:3;
x2=-3:delta:3;
for i=1:length(x1),
    for j=1:length(x2),
        f(i,j)=(1/((2*pi)*det(Cx)^1/2))*exp((-1/2)*(([x1(i) x2(j)]-mx')*inv(Cx)*([x1(i);x2(j)]-mx)));
        echo off ;
    end;
end;
echo on ;
% plotting command for pdf follows
mesh(x1,x2,f);
```

M-FILE

```
function [x] = multi_gp(m,C)
%   [x]=multi_gp(m,C)
%                MULTI_GP   generates a multivariate Gaussian random
%                process with mean vector m (column vector) and covariance matrix C.
N=length(m);
for i=1:N,
   y(i)=gngauss;
end;
y=y.';
x=sqrtm(C)*y+m;
```

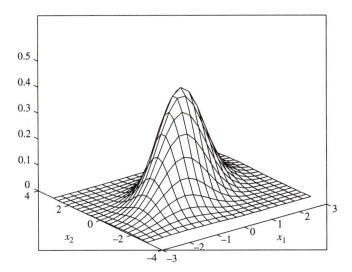

Figure 2.6 Joint probability density function of x_1 and x_2

Figure 2.6 illustrates the joint pdf $f(x_1, x_2)$ for the covariance matrix C given by (2.3.6).

As indicated, the most difficult step in the computation is determining $C^{1/2}$. Given the desired covariance matrix, we may determine the eigenvalues $\{\lambda_k, 1 \leq k \leq n\}$ and the corresponding eigenvectors $\{v_k, 1 \leq k \leq n\}$. Then the covariance matrix C can be expressed as

$$C = \sum_{k=1}^{n} \lambda_k v_k v_k^t \tag{2.3.10}$$

and since $C = C^{1/2}(C^{1/2})^t$, it follows that

$$C^{1/2} = \sum_{k=1}^{n} \lambda_k^{1/2} v_k v_k^t \tag{2.3.11}$$

Definition: A *Markov process* $X(t)$ is a random process whose past has no influence on the future if its present is specified; that is, if $t_n > t_{n-1}$, then

$$P\left[X(t_n) \leq x_n \mid X(t), \quad t \leq t_{n-1}\right] = P\left[X(t_n) \leq x_n \mid X(t_{n-1})\right] \tag{2.3.12}$$

From this definition, it follows that if $t_1 < t_2 < \cdots < t_n$, then

$$P\left[X(t_n) \leq x_n \mid X(t_{n-1}), X(t_{n-2}), \ldots, X(t_1)\right] = P\left[X(t_n) \leq x_n \mid X(t_{n-1})\right] \tag{2.3.13}$$

Definition: A *Gauss-Markov process* $X(t)$ is a Markov process whose probability density function is Gaussian.

The simplest method for generating a Markov process is by means of the simple recursive formula

$$X_n = \rho X_{n-1} + w_n \tag{2.3.14}$$

when w_n is a sequence of zero-mean i.i.d. (white) random variables and ρ is a parameter that determines the degree of correlation between X_n and X_{n-1}; that is,

$$E(X_n X_{n-1}) = \rho E(X_{n-1}^2) = \rho \sigma_{n-1}^2 \tag{2.3.15}$$

If the sequence $\{w_n\}$ is Gaussian, then the resulting process $X(t)$ is a Gauss-Markov process.

ILLUSTRATIVE PROBLEM

Illustrative Problem 2.3 [The Gauss-Markov Process] Generate a sequence of 1000 (equally spaced) samples of a Gauss-Markov process from the recursive relation

$$X_n = 0.95 X_{n-1} + w_n, \qquad n = 1, 2, \ldots, 1000 \tag{2.3.16}$$

where $X_0 = 0$ and $\{w_n\}$ is a sequence of zero-mean and unit-variance i.i.d. Gaussian random variables. Plot the sequence $\{X_n, \quad 1 \le n \le 1000\}$ as a function of the time index n and the autocorrelation

$$R_x(m) = \frac{1}{N-m} \sum_{n=1}^{N-m} X_n X_{n+m}, \qquad m = 0, 1, \ldots, 50 \tag{2.3.17}$$

where $N = 1000$.

SOLUTION

The MATLAB scripts for this computation are given next. Figures 2.7 and 2.8 illustrate the sequence (X_n) and the autocorrelation function $\hat{R}_x(m)$, respectively.

M-FILE

```
% MATLAB script for Illustrative Problem 2.3.
echo on
rho=0.95;
X0=0;
N=1000;
X=gaus_mar(X0,rho,N);
M=50;
Rx=Rx_est(X,M);
% plotting commands follow
```

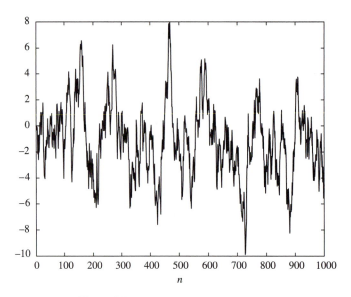

Figure 2.7   The Gauss-Markov sequence

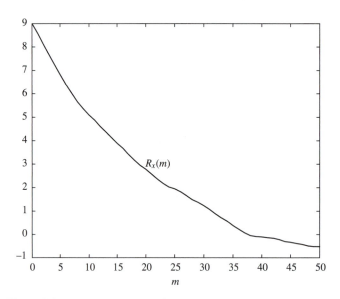

Figure 2.8   The autocorrelation function of the Gauss-Markov process

M-FILE

```
function [X]=gaus_mar(X0,rho,N)
%  [X]=gaus_mar(X0,rho,N)
%
%              GAUS_MAR  generates a Gauss-Markov process of length N.
%              The noise process is taken to be white Gaussian
%              noise with zero mean and unit variance.
for  i=1:2:N,
   [Ws(i) Ws(i+1)]=gngauss;        % Generate the noise process
end;
X(1)=rho*X0+Ws(1);                 % First element in the Gauss-Markov process
for  i=2:N,
   X(i)=rho*X(i−1)+Ws(i);          % The remaining elements
end;
```

2.4   Power Spectrum of Random Processes and White Processes

A stationary random process $X(t)$ is characterized in the frequency domain by its power spectrum $S_x(f)$, which is the Fourier transform of the autocorrelation function $R_x(\tau)$ of the random process; that is,

$$S_x(f) = \int_{-\infty}^{\infty} R_x(\tau)e^{-j2\pi f \tau}\, dt \qquad (2.4.1)$$

Conversely, the autocorrelation function $R_x(\tau)$ of a stationary random process $X(t)$ is obtained from the power spectrum $S_x(f)$ by means of the inverse Fourier transform; that is,

$$R_x(\tau) = \int_{-\infty}^{\infty} S_x(f)e^{j2\pi f \tau}\, df \qquad (2.4.2)$$

In modeling thermal noise that is generated in electronic devices used in the implementation of communication systems, we often assume that such noise is a *white random process*. Such a process is defined as follows.

Definition: A random process $X(t)$ is called a *white process* if it has a flat power spectrum— that is, if $S_x(f)$ is a constant for all f.

As indicated, the importance of white processes stems from the fact that thermal noise can be closely modeled as spectrally constant over a wide range of frequencies. Also, a number of processes that are used to describe a variety of information sources are modeled as the output of LTI systems driven by a white process.

We observe, however, that if $S_x(f) = C$ for all f, then

$$\int_{-\infty}^{\infty} S_x(f)\, df = \int_{-\infty}^{\infty} C\, df = \infty \qquad (2.4.3)$$

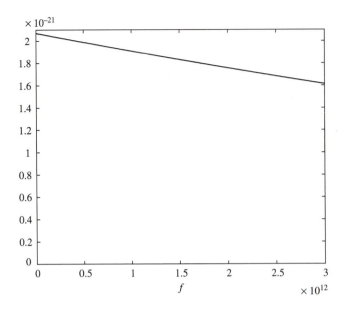

Figure 2.9   Plot of $\mathcal{S}_n(f)$ in (2.4.4)

so that the total power is infinite. Obviously, no real physical process can have infinite power and, therefore, a white process may not be a meaningful physical process. However, quantum mechanical analysis of the thermal noise shows that it has a power-spectral density given by

$$\mathcal{S}_n(f) = \frac{hf}{2(e^{hf/kT} - 1)} \tag{2.4.4}$$

in which h denotes *Planck's constant* (equal to 6.6×10^{-34}J $\times$ s) and k is *Boltzmann's constant* (equal to 1.38×10^{-23}J/K). T denotes the temperature in kelvins. This power spectrum is shown in Figure 2.9.

   This spectrum achieves its maximum at $f = 0$, and the value of this maximum is $kT/2$. The spectrum goes to 0 as f goes to infinity, but the rate of convergence to 0 is very low. For instance, at room temperature ($T = 300$K), $\mathcal{S}_n(f)$ drops to 90% of its maximum at about $f \approx 2 \times 10^{12}$ Hz, which is beyond the frequencies employed in conventional communication systems. From this we conclude that thermal noise, although not precisely white, can be modeled for all practical purposes as a white process with the power spectrum equaling $kT/2$. The value kT is usually denoted by N_0; therefore, the power-spectral density of thermal noise is usually given as $\mathcal{S}_n(f) = N_0/2$ and is sometimes referred to as the *two-sided power-spectral density,* emphasizing that this spectrum extends to both positive and negative frequencies. We will avoid this terminology throughout and simply use *power spectrum* or *power-spectral density.*

For a white random process $X(t)$ with power spectrum $S_x(f) = N_0/2$, the autocorrelation function $R_x(\tau)$ is

$$R_x(\tau) = \int_{-\infty}^{\infty} S_x(f)e^{j2\pi f\tau}\,df = \frac{N_0}{2}\int_{-\infty}^{\infty} e^{j2\pi f\tau}\,df = \frac{N_0}{2}\delta(\tau) \qquad (2.4.5)$$

where $\delta(\tau)$ is the unit impulse. Consequently for all $\tau \neq 0$, we have $R_x(\tau) = 0$; that is, if we sample a white process at two points t_1 and t_2 $(t_1 \neq t_2)$, the resulting random variables will be uncorrelated. If, in addition to being white, the random process is Gaussian, the sampled random variables will be statistically independent Gaussian random variables.

ILLUSTRATIVE PROBLEM

Illustrative Problem 2.4 [Autocorrelation and Power Spectrum] Generate a discrete-time sequence of $N = 1000$ i.i.d. uniformly distributed random numbers in the interval $(-\frac{1}{2}, \frac{1}{2})$ and compute the autocorrelation of the sequence $\{X_n\}$, defined as

$$R_x(m) = \frac{1}{N-m}\sum_{n=1}^{N-m} X_n X_{n+m}, \qquad m = 0, 1, \ldots, M$$

$$= \frac{1}{N-|m|}\sum_{n=|m|}^{N} X_n X_{n+m}, \qquad m = -1, -2, \ldots, -M \qquad (2.4.6)$$

Also, determine the power spectrum of the sequence $\{X_n\}$ by computing the discrete Fourier transform (DFT) of $R_x(m)$. The DFT, which is efficiently computed by use of the fast Fourier transform (FFT) algorithm, is defined as

$$S_x(f) = \sum_{m=-M}^{M} R_x(m)e^{-j2\pi fm/(2M+1)} \qquad (2.4.7)$$

SOLUTION

The MATLAB script that implements the generation of the sequence $\{X_n\}$, the computation of the autocorrelation, and the computation of the power spectrum $S_x(f)$ is given next. We should note that the autocorrelation function and the power spectrum exhibit a significant variability. Therefore, it is necessary to average the sample autocorrelation over several realizations. Figures 2.10 and 2.11 illustrate $R_x(m)$ and $S_x(f)$ obtained by running this program using average autocorrelation over ten realizations of the random process.

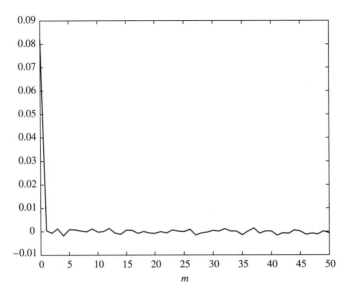

Figure 2.10   The autocorrelation function in Illustrative Problem 2.4

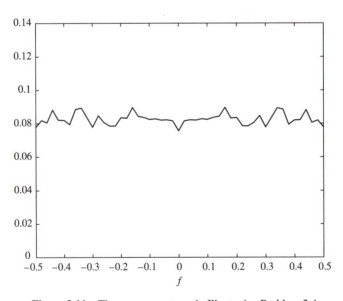

Figure 2.11   The power spectrum in Illustrative Problem 2.4

◖ **M-FILE** ◗

```
% MATLAB script for Illustrative Problem 2.4.
echo on
N=1000;
M=50;
Rx_av=zeros(1,M+1);
Sx_av=zeros(1,M+1);
for j=1:10,                          % take the ensemble average over ten realizations
    X=rand(1,N)−1/2;                 % N i.i.d. uniformly distributed random variables
                                     % between -1/2 and 1/2
    Rx=Rx_est(X,M);                  % autocorrelation of the realization
    Sx=fftshift(abs(fft(Rx)));       % power spectrum of the realization
    Rx_av=Rx_av+Rx;                  % sum of the autocorrelations
    Sx_av=Sx_av+Sx;                  % sum of the spectrums
    echo off ;
end;
echo on ;
Rx_av=Rx_av/10;                      % ensemble average autocorrelation
Sx_av=Sx_av/10;                      % ensemble average spectrum
% Plotting comments follow
```

A bandlimited random process $X(t)$ has the power spectrum

$$S_x(f) = \begin{cases} \frac{N_0}{2}, & |f| \le B \\ 0, & |f| > B \end{cases} \tag{2.4.8}$$

Let us determine its autocorrelation function. From (2.4.1), we have

$$R_x(\tau) = \int_{-B}^{B} \frac{N_0}{2} e^{j2\pi f\tau} \, df$$

$$= N_0 B \left(\frac{\sin 2\pi B\tau}{2\pi B\tau} \right) \tag{2.4.9}$$

Figure 2.12 illustrates $R_x(\tau)$.

MATLAB may be used to compute $R_x(\tau)$ from $S_x(f)$ and vice versa. The fast Fourier transform (FFT) algorithm may be used for this computation.

◖ **ILLUSTRATIVE PROBLEM** ◗

Illustrative Problem 2.5 [Autocorrelation and Power Spectrum] Compute the autocorrelation $R_x(\tau)$ for the random process whose power spectrum is given by (2.4.8).

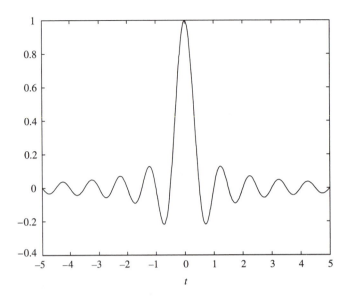

Figure 2.12   Plot of autocorrelation function $R_x(\tau)$ given by (2.4.9) for $B = N_0 = 1$

SOLUTION

To perform the computation, we represent $S_x(f)$ by N samples in the frequency range $|f| \leq B$, with each sample normalized to unity. The result of computing the inverse FFT with $N = 32$ is illustrated in Figure 2.13. Note that we obtain only a coarse representation of the autocorrelation function $R_x(\tau)$, because we sampled $S_x(f)$ only in the frequency range $|f| \leq B$. The frequency separation in this example is $\Delta f = 2B/N$. If we keep Δf fixed and increase the number of samples by including samples for $|f| > B$, we obtain intermediate values of $R_x(\tau)$. Figure 2.14 illustrates the result of computing the inverse FFT with $N_1 = 256$ samples, of which $N = 32$ are unity.

2.5   Linear Filtering of Random Processes

Suppose that a stationary random process $X(t)$ is passed through a linear time-invariant filter that is characterized in the time domain by its impulse response $h(t)$ and in the frequency domain by its frequency response

$$H(f) = \int_{-\infty}^{\infty} h(t)e^{-j2\pi ft}\, dt \tag{2.5.1}$$

It follows that the output of the linear filter is the random process

$$Y(t) = \int_{-\infty}^{\infty} X(t)h(t - \tau)\, d\tau \tag{2.5.2}$$

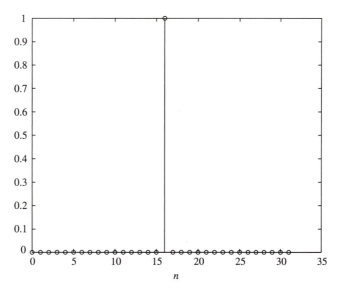

Figure 2.13 Inverse FFT of the power spectrum of the bandlimited random process in Illustrative Problem 2.5 with 32 samples

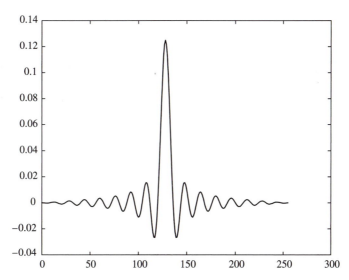

Figure 2.14 Inverse FFT of the power spectrum of the bandlimited random process in Illustrative Problem 2.5 with 256 samples

The mean value of $Y(t)$ is

$$m_y \equiv E[Y(t)]$$

$$= \int_{-\infty}^{\infty} E[X(\tau)]h(t - \tau)\,d\tau$$

$$= m_x \int_{-\infty}^{\infty} h(t - \tau)\,d\tau$$

$$= m_x \int_{-\infty}^{\infty} h(\tau)\,d\tau$$

$$= m_x H(0) \qquad (2.5.3)$$

where $H(0)$ is the frequency response $H(f)$ of the filter evaluated at $f = 0$.

The autocorrelation function of $Y(t)$ is

$$R_y(\tau) = E[Y(\tau)Y(t + \tau)]$$

$$= \int_{-\infty}^{\infty}\int_{-\infty}^{\infty} E[X(\tau)X(\alpha)]h(t - \tau)h(t + \tau - \alpha)\,d\tau\,d\alpha$$

$$= \int_{-\infty}^{\infty}\int_{-\infty}^{\infty} R_x(\tau - \alpha)h(t - \tau)h(t + \tau - \alpha)\,d\tau\,d\alpha \qquad (2.5.4)$$

In the frequency domain, the power spectrum of the output process $Y(t)$ is related to the power spectrum of the input process $X(t)$ and the frequency response of the linear filter by the expression

$$S_y(f) = S_x(f)|H(f)|^2 \qquad (2.5.5)$$

This is easily shown by taking the Fourier transform of (2.5.4).

─────**ILLUSTRATIVE PROBLEM**─────────────────────

Illustrative Problem 2.6 [Filtered Noise] Suppose that a white random process $X(t)$ with power spectrum $S_x(f) = 1$ for all f excites a linear filter with impulse response

$$h(t) = \begin{cases} e^{-t}, & t \geq 0 \\ 0, & t < 0 \end{cases} \qquad (2.5.6)$$

Determine the power spectrum $S_y(f)$ of the filter output.

─────**SOLUTION**─────────────────────────

The frequency response of the filter is easily shown to be

$$H(f) = \frac{1}{1 + j2\pi f} \qquad (2.5.7)$$

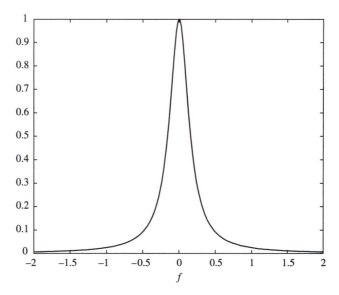

Figure 2.15   Plot of $\mathcal{S}_y(f)$ given by (2.5.8)

Hence,

$$\mathcal{S}_y(f) = |H(f)|^2$$

$$= \frac{1}{1 + (2\pi f)^2} \tag{2.5.8}$$

The graph of $\mathcal{S}_y(f)$ is illustrated in Figure 2.15. The MATLAB script for computing $\mathcal{S}_y(f)$ for a specified $\mathcal{S}_x(f)$ and $H(f)$ is given next.

M-FILE

```
% MATLAB script for Illustrative Problem 2.6.
echo on
delta=0.01;
F_min=-2;
F_max=2;
f=F_min:delta:F_max;
Sx=ones(1,length(f));
H=1./(1+(2*pi*f).^2);
Sy=Sx.*H.^2;
```

---**(ILLUSTRATIVE PROBLEM)**-------------------

Illustrative Problem 2.7 [Autocorrelation and Power Spectrum] Compute the autocorrelation function $R_y(\tau)$ corresponding to $S_y(f)$ in Illustrative Problem 2.6 for the specified $S_x(f) = 1$.

---**(SOLUTION)**-------------------

In this case, we may use the inverse FFT algorithm on samples of $S_y(f)$ given by (2.5.8). Figure 2.16 illustrates this computation with $N = 256$ frequency samples and a frequency separation $\Delta f = 0.1$. The MATLAB script for this computation is given next.

---**(M-FILE)**-------------------

```
% MATLAB script for Illustrative Problem 2.7.
echo on
N=256;                              % number of samples
deltaf=0.1;                         % frequency separation
f=[0:deltaf:(N/2)*deltaf, −(N/2−1)*deltaf:deltaf:−deltaf];
                                    % swap the first half
Sy=1./(1+(2*pi*f).^2);             % sampled spectrum
Ry=ifft(Sy);                        % autocorrelation of Y
% plotting command follows
plot(fftshift(real(Ry)));
```

Let us now consider the equivalent discrete-time problem. Suppose that a stationary random process $X(t)$ is sampled and the samples are passed through a discrete-time linear filter with impulse response $h(n)$. The output of the linear filter is given by the convolution sum formula

$$Y(n) = \sum_{k=0}^{\infty} h(k)X(n-k) \tag{2.5.9}$$

where $X(n) \equiv X(t_n)$ are discrete-time values of the input random process and $Y(n)$ is the output of the discrete-time filter. The mean value of the output process is

$$m_y = E[Y(n)]$$
$$= \sum_{k=0}^{\infty} h(k)E[X(n-k)]$$
$$= m_x \sum_{k=0}^{\infty} h(k)$$
$$= m_x H(0) \tag{2.5.10}$$

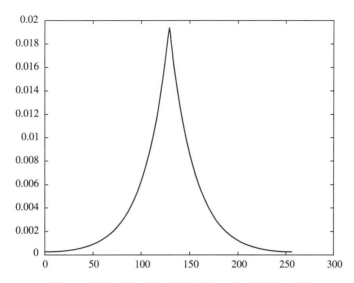

Figure 2.16 Plot of $R_y(\tau)$ in Illustrative Problem 2.7

where $H(0)$ is the frequency response $H(f)$ of the filter evaluated at $f = 0$ and

$$H(f) = \sum_{n=0}^{\infty} h(n)e^{-j2\pi fn} \tag{2.5.11}$$

The autocorrelation function of the output process is

$$
\begin{aligned}
R_y(m) &= E[Y(n)Y(n+m)] \\
&= \sum_{k=0}^{\infty}\sum_{l=0}^{\infty} h(k)h(l)E[X(n-k)X(n+m-l)] \\
&= \sum_{k=0}^{\infty}\sum_{l=0}^{\infty} h(k)h(l)R_x(m-l+k) \tag{2.5.12}
\end{aligned}
$$

The corresponding expression in the frequency domain is

$$S_y(f) = S_x(f)|H(f)|^2 \tag{2.5.13}$$

where the power spectra are defined as

$$S_x(f) = \sum_{m=-\infty}^{\infty} R_x(m)e^{-j2\pi fm} \tag{2.5.14}$$

and

$$S_y(f) = \sum_{m=-\infty}^{\infty} R_y(m)e^{-j2\pi fm} \tag{2.5.15}$$

─────(ILLUSTRATIVE PROBLEM)─────────────────────────────────

Illustrative Problem 2.8 [Filtered White Noise] Suppose that a white random process
with samples $\{X(n)\}$ is passed through a linear filter with impulse response

$$h(n) = \begin{cases} (0.95)^n, & n \geq 0 \\ 0, & n < 0 \end{cases}$$

Determine the power spectrum of the output process $\{Y(n)\}$.

─────(SOLUTION)───

It is easily seen that

$$\begin{aligned} H(f) &= \sum_{n=0}^{\infty} h(n) e^{-j2\pi f n} \\ &= \sum_{n=0}^{\infty} (0.95 e^{-j2\pi f})^n \\ &= \frac{1}{1 - 0.95 e^{-j2\pi f}} \end{aligned} \tag{2.5.16}$$

and

$$\begin{aligned} |H(f)|^2 &= \frac{1}{\left| 1 - 0.95 e^{-j2\pi f} \right|^2} \\ &= \frac{1}{1.9025 - 1.9\cos(2\pi f)} \end{aligned} \tag{2.5.17}$$

Therefore, the power spectrum of the output process is

$$S_y(f) = |H(f)|^2 S_x(f) \tag{2.5.18}$$

$$= \frac{1}{1.9025 - 1.9\cos(2\pi f)} \tag{2.5.19}$$

where we assumed $S_x(f)$ is normalized to unity. Figure 2.17 illustrates $S_y(f)$. Note that
$S_y(f)$ is periodic with period 2π. The MATLAB script for this computation is given next.

─────(M-FILE)───

```
% MATLAB script for Illustrative Problem 2.8.
delta_w=2*pi/100;
w=-pi:delta_w:pi;                    % one period of Sy
```

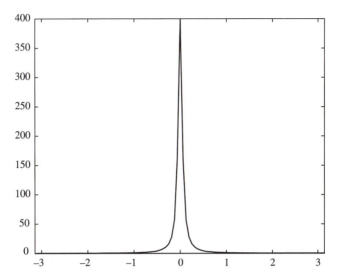

Figure 2.17   Plot of $S_y(f)$ in Illustrative Problem 2.8

```
Sy=1./(1.9025−1.9*cos(w));
% plotting command follows
plot(w,Sy);
```

The autocorrelation of the output process $\{Y(n)\}$ may be determined by taking the inverse FFT of $S_y(f)$. The student will find it interesting to compare this autocorrelation with that obtained in Illustrative Problem 2.3.

2.6   Lowpass and Bandpass Processes

Just as in the case of deterministic signals, random signals can also be characterized as lowpass and bandpass random processes.

Definition: A random process is called *lowpass* if its power spectrum is large in the vicinity of $f = 0$ and small (approaching 0) at high frequencies. In other words, a lowpass random process has most of its power concentrated at low frequencies.

Definition: A lowpass random process $X(t)$ is *bandlimited* if the power spectrum $S_x(f) = 0$ for $|f| > B$. The parameter B is called the *bandwidth* of the random process.

⬤ILLUSTRATIVE PROBLEM

Illustrative Problem 2.9 [Lowpass Processes] Consider the problem of generating samples of a lowpass random process by passing a white noise sequence $\{X_n\}$ through a lowpass filter. The input sequence is an i.i.d. sequence of uniformly distributed random variables on the interval $(-\frac{1}{2}, \frac{1}{2})$. The lowpass filter has the impulse response

$$h(n) = \begin{cases} (0.9)^n, & n \geq 0 \\ 0, & n < 0 \end{cases}$$

and is characterized by the input-output recursive (difference) equation

$$y_n = 0.9 y_{n-1} + x_n, \qquad n \geq 1, \qquad y_{-1} = 0$$

Compute the output sequence $\{y_n\}$ and determine the autocorrelation functions $R_x(m)$ and $R_y(m)$, as indicated in (2.4.6). Determine the power spectra $S_x(f)$ and $S_y(f)$ by computing the DFT of $R_x(m)$ and $R_y(m)$.

SOLUTION

The MATLAB scripts for these computations are given next. Figures 2.18 (p. 74) and 2.19 (p. 75) illustrate the autocorrelation functions and the power spectra. We note that the plots of the autocorrelation function and the power spectra are averages over ten realizations of the random process.

M-FILE

```
% MATLAB script for Illustrative Problem 2.9.
N=1000;                              % The maximum value of n
M=50;
Rxav=zeros(1,M+1);
Ryav=zeros(1,M+1);
Sxav=zeros(1,M+1);
Syav=zeros(1,M+1);
for i=1:10,                          % Take the ensemble average over ten realizations
    X=rand(1,N)−(1/2);               % Generate a uniform number sequence on (-1/2,1/2)
    Y(1)=0;
    for n=2:N, Y(n)=0.9*Y(n−1)+X(n); end; % Note that Y(n) means Y(n-1)
    Rx=Rx_est(X,M);                  % Autocorrelation of {Xn}
    Ry=Rx_est(Y,M);                  % Autocorrelation of {Yn}
    Sx=fftshift(abs(fft(Rx)));       % Power spectrum of {Xn}
    Sy=fftshift(abs(fft(Ry)));       % Power spectrum of {Yn}
    Rxav=Rxav+Rx;
    Ryav=Ryav+Ry;
    Sxav=Sxav+Sx;
    Syav=Syav+Sy;
    echo off ;
end;
echo on ;
Rxav=Rxav/10;
```

Ryav=Ryav/10;
Sxav=Sxav/10;
Syav=Syav/10;
% Plotting commands follow

● **M-FILE**

```
function [Rx]=Rx_est(X,M)
% [Rx]=Rx_est(X,M)
%              RX_EST   Estimates the autocorrelation of the sequence of random
%              variables given in X. Only Rx(0), Rx(1), ... , Rx(M) are computed.
%              Note that Rx(m) actually means Rx(m-1).
N=length(X);
Rx=zeros(1,M+1);
for m=1:M+1,
   for n=1:N−m+1,
     Rx(m)=Rx(m)+X(n)*X(n+m−1);
   end;
   Rx(m)=Rx(m)/(N−m+1);
end;
```

Definition: A random process is called *bandpass* if its power spectrum is large in a band of frequencies centered in the neighborhood of a central frequency $\pm f_0$ and relatively small outside of this band of frequencies. A random process is called *narrowband* if its bandwidth $B \ll f_0$.

Bandpass processes are suitable for representing modulated signals. In a communication system, the information-bearing signal is usually a lowpass random process that modulates a carrier for transmission over a bandpass (narrowband) communication channel. Thus, the modulated signal is a bandpass random process.

As in the case of deterministic signals, a bandpass random process $X(t)$ can be represented as

$$X(t) = X_c(t) \cos 2\pi f_0 t - X_s(t) \sin 2\pi f_0 t \tag{2.6.1}$$

where $X_c(t)$ and $X_s(t)$ are called the in-phase and quadratic components of $X(t)$. The random processes $X_c(t)$ and $X_s(t)$ are lowpass processes. The following theorem, stated without proof, provides an important relationship among $X(t)$, $X_c(t)$, and $X_s(t)$.

Theorem: If $X(t)$ is a zero-mean, stationary random process, the processes $X_c(t)$ and $X_s(t)$ are also zero-mean, jointly stationary processes.

In fact, it can be easily proved (see [1]) that the autocorrelation functions of $X_c(t)$ and $X_s(t)$ are identical and may be expressed as

$$R_c(\tau) = R_s(\tau) = R_x(\tau) \cos 2\pi f_0 \tau + \hat{R}_x(\tau) \sin 2\pi f_0 \tau \tag{2.6.2}$$

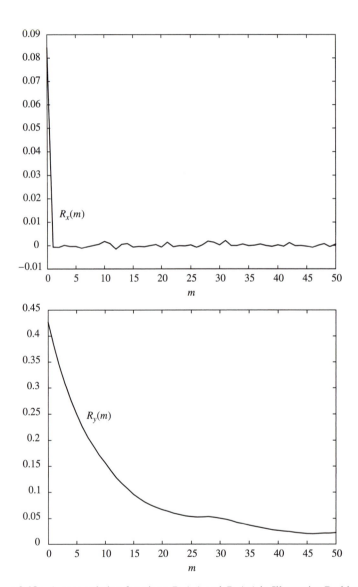

Figure 2.18   Autocorrelation functions $R_x(m)$ and $R_y(m)$ in Illustrative Problem 2.9

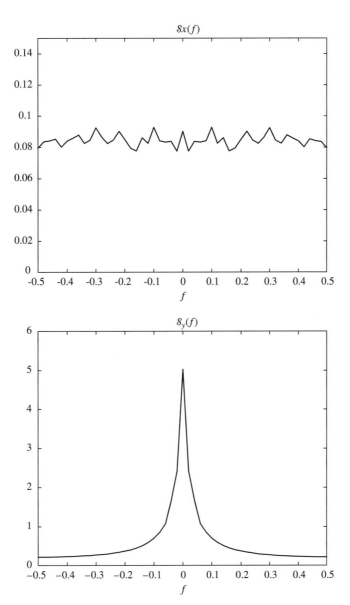

Figure 2.19   Power spectra $\mathcal{S}_x(f)$ and $\mathcal{S}_y(f)$ in Illustrative Problem 2.9

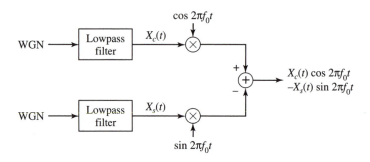

Figure 2.20   Generation of a bandpass random process

where $R_x(\tau)$ is the autocorrelation function of the bandpass process $X(t)$ and $\hat{R}_x(\tau)$ is the Hilbert transform of $R_x(\tau)$, which is defined as

$$\hat{R}_x(\tau) = \frac{1}{\pi} \int_{-\infty}^{\infty} \frac{R(t)}{\tau - t} \, dt \qquad (2.6.3)$$

Also, the cross-correlation function of $X_c(t)$ and $X_s(t)$ is expressed as

$$R_{cs}(\tau) = R_x(\tau) \sin 2\pi f_0 \tau - \hat{R}_x(\tau) \cos 2\pi f_0 \tau \qquad (2.6.4)$$

Finally, the autocorrelation function of the bandpass process $X(t)$ is expressed in terms of the autocorrelation function $R_c(\tau)$ and the cross-correlation function $R_{cs}(\tau)$ as

$$R_x(\tau) = R_c(\tau) \cos 2\pi f_0 \tau - R_{cs}(\tau) \sin 2\pi f_0 \tau \qquad (2.6.5)$$

──(ILLUSTRATIVE PROBLEM)──────────────

Illustrative Problem 2.10 [Generation of samples of a bandpass random process] Generate samples of a bandpass random process by first generating samples of two statistically independent random processes $X_c(t)$ and $X_s(t)$ and then using these to modulate the quadrature carriers $\cos 2\pi f_0 t$ and $\sin 2\pi f_0 t$, as shown in Figure 2.20.

──(SOLUTION)──────────────

On a digital computer samples of the lowpass processes $X_c(t)$ and $X_s(t)$ are generated by filtering two independent white noise processes by two identical lowpass filters. Thus, we obtain the samples $X_c(n)$ and $X_s(n)$, corresponding to the sampled values of $X_c(t)$ and $X_s(t)$. Then $X_c(n)$ modulates the sampled carrier $\cos 2\pi f_0 nT$ and $X_s(n)$ modulates the quadrature carrier $\sin 2\pi f_0 nT$, where T is the appropriate sampling interval.

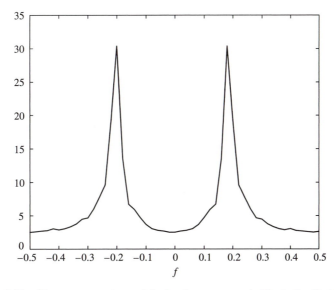

Figure 2.21   The power spectrum of the bandpass process in Illustrative Problem 2.10

The MATLAB script for these computations is given next. For illustrative purposes, we have selected the lowpass filter to have a transfer function

$$H(z) = \frac{1}{1 - 0.9z^{-1}}$$

Also, we selected $T = 1$ and $f_0 = 1000/\pi$. The resulting power spectrum of the bandpass process is shown in Figure 2.21.

M-FILE

```
% MATLAB script for Illustrative Problem 2.10.
N=1000;                              % number of samples
for i=1:2:N,
    [X1(i)  X1(i+1)]=gngauss;
    [X2(i)  X2(i+1)]=gngauss;
  echo off ;
end;                                 % standard Gaussian input noise processes
echo on ;
A=[1  −0.9];                         % lowpass filter parameters
B=1;
Xc=filter(B,A,X1);
Xs=filter(B,A,X2);
fc=1000/pi;                          % carrier frequency
for i=1:N,
    band_pass_process(i)=Xc(i)*cos(2*pi*fc*i)−Xs(i)*sin(2*pi*fc*i);
    echo off ;
```

```
end;                                      % T=1 is assumed
echo on;
% Determine the autocorrelation and the spectrum of the bandpass process
M=50;
bpp_autocorr=Rx_est(band_pass_process,M);
bpp_spectrum=fftshift(abs(fft(bpp_autocorr)));
% plotting commands follow
```

Problems

2.1 Generate a set of 1000 uniform random numbers in the interval $[0, 1]$ using the MATLAB function rand$(1, N)$. Plot the histogram and the probability distribution function for the sequence. The histogram may be determined by quantizing the interval in ten equal-width subintervals covering the range $[0, 1]$ and counting the numbers in each subinterval.

2.2 Generate a set of 1000 uniform random numbers in the interval $[-\frac{1}{2}, \frac{1}{2}]$ using the MATLAB function rand$(1, N)$. Plot the histogram and the probability distribution function for the sequence.

2.3 Generate a set of 1000 uniform random numbers in the interval $[-2, 2]$ by using the MATLAB function rand$(1, N)$. Plot the histogram and the probability distribution function for the sequence.

2.4 Generate a set of 1000 random numbers having the linear probability density function

$$f(x) = \begin{cases} \frac{x}{2}, & 0 \le x \le 2 \\ 0, & \text{otherwise} \end{cases}$$

Plot the histogram and the probability distribution function.

2.5 Generate a set of 1000 Gaussian random numbers having zero mean and unit variance using the method described in Section 2.2. Plot the histogram and the probability distribution function for the sequence. In determining the histogram, the range of the random numbers may be subdivided into subintervals of width $\sigma^2/5$, beginning with the first interval covering the range $-\sigma^2/10 < x < \sigma^2/10$, where σ^2 is the variance.

2.6 Generate a set of 1000 Gaussian random numbers having zero mean and unit variance by using the MATLAB function randn$(1, N)$. Plot the histogram and the probability distribution function for the sequence. Compare these results with the results obtained in Problem 2.5.

2.7 Generate 1000 pairs of Gaussian random numbers (x_1, x_2) that have mean vector

$$\boldsymbol{m} = E\,[x_1 \quad x_2] = \begin{bmatrix} \frac{1}{2} & \frac{1}{2} \end{bmatrix}$$

and covariance matrix

$$C = \begin{bmatrix} 1 & \frac{1}{2} \\ \frac{1}{2} & 1 \end{bmatrix}$$

a. Determine the means of the samples (x_{1i}, x_{2i}), $i = 1, 2, \ldots, 1000$, defined as

$$\hat{m}_1 = \frac{1}{1000} \sum_{i=1}^{1000} x_{1i}$$

$$\hat{m}_2 = \frac{1}{1000} \sum_{i=1}^{1000} x_{2i}$$

Also, determine their variances,

$$\hat{\sigma}_1^2 = \frac{1}{1000} \sum_{i=1}^{1000} (x_{1i} - \hat{m}_1)^2$$

$$\hat{\sigma}_2^2 = \frac{1}{1000} \sum_{i=1}^{1000} (x_{2i} - \hat{m}_2)^2$$

and their covariance,

$$\hat{c}_{ij} = \frac{1}{1000} \sum_{i=1}^{1000} (x_{1i} - \hat{m}_1)(x_{2i} - \hat{m}_2)$$

b. Compare the values obtained from the samples with the theoretical values.

2.8 Generate a sequence of 1000 samples of a Gauss-Markov process described by the recursive relation

$$X_n = \rho X_{n-1} + W_n, \qquad n = 1, 2, \ldots, 1000$$

where $X_0 = 0$, $\rho = 0.9$, and $\{W_n\}$ is a sequence of zero-mean and unit-variance i.i.d. Gaussian random variables.

2.9 Repeat Illustrative Problem 2.4 with an i.i.d. sequence of zero-mean, unit-variance, Gaussian random variables.

2.10 Repeat Illustrative Problem 2.5 when the power spectrum of a bandlimited random process is

$$S_x(f) = \begin{cases} 1 - \frac{|f|}{B}, & |f| \leq B \\ 0, & \text{otherwise} \end{cases}$$

2.11 Repeat Illustrative Problem 2.6 with a linear filter that has impulse response

$$h(t) = \begin{cases} e^{-3t}, & t \geq 0 \\ 0, & t < 0 \end{cases}$$

2.12 Determine numerically the autocorrelation function of the random process at the output of the linear filter in Problem 2.11.

2.13 Repeat Illustrative Problem 2.8 when

$$h(n) = \begin{cases} (0.8)^n, & n \geq 0 \\ 0, & n < 0 \end{cases}$$

2.14 Generate an i.i.d. sequence $\{x_n\}$ of $N = 1000$ uniformly distributed random numbers in the interval $[-\frac{1}{2}, \frac{1}{2}]$. This sequence is passed through a linear filter with impulse response

$$h(n) = \begin{cases} (0.95)^n, & n \geq 0 \\ 0, & n < 0 \end{cases}$$

The recursive equation that describes the output of this filter as a function of the input is

$$y_n = 0.95 y_{n-1} + x_n, \qquad n \geq 0, \quad y_{-1} = 0$$

Compute the autocorrelation functions $R_x(m)$ and $R_y(m)$ of the sequences $\{x_n\}$ and $\{y_n\}$ and the corresponding power spectra $S_x(f)$ and $S_y(f)$ using the relations given in (2.4.6) and (2.4.7). Compare this result for $S_y(f)$ with that obtained in Illustrative Problem 2.8.

2.15 Generate two i.i.d. sequences $\{w_{cn}\}$ and $\{w_{sn}\}$ of $N = 1000$ uniformly distributed random numbers in the interval $[-\frac{1}{2}, \frac{1}{2}]$. Each of these sequences is passed through a linear filter with impulse response

$$h(n) = \begin{cases} \left(\frac{1}{2}\right)^n, & n \geq 0 \\ 0, & n < 0 \end{cases}$$

whose input-output characteristic is given by the recursive relation

$$x_n = \frac{1}{2} x_{n-1} + w_n, \qquad n \geq 1, \quad x_0 = 0$$

Thus, we obtain two sequences, $\{x_{cn}\}$ and $\{x_{sn}\}$. The output sequence $\{x_{cn}\}$ modulates the carrier $\cos(\pi/2)n$, and the output sequence $\{x_{sn}\}$ modulates the quadrature carrier $\sin(\pi/2)n$. The bandpass signal is formed by combining the modulated components as in (2.6.1).

Compute and plot the autocorrelation components $R_c(m)$ and $R_s(m)$ for $|m| \leq 10$ for the sequences $\{x_{cn}\}$ and $\{x_{sn}\}$, respectively. Compute the autocorrelation function $R_x(m)$ of the bandpass signal for $|m| \leq 10$. Use the DFT (or the FFT algorithm) to compute the power spectra $S_c(f)$, $S_s(f)$, and $S_x(f)$. Plot these power spectra and comment on the results.

Chapter 3

Analog Modulation

3.1 Preview

In this chapter, we study the performance of various analog modulation-demodulation schemes, both in the presence and in the absence of additive noise. Systems studied in this chapter include amplitude-modulation (AM) schemes, such as DSB-AM, SSB-AM, and conventional AM, and angle-modulation schemes, such as frequency and phase modulation. Each member of the class of analog modulation systems is characterized by five basic properties:

1. Time-domain representation of the modulated signal

2. Frequency-domain representation of the modulated signal

3. Bandwidth of the modulated signal

4. Power content of the modulated signal

5. Signal-to-noise ratio (SNR) after demodulation

These properties are obviously not independent of one another. There exists a close relationship between time- and frequency-domain representations of signals expressed through the Fourier transform relation. Also, the bandwidth of a signal is defined in terms of its frequency characteristics.

Due to the fundamental difference between amplitude- and angle-modulation schemes, these schemes are treated separately and in different sections. We begin this chapter with the study of the simplest modulation scheme, amplitude modulation.

3.2 Amplitude Modulation (AM)

Amplitude modulation (AM), which is frequently referred to as *linear modulation*, is the family of modulation schemes in which the amplitude of a sinusoidal carrier is changed as a function of the modulating signal. This class of modulation schemes consists of DSB-AM (double-sideband amplitude modulation), conventional amplitude modulation, SSB-AM (single-sideband amplitude modulation), and VSB-AM (vestigial-sideband amplitude modulation). The dependence between the modulating signal and the amplitude of the modulated carrier can be very simple, as, for example, in the DSB-AM case, or much more complex, as in SSB-AM or VSB-AM. Amplitude-modulation systems are usually characterized by a relatively low bandwidth requirement and power inefficiency in comparison to the angle-modulation schemes. The bandwidth requirement for AM systems varies between W and $2W$, where W denotes the bandwidth of the message signal. For SSB-AM the bandwidth is W, for DSB-AM and conventional AM the bandwidth is $2W$, and for VSB-AM the bandwidth is between W and $2W$. These systems are widely used in broadcasting (AM radio and TV video broadcasting), point-to-point communication (SSB), and multiplexing applications (for example, transmission of many telephone channels over microwave links).

3.2.1 DSB-AM

In DSB-AM, the amplitude of the modulated signal is proportional to the message signal. This means that the time-domain representation of the modulated signal is given by

$$u(t) = A_c m(t) \cos(2\pi f_c t) \tag{3.2.1}$$

where

$$c(t) = A_c \cos(2\pi f_c t) \tag{3.2.2}$$

is the carrier and $m(t)$ is the message signal. The frequency-domain representation of the DSB-AM signal is obtained by taking the Fourier transform of $u(t)$ and results in

$$U(f) = \frac{A_c}{2} M(f - f_c) + \frac{A_c}{2} M(f + f_c) \tag{3.2.3}$$

where $M(f)$ is the Fourier transform of $m(t)$. Obviously, this type of modulation results in a shift of $\pm f_c$ and a scaling of $A_c/2$ in the spectrum of the message signal. The transmission bandwidth denoted by B_T is twice the bandwidth of the message signal:

$$B_T = 2W \tag{3.2.4}$$

A typical message spectrum and the spectrum of the corresponding DSB-AM modulated signal are shown in Figure 3.1.

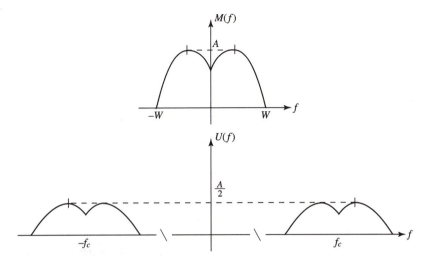

Figure 3.1   Spectra of the message and the DSB-AM modulated signal

The power content of the modulated signal is given by

$$P_u = \lim_{T \to \infty} \frac{1}{T} \int_{-T/2}^{T/2} u^2(t)\, dt$$

$$= \lim_{T \to \infty} \frac{1}{T} \int_{-T/2}^{T/2} A_c^2 m^2(t) \cos^2(2\pi f_c t)\, dt$$

$$= \lim_{T \to \infty} \frac{1}{T} \int_{-T/2}^{T/2} A_c^2 m^2(t) \frac{1 + \cos(4\pi f_c t)}{2}\, dt$$

$$= A_c^2 \left\{ \lim_{T \to \infty} \frac{1}{T} \int_{-T/2}^{T/2} \frac{m^2(t)}{2}\, dt + \lim_{T \to \infty} \frac{1}{T} \int_{-T/2}^{T/2} m^2(t) \frac{\cos(4\pi f_c t)}{2}\, dt \right\}$$

$$= A_c^2 \lim_{T \to \infty} \frac{1}{T} \int_{-T/2}^{T/2} \frac{m^2(t)}{2}\, dt \tag{3.2.5}$$

$$= \frac{A_c^2}{2} P_m \tag{3.2.6}$$

where P_m is the power content of the message signal. Equation (3.2.5) follows from the fact that $m(t)$ is a lowpass signal with frequency content much less than $2f_c$, the frequency content of $\cos(4\pi f_c t)$. Therefore, the integral

$$\int_{-T/2}^{T/2} m^2(t) \frac{\cos(4\pi f_c t)}{2}\, dt \tag{3.2.7}$$

goes to zero as $T \to \infty$. Finally, the SNR for a DSB-AM system is equal to the baseband SNR; that is, it is given by

$$\left(\frac{S}{N} \right)_o = \frac{P_R}{N_0 W} \tag{3.2.8}$$

where P_R is the received power (the power in the modulated signal at the receiver), $N_0/2$ is the noise power spectral density (assuming white noise), and W is the message bandwidth.

─(ILLUSTRATIVE PROBLEM)──────────

Illustrative Problem 3.1 [DSB-AM modulation] The message signal $m(t)$ is defined as

$$m(t) = \begin{cases} 1, & 0 \le t \le \frac{t_0}{3} \\ -2, & \frac{t_0}{3} < t \le \frac{2t_0}{3} \\ 0, & \text{otherwise} \end{cases}$$

This message DSB-AM modulates the carrier $c(t) = \cos 2\pi f_c t$, and the resulting modulated signal is denoted by $u(t)$. It is assumed that $t_0 = 0.15$ s and $f_c = 250$ Hz.

1. Obtain the expression for $u(t)$.

2. Derive the spectra of $m(t)$ and $u(t)$.

3. Assuming that the message signal is periodic with period $T_0 = t_0$, determine the power in the modulated signal.

4. If a noise is added to the modulated signal in part 3 such that the resulting SNR is 10 dB, find the noise power.

──(SOLUTION)──────────

1. $m(t)$ can be written as

$$m(t) = \Pi\left(\frac{t - t_0/6}{t_0/3}\right) - 2\Pi\left(\frac{t - t_0/2}{t_0/3}\right)$$

Therefore,

$$u(t) = \left[\Pi\left(\frac{t - 0.025}{0.05}\right) - 2\Pi\left(\frac{t - 0.075}{0.05}\right)\right]\cos(500\pi t) \qquad (3.2.9)$$

2. Using the standard Fourier transform relation $\mathcal{F}[\Pi(t)] = \text{sinc}(t)$ together with the shifting and the scaling theorems of the Fourier transform, we obtain

$$\mathcal{F}[m(t)] = \frac{t_0}{3}e^{-j\pi f t_0/3}\text{sinc}\left(\frac{t_0 f}{3}\right) - 2\frac{t_0}{3}e^{-j\pi f t_0}\text{sinc}\left(\frac{t_0 f}{3}\right)$$

$$= \frac{t_0}{3}e^{-j\pi f t_0/3}\text{sinc}\left(\frac{t_0 f}{3}\right)\left(1 - 2e^{-j2\pi t_0 f/3}\right) \qquad (3.2.10)$$

Substituting $t_0 = 0.15$ s gives

$$\mathcal{F}[m(t)] = 0.05e^{-0.05j\pi f}\text{sinc}(0.05 f)\left(1 - 2e^{-0.1j\pi f}\right) \qquad (3.2.11)$$

For the modulated signal $u(t)$, we have

$$U(f) = 0.025e^{-0.05j\pi(f-f_c)}\text{sinc}\,(0.05(f - f_c))\left(1 - 2e^{-0.1j\pi(f-f_c)}\right)$$
$$+ 0.025e^{-0.05j\pi(f+f_c)}\text{sinc}\,(0.05(f + f_c))\left(1 - 2e^{-0.1j\pi(f+f_c)}\right)$$

Plots of the magnitude spectra of the message and the modulated signals are shown in Figure 3.2.

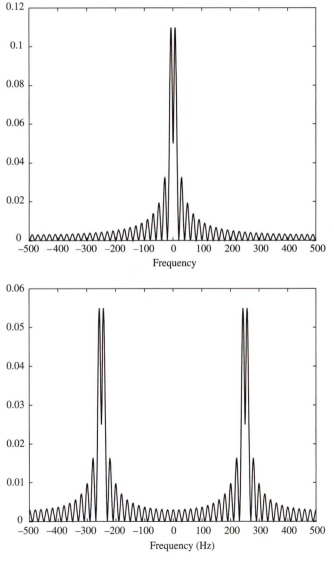

Figure 3.2   Magnitude spectra of the message and the modulated signals in
Illustrative Problem 3.1

3. The power in the modulated signal is given by

$$P_u = \frac{A_c^2}{2} P_m = \frac{1}{2} P_m$$

where P_m is the power in the message signal;

$$P_m = \frac{1}{t_0} \int_0^{2t_0/3} m^2(t)\, dt = \frac{1}{t_0} \left(\frac{t_0}{3} + \frac{4t_0}{3} \right) = \frac{5}{3} = 1.666$$

and

$$P_u = \frac{1.666}{2} = 0.833$$

4. Here

$$10 \log_{10} \left(\frac{P_R}{P_n} \right) = 10$$

or $P_R = P_u = 10 P_n$, which results in $P_n = P_u/10 = 0.0833$.

The MATLAB script for the preceding problem follows.

M-FILE

```
% Matlab script for Illustrative Problem 3.1.
% Matlab demonstration script for DSB-AM modulation. The message signal
% is +1 for 0 < t < t0/3, -2 for t0/3 < t < 2t0/3, and zero otherwise.
echo on
t0=.15;                              % signal duration
ts=0.001;                            % sampling interval
fc=250;                              % carrier frequency
snr=20;                              % SNR in dB (logarithmic)
fs=1/ts;                             % sampling frequency
df=0.3;                              % desired freq. resolution
t=[0:ts:t0];                         % time vector
snr_lin=10^(snr/10);                 % linear SNR
% message signal
m=[ones(1,t0/(3*ts)),−2*ones(1,t0/(3*ts)),zeros(1,t0/(3*ts)+1)];
c=cos(2*pi*fc.*t);                   % carrier signal
u=m.*c;                              % modulated signal
[M,m,df1]=fftseq(m,ts,df);           % Fourier transform
M=M/fs;                              % scaling
[U,u,df1]=fftseq(u,ts,df);           % Fourier transform
U=U/fs;                              % scaling
[C,c,df1]=fftseq(c,ts,df);           % Fourier transform
f=[0:df1:df1*(length(m)−1)]−fs/2;    % freq. vector
signal_power=spower(u(1:length(t))); % power in modulated signal
noise_power=signal_power/snr_lin;    % compute noise power
noise_std=sqrt(noise_power);         % compute noise standard deviation
```

```
noise=noise_std*randn(1,length(u));        % generate noise
r=u+noise;                                 % add noise to the modulated signal
[R,r,df1]=fftseq(r,ts,df);                 % spectrum of the signal+noise
R=R/fs;                                    % scaling
pause  % Press a key to show the modulated signal power
signal_power
pause  % Press any key to see plot of the message
clf
subplot(2,2,1)
plot(t,m(1:length(t)))
xlabel('Time')
title('The message signal')
pause  % Press any key to see a plot of the carrier
subplot(2,2,2)
plot(t,c(1:length(t)))
xlabel('Time')
title('The carrier')
pause  % Press any key to see a plot of the modulated signal
subplot(2,2,3)
plot(t,u(1:length(t)))
xlabel('Time')
title('The modulated signal')
pause    % Press any key to see plots of the magnitude of the message and the
         % modulated signal in the frequency domain.
subplot(2,1,1)
plot(f,abs(fftshift(M)))
xlabel('Frequency')
title('Spectrum of the message signal')
subplot(2,1,2)
plot(f,abs(fftshift(U)))
title('Spectrum of the modulated signal')
xlabel('Frequency')
pause  % Press a key to see a noise sample
subplot(2,1,1)
plot(t,noise(1:length(t)))
title('Noise sample')
xlabel('Time')
pause  % Press a key to see the modulated signal and noise
subplot(2,1,2)
plot(t,r(1:length(t)))
title('Signal and noise')
xlabel('Time')
pause  % Press a key to see the modulated signal and noise in freq. domain
subplot(2,1,1)
plot(f,abs(fftshift(U)))
title('Signal spectrum')
xlabel('Frequency')
subplot(2,1,2)
plot(f,abs(fftshift(R)))
title('Signal and noise spectrum')
xlabel('Frequency')
```

Illustrative Problem 3.2 [DSB modulation for an almost bandlimited signal] The message signal $m(t)$ is given by

$$m(t) = \begin{cases} \text{sinc}(100t), & |t| \leq t_0 \\ 0, & \text{otherwise} \end{cases}$$

where $t_0 = 0.1$. This message modulates the carrier $c(t) = \cos(2\pi f_c t)$, where $f_c = 250$ Hz.

1. Determine the modulated signal $u(t)$.

2. Determine the spectra of $m(t)$ and $u(t)$.

3. If the message signal is periodic with period $T_0 = 0.2$ s, determine the power in the modulated signal.

4. If a Gaussian noise is added to the modulated signal such that the resulting SNR is 10 dB, find the noise power.

SOLUTION

---

1. We have

$$u(t) = m(t)c(t)$$

$$= \begin{cases} \text{sinc}(100t)\cos(500t), & |t| \leq 0.1 \\ 0, & \text{otherwise} \end{cases} \qquad (3.2.12)$$

$$= \text{sinc}(100t)\Pi(5t)\cos(500t) \qquad (3.2.13)$$

2. A plot of the spectra of $m(t)$ and $u(t)$ is shown in Figure 3.3. As the figure shows, the message signal is an almost bandlimited signal with a bandwidth of 50 Hz.

3. The power in the modulated signal is half of the power in the message signal. The power in the message signal is given by

$$P_m = \frac{1}{0.2} \int_{-0.1}^{0.1} \text{sinc}^2(100t)\,dt$$

The integral can be computed using MATLAB's quad8.m m-file, which results in $P_m = 0.0495$ and hence, $P_u = 0.0247$.

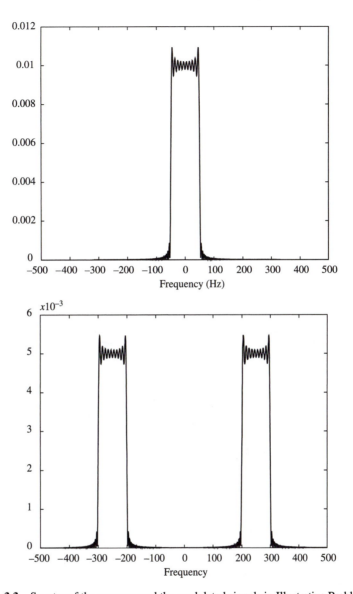

Figure 3.3   Spectra of the message and the modulated signals in Illustrative Problem 3.2

4. Here

$$10 \log_{10} \left(\frac{P_R}{P_n} \right) = 10 \implies P_n = 0.1 P_R = 0.1 P_U = 0.00247$$

The MATLAB script for the problem follows.

M-FILE

```
% Matlab script for Illustrative Problem 3.2.
% Matlab demonstration script for DSB-AM modulation. The message signal
% is m(t)=sinc(100t).
echo on
t0=.2;                                    % signal duration
ts=0.001;                                 % sampling interval
fc=250;                                   % carrier frequency
snr=20;                                   % SNR in dB (logarithmic)
fs=1/ts;                                  % sampling frequency
df=0.3;                                   % required freq. resolution
t=[−t0/2:ts:t0/2];                        % time vector
snr_lin=10^(snr/10);                      % linear SNR
m=sinc(100*t);                            % the message signal
c=cos(2*pi*fc.*t);                        % the carrier signal
u=m.*c;                                   % the DSB-AM modulated signal
[M,m,df1]=fftseq(m,ts,df);                % Fourier transform
M=M/fs;                                   % scaling
[U,u,df1]=fftseq(u,ts,df);                % Fourier transform
U=U/fs;                                   % scaling
f=[0:df1:df1*(length(m)−1)]−fs/2;         % frequency vector
signal_power=spower(u(1:length(t)));      % compute modulated signal power
noise_power=signal_power/snr_lin;         % compute noise power
noise_std=sqrt(noise_power);              % compute noise standard deviation
noise=noise_std*randn(1,length(u));       % generate noise sequence
r=u+noise;                                % add noise to the modulated signal
[R,r,df1]=fftseq(r,ts,df);                % Fourier transform
R=R/fs;                                   % scaling
pause   % Press a key to show the modulated signal power
signal_power
pause   %Press any key to see a plot of the message
clf
subplot(2,2,1)
plot(t,m(1:length(t)))
xlabel('Time')
title('The message signal')
pause % Press any key to see a plot of the carrier
subplot(2,2,2)
plot(t,c(1:length(t)))
xlabel('Time')
title('The carrier')
pause   % Press any key to see a plot of the modulated signal
subplot(2,2,3)
plot(t,u(1:length(t)))
xlabel('Time')
title('The modulated signal')
pause   % Press any key to see a plot  of the magnitude of the message and the
        % modulated signal in the frequency domain.
subplot(2,1,1)
plot(f,abs(fftshift(M)))
xlabel('Frequency')
title('Spectrum of the message signal')
```

```
subplot(2,1,2)
plot(f,abs(fftshift(U)))
title('Spectrum of the modulated signal')
xlabel('Frequency')
pause  % Press a key to see a noise sample
subplot(2,1,1)
plot(t,noise(1:length(t)))
title('Noise sample')
xlabel('Time')
pause  % Press a key to see the modulated signal and noise
subplot(2,1,2)
plot(t,r(1:length(t)))
title('Signal and noise')
xlabel('Time')
pause  % Press a key to see the modulated signal and noise in freq. domain
subplot(2,1,1)
plot(f,abs(fftshift(U)))
title('Signal spectrum')
xlabel('Frequency')
subplot(2,1,2)
plot(f,abs(fftshift(R)))
title('Signal and noise spectrum')
xlabel('Frequency')
```

WHAT IF?

What happens if the duration of the message signal t_0 changes; in particular, what is the effect of having large t_0's and small t_0's? What is the effect on the bandwidth and signal power?

The m-file dsb_mod.m given next is a general DSB modulator of the message signal given in vector m on a carrier of frequency f_c.

M-FILE

```
function u=dsb_mod(m,ts,fc)
%                 u=dsb_mod(m,ts,fc)
%DSB_MOD          takes signal m sampled at ts and carrier
%                 freq. fc as input and returns the DSB modulated
%                 signal. ts << 1/2fc. The modulated signal
%                 is normalized to have half the message power.
%                 The message signal starts at 0.

t=[0:length(m)−1]*ts;
u=m.*cos(2*pi*t);
```

3.2.2   Conventional AM

In many respects, conventional AM is quite similar to DSB-AM; the only difference is that in conventional AM, $m(t)$ is substituted with $[1 + am_n(t)]$, where $m_n(t)$ is the normalized message signal (i.e., $|m_n(t)| \leq 1$) and a is the *index of modulation*, which is a positive constant between 0 and 1. Thus, we have

$$u(t) = A_c [1 + am_n(t)] \cos(2\pi f_c t) \qquad (3.2.14)$$

and

$$U(f) = \frac{A_c}{2} [\delta(f - f_c) + aM_n(f - f_c) + \delta(f + f_c) + aM_n(f + f_c)] \qquad (3.2.15)$$

The net effect of scaling the message signal and adding a constant to it is that the term $[1 + am_n(t)]$ is always positive. This makes the demodulation of these signals much easier by employing envelope detectors. Note the existence of the sinusoidal component at the frequency f_c in $U(f)$. This means that a (usually substantial) fraction of the transmitted power is in the signal carrier that does not really serve the transmission of information. This fact shows that compared to DSB-AM, conventional AM is a less economical modulation scheme in terms of power utilization. The bandwidth, of course, is equal to the bandwidth of DSB-AM and is given by

$$B_T = 2W \qquad (3.2.16)$$

Typical frequency-domain plots of the message and the corresponding conventional AM signal are shown in Figure 3.4.

The power content of the modulated signal, assuming that the message signal is a zero-mean signal, is given by

$$P_u = \frac{A_c^2}{2} \left[1 + a^2 P_{m_n} \right] \qquad (3.2.17)$$

which comprises two parts: $A_c^2/2$, which denotes the power in the carrier, and $(A_c^2/2)a^2 P_{m_n}$, which is the power in the message-bearing part of the modulated signal. This is the power that is really used to transmit the message. The ratio of the power that is used to transmit the message to the total power in the modulated signal is called the *modulation efficiency* and is defined by

$$\eta = \frac{a^2 P_{m_n}}{1 + a^2 P_{m_n}} \qquad (3.2.18)$$

Since $|m_n(t)| \leq 1$ and $a \leq 1$, we always have $\eta \leq 0.5$. In practice, however, the value of η is around 0.1. The signal-to-noise ratio is given by

$$\left(\frac{S}{N} \right)_o = \eta \frac{P_R}{N_0 W} \qquad (3.2.19)$$

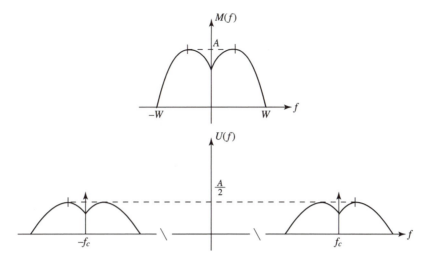

Figure 3.4   Spectra of the message and the conventional AM signal

where η is the modulation efficiency. We see that the SNR is reduced by a factor equal to η compared to a DSB-AM system. This reduction of performance is a direct consequence of the fact that a significant part of the total power is in the carrier (the deltas in the spectrum), which does not carry any information and is filtered out at the receiver.

ILLUSTRATIVE PROBLEM

Illustrative Problem 3.3 [Conventional AM] The message signal

$$m(t) = \begin{cases} 1, & 0 \leq t < \frac{t_0}{3} \\ -2, & \frac{t_0}{3} \leq t < \frac{2t_0}{3} \\ 0, & \text{otherwise} \end{cases}$$

modulates the carrier $c(t) = \cos(2\pi f_c t)$ using a conventional AM scheme. It is assumed that $f_c = 250$ Hz and $t_0 = 0.15$; the modulation index is $a = 0.85$.

1. Derive an expression for the modulated signal.

2. Determine the spectra of the message and the modulated signals.

3. If the message signal is periodic with a period equal to t_0, determine the power in the modulated signal and the modulation efficiency.

4. If a noise signal is added to the message signal such that the SNR at the output of the demodulator is 10 dB, find the power content of the noise signal.

SOLUTION

1. First note that max $|m(t)| = 2$; therefore, $m_n(t) = m(t)/2$. From this, we have

$$u(t) = \left[1 + 0.85\frac{m(t)}{2}\right]\cos(2\pi f_c t)$$

$$= \left[1 + 0.425\Pi\left(\frac{t - 0.025}{0.05}\right) - 0.85\Pi\left(\frac{t - 0.075}{0.05}\right)\right]\cos(500\pi t)$$

A plot of the message and the modulated signal is shown in Figure 3.5.

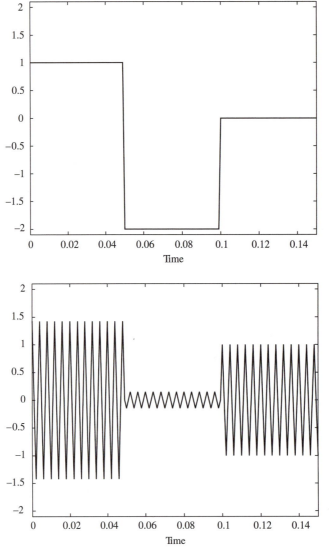

Figure 3.5 The message and the modulated signals in Illustrative Problem 3.3

2. For the message signal, we have

$$\mathcal{F}[m(t)] = 0.05e^{-0.05j\pi f}\,\text{sinc}(0.05f)\left(1 - 2e^{-0.1j\pi f}\right) \qquad (3.2.20)$$

and for the modulated signal,

$$U(f) = 0.010625e^{-0.05j\pi(f-250)}\,\text{sinc}(0.05(f-250))\left(1 - 2e^{-0.1j\pi(f-250)}\right)$$
$$+ 0.010625e^{-0.05j\pi(f+250)}\,\text{sinc}(0.05(f+250))\left(1 - 2e^{-0.1j\pi(f+250)}\right)$$

Plots of the spectra of the message and the modulated signal are shown in Figure 3.6.

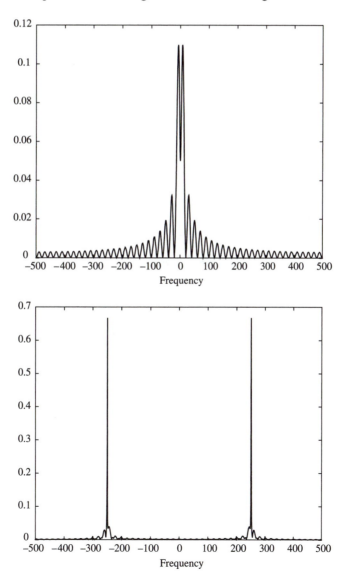

Figure 3.6   Spectra of the message and the modulated signal in Illustrative Problem 3.3

Note that the scales on the two spectra plots are different. The presence of the two delta functions in the spectrum of the modulated signal is apparent.

3. The power in the message signal can be obtained as

$$P_m = \frac{1}{0.15} \left[\int_0^{0.05} dt + 4 \int_{0.05}^{0.1} dt \right] = 1.667$$

The power in the normalized message signal P_{m_n} is given by

$$P_{m_n} = \frac{1}{4} P_m = \frac{1.66}{4} = 0.4167$$

and the modulation efficiency is

$$\eta = \frac{a^2 P_{m_n}}{1 + a^2 P_{m_n}} = \frac{0.85^2 \times 0.4167}{1 + 0.85^2 \times 0.4167} = 0.2314$$

The power in the modulated signal is given by (E denotes the expected value)

$$P_u = \frac{A_c^2}{2} E \left[1 + a m_n(t) \right]^2$$
$$= \frac{1}{2} \left(1 + 0.3010 - 1.7 \times \frac{0.025}{0.15} \right)$$
$$= 0.5088$$

4. In this case,

$$10 \log_{10} \left[\eta \left(\frac{P_R}{N_0 W} \right) \right] = 10$$

or

$$\eta \left(\frac{P_R}{P_n} \right) = 10$$

Substituting $\eta = 0.2314$ and $P_R = P_u = 0.5088$ yields

$$P_n = \frac{\eta P_u}{10} = 0.0118$$

COMMENT

In finding the power in the modulated signal in this problem, we could not use the relation

$$P_u = \frac{A_c^2}{2} \left[1 + a^2 P_{m_n} \right]$$

because in this problem $m(t)$ is not a zero-mean signal.

The MATLAB script for this problem is given next.

M-FILE

```
% am.m
% Matlab demonstration script for DSB-AM modulation. The message signal
% is +1 for 0 < t < t0/3, -2 for t0/3 < t < 2t0/3, and zero otherwise.
echo on
t0=.15;                              % signal duration
ts=0.001;                            % sampling interval
fc=250;                              % carrier frequency
snr=10;                              % SNR in dB (logarithmic)
a=0.85;                              % modulation index
fs=1/ts;                             % sampling frequency
t=[0:ts:t0];                         % time vector
df=0.2;                              % required frequency resolution
snr_lin=10^(snr/10);                 % SNR
% message signal
m=[ones(1,t0/(3*ts)),−2*ones(1,t0/(3*ts)),zeros(1,t0/(3*ts)+1)];
c=cos(2*pi*fc.*t);                   % carrier signal
m_n=m/max(abs(m));                   % normalized message signal
[M,m,df1]=fftseq(m,ts,df);           % Fourier transform
M=M/fs;                              % scaling
f=[0:df1:df1*(length(m)−1)]−fs/2;    % frequency vector
u=(1+a*m_n).*c;                      % modulated signal
[U,u,df1]=fftseq(u,ts,df);           % Fourier transform
U=U/fs;                              % scaling
signal_power=spower(u(1:length(t))); % power in modulated signal
% power in normalized message
pmn=spower(m(1:length(t)))/(max(abs(m)))^2;
eta=(a^2*pmn)/(1+a^2*pmn);           % modulation efficiency
noise_power=eta*signal_power/snr_lin;% noise power
noise_std=sqrt(noise_power);         % noise standard deviation
noise=noise_std*randn(1,length(u));  % generate noise
r=u+noise;                           % add noise to the modulated signal
[R,r,df1]=fftseq(r,ts,df);           % Fourier transform
R=R/fs;                              % scaling
pause  % Press a key to show the modulated signal power
signal_power
pause  % Press a key to show the modulation efficiency
eta
pause  % Press any key to see plot of the message
subplot(2,2,1)
plot(t,m(1:length(t)))
axis([0 0.15 −2.1 2.1])
xlabel('Time')
title('The message signal')
pause
pause  % Press any key to see a plot of the carrier
subplot(2,2,2)
plot(t,c(1:length(t)))
```

```
axis([0 0.15 −2.1 2.1])
xlabel('Time')
title('The carrier')
pause    % Press any key to see a plot of the modulated signal
subplot(2,2,3)
plot(t,u(1:length(t)))
axis([0 0.15 −2.1 2.1])
xlabel('Time')
title('The modulated signal')
pause    % Press any key to see a plot of the magnitude of the message and the
         % modulated signal in the frequency domain.
subplot(2,1,1)
plot(f,abs(fftshift(M)))
xlabel('Frequency')
title('Spectrum of the message signal')
subplot(2,1,2)
plot(f,abs(fftshift(U)))
title('Spectrum of the modulated signal')
xlabel('Frequency')
pause    % Press a key to see a noise sample
subplot(2,1,1)
plot(t,noise(1:length(t)))
title('Noise sample')
xlabel('Time')
pause    % Press a key to see the modulated signal and noise
subplot(2,1,2)
plot(t,r(1:length(t)))
title('Signal and noise')
xlabel('Time')
pause    % Press a key to see the modulated signal and noise in freq. domain
subplot(2,1,1)
plot(f,abs(fftshift(U)))
title('Signal spectrum')
xlabel('Frequency')
subplot(2,1,2)
plot(f,abs(fftshift(R)))
title('Signal and noise spectrum')
xlabel('Frequency')
```

The MATLAB m-file am_mode.m given next is a general conventional AM modulator.

M-FILE

```
function u=am_mod(a,m,ts,fc)
%               u=am_mod(a,m,ts,fc)
%AM_MOD     takes signal m sampled at ts and carrier
%           freq. fc as input and returns the AM modulated
%           signal. "a" is the modulation index.
%           and ts << 1/2fc.

t=[0:length(m)−1]*ts;
```

```
c=cos(2*pi*fc.*t);
m_n=m/max(abs(m));
u=(1+a*m_n).*c;
```

3.2.3 SSB-AM

SSB-AM is derived from DSB-AM by eliminating one of the sidebands. Therefore, it occupies half the bandwidth of DSB-AM. Depending on the sideband that remains, either the upper or the lower sideband, there exist two types of SSB-AM: USSB-AM and LSSB-AM. The time representation for these signals is given by

$$u(t) = \frac{A_c}{2} m(t) \cos(2\pi f_c t) \mp \frac{A_c}{2} \hat{m}(t) \sin(2\pi f_c t) \tag{3.2.21}$$

where the minus sign corresponds to USSB-AM and the plus sign corresponds to LSSB-AM. The signal denoted by $\hat{m}(t)$ is the Hilbert transform of $m(t)$, defined by $\hat{m}(t) = m(t) \star 1/\pi t$ or, in the frequency domain, by $\hat{M}(f) = -j \operatorname{sgn}(f) M(f)$. In other words, the Hilbert transform of a signal represents a $\pi/2$ phase shift in all signal components. In the frequency domain, we have

$$U_{\text{USSB}}(f) = \begin{cases} [M(f - f_c) + M(f + f_c)], & f_c \leq |f| \\ 0, & \text{otherwise} \end{cases} \tag{3.2.22}$$

and

$$U_{\text{LSSB}}(f) = \begin{cases} [M(f - f_c) + M(f + f_c)], & |f| \leq f_c \\ 0, & \text{otherwise} \end{cases} \tag{3.2.23}$$

Typical plots of the spectra of a message signal and the corresponding USSB-AM modulated signal are shown in Figure 3.7.

The bandwidth of the SSB signal is half the bandwidth of DSB and conventional AM and so is equal to the bandwidth of the message signal; that is,

$$B_T = W \tag{3.2.24}$$

The power in the SSB signal is given by

$$P_u = \frac{A_c^2}{4} P_m \tag{3.2.25}$$

Note that the power is half of the power in the corresponding DSB-AM signal because one of the sidebands has been removed. On the other hand, since the modulated signal has half the bandwidth of the corresponding DSB-AM signal, the noise power at the front end of the receiver is also half of a comparable DSB-AM signal, and therefore the SNR in both systems is the same; that is,

$$\left(\frac{S}{N}\right)_o = \frac{P_R}{N_0 W} \tag{3.2.26}$$

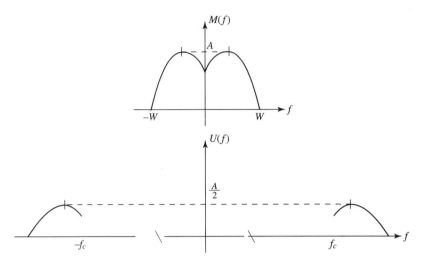

Figure 3.7   Spectra of the message and the USSB-AM signal

ILLUSTRATIVE PROBLEM

Illustrative Problem 3.4 [Single-sideband example] The message signal

$$m(t) = \begin{cases} 1, & 0 \le t < \frac{t_0}{3} \\ -2, & \frac{t_0}{3} \le t < \frac{2t_0}{3} \\ 0, & \text{otherwise} \end{cases}$$

modulates the carrier $c(t) = \cos(2\pi f_c t)$ using an LSSB-AM scheme. It is assumed that $t_0 = 0.15$ s and $f_c = 250$ Hz.

1. Plot the Hilbert transform of the message signal and the modulated signal $u(t)$. Also plot the spectrum of the modulated signal.

2. Assuming the message signal is periodic with period t_0, determine the power in the modulated signal.

3. If a noise is added to the modulated signal such that the SNR after demodulation is 10 dB, determine the power in the noise.

SOLUTION

1. The Hilbert transform of the message signal can be computed using the Hilbert transform m-file of MATLAB—that is, hilbert.m. It should be noted, however, that this function returns a complex sequence whose real part is the original signal and whose

imaginary part is the desired Hilbert transform. Therefore, the Hilbert transform of the sequence m is obtained by using the command imag(hilbert(m)). Now, using the relation

$$u(t) = m(t)\cos(2\pi f_c t) + \hat{m}(t)\sin(2\pi f_c t) \qquad (3.2.27)$$

we can find the modulated signal. Plots of $\hat{m}(t)$ and the spectrum of the LSSB-AM modulated signal $u(t)$ are shown in Figure 3.8.

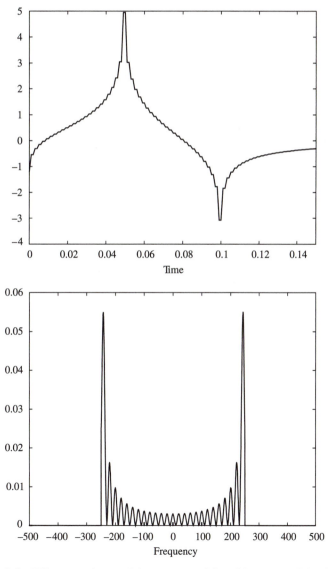

Figure 3.8   Hilbert transform and the spectrum of the LSSB-AM modulated signal for $m(t)$

2. The power in the message signal is

$$P_m = \frac{1}{0.15} \int_0^{0.15} m^2(t) \, dt = 1.667$$

and therefore

$$P_u = \frac{A_c^2}{4} P_m = 0.416$$

3. The post-demodulation SNR is given by

$$10 \log_{10} \left(\frac{P_R}{P_n} \right)_o = 10$$

Hence, $P_n = 0.1 P_R = 0.1 P_u = 0.0416$.

The MATLAB script for this problem follows.

M-FILE

```
% lssb.m
% Matlab demonstration script for LSSB-AM modulation. The message signal
% is +1 for 0 < t < t0/3, -2 for t0/3 < t < 2t0/3, and zero otherwise.
echo on
t0=.15;                                 % signal duration
ts=0.001;                               % sampling interval
fc=250;                                 % carrier frequency
snr=10;                                 % SNR in dB (logarithmic)
fs=1/ts;                                % sampling frequency
df=0.25;                                % desired freq. resolution
t=[0:ts:t0];                            % time vector
snr_lin=10^(snr/10);                    % SNR
% the message vector
m=[ones(1,t0/(3*ts)),-2*ones(1,t0/(3*ts)),zeros(1,t0/(3*ts)+1)];
c=cos(2*pi*fc.*t);                      % carrier vector
udsb=m.*c;                              % DSB modulated signal
[UDSB,udssb,df1]=fftseq(udsb,ts,df);    % Fourier transform
UDSB=UDSB/fs;                           % scaling
f=[0:df1:df1*(length(udssb)-1)]-fs/2;   % frequency vector
n2=ceil(fc/df1);                        % location of carrier in freq. vector
% remove the upper sideband from DSB
UDSB(n2:length(UDSB)-n2)=zeros(size(UDSB(n2:length(UDSB)-n2)));
ULSSB=UDSB;                             % generate LSSB-AM spectrum
[M,m,df1]=fftseq(m,ts,df);             % Fourier transform
M=M/fs;                                 % scaling
u=real(ifft(ULSSB))*fs;                 % generate LSSB signal from spectrum
signal_power=spower(udsb(1:length(t)))/2;
%                                       % compute signal power
```

```
noise_power=signal_power/snr_lin;        % compute noise power
noise_std=sqrt(noise_power);             % compute noise standard deviation
noise=noise_std*randn(1,length(u));      % generate noise vector
r=u+noise;                               % add the signal to noise
[R,r,df1]=fftseq(r,ts,df);               % Fourier transform
R=R/fs;                                  % scaling
pause  % Press a key to show the modulated signal power
signal_power
pause  % Press any key to see a plot of the message signal
clf
subplot(2,1,1)
plot(t,m(1:length(t)))
axis([0,0.15,−2.1,2.1])
xlabel('Time')
title('The message signal')
pause  % Press any key to see a plot of the carrier
subplot(2,1,2)
plot(t,c(1:length(t)))
xlabel('Time')
title('The carrier')
pause  % Press any key to see a plot of the modulated signal and its spectrum
clf
subplot(2,1,1)
plot([0:ts:ts*(length(u)−1)/8],u(1:length(u)/8))
xlabel('Time')
title('The LSSB-AM modulated signal')
subplot(2,1,2)
plot(f,abs(fftshift(ULSSB)))
xlabel('Frequency')
title('Spectrum of the LSSB-AM modulated signal')
pause    % Press any key to see the spectra of the message and the modulated signals
clf
subplot(2,1,1)
plot(f,abs(fftshift(M)))
xlabel('Frequency')
title('Spectrum of the message signal')
subplot(2,1,2)
plot(f,abs(fftshift(ULSSB)))
xlabel('Frequency')
title('Spectrum of the LSSB-AM modulated signal')

pause  % Press any key to see a noise sample
subplot(2,1,1)
plot(t,noise(1:length(t)))
title('Noise sample')
xlabel('Time')
pause  % Press a key to see the modulated signal and noise
subplot(2,1,2)
plot(t,r(1:length(t)))
title('Modulated signal and noise')
xlabel('Time')
subplot(2,1,1)
pause % Press any key to see the spectrum of the modulated signal
plot(f,abs(fftshift(ULSSB)))
```

```
title('Modulated signal spectrum')
xlabel('Frequency')
subplot(2,1,2)


pause   % Press a key to see the modulated signal noise in freq. domain
plot(f,abs(fftshift(R)))
title('Modulated signal noise spectrum')
xlabel('Frequency')
```

The m-files ussb_mod.m and lssb_mod.m given next modulate the message signal given in vector m using USSB and LSSB modulation schemes.

M-FILE

```
function u=ussb_mod(m,ts,fc)
%                   u=ussb_mod(m,ts,fc)
%USSB_MOD    takes signal m sampled at ts and carrier
%                   freq. fc as input and returns the USSB modulated
%                   signal. ts << 1/2fc.
t=[0:length(m)−1]*ts;
u=m.*cos(2*pi*t)−imag(hilbert(m)).*sin(2*pi*t);
```

M-FILE

```
function u=lssb_mod(m,ts,fc)
%                   u=lssb_mod(m,ts,fc)
%LSSB_MOD    takes signal m sampled at ts and carrier
%                   freq. fc as input and returns the LSSB modulated
%                   signal. ts << 1/2fc.
t=[0:length(m)−1]*ts;
u=m.*cos(2*pi*t)+imag(hilbert(m)).*sin(2*pi*t);
```

3.3   Demodulation of AM Signals

Demodulation is the process of extracting the message signal from the modulated signal. The demodulation process depends on the type of modulation employed. For DSB-AM and SSB-AM, the demodulation method is coherent demodulation, which requires the existence of a signal with the same frequency and phase of the carrier at the receiver. For conventional AM, envelope detectors are used for demodulation. In this case precise knowledge of the frequency and the phase of the carrier at the receiver is not crucial, so the demodulation process is much easier. Coherent demodulation for DSB-AM and SSB-AM

Figure 3.9   Demodulation of DSB-AM signals

consists of multiplying (mixing) the modulated signal by a sinusoidal with the same frequency and phase of the carrier and then passing the product through a lowpass filter. The oscillator that generates the required sinusoidal at the receiver is called the *local oscillator.*

3.3.1   DSB-AM Demodulation

In the DSB case the modulated signal is given by $A_c m(t) \cos(2\pi f_c t)$, which when multiplied by $\cos(2\pi f_c t)$, or mixed with $\cos(2\pi f_c t)$, results in

$$y(t) = A_c m(t) \cos(2\pi f_c t) \cos(2\pi f_c t) = \frac{A_c}{2} m(t) + \frac{A_c}{2} m(t) \cos(4\pi f_c t) \qquad (3.3.1)$$

where $y(t)$ denotes the mixer output, and its Fourier transform is given by

$$Y(f) = \frac{A_c}{2} M(f) + \frac{A_c}{4} M(f - 2f_c) + \frac{A_c}{4} M(f + 2f_c) \qquad (3.3.2)$$

As it is seen, the mixer output has a lowpass component of $(A_c/2)M(f)$ and high-frequency components in the neighborhood of $\pm 2f_c$. When $y(t)$ passes through a lowpass filter with bandwidth W, the high-frequency components will be filtered out and the lowpass component, $(A_c/2)m(t)$, which is proportional to the message signal, will be demodulated. This process is shown in Figure 3.9.

───────(ILLUSTRATIVE PROBLEM)───────────────────────────

Illustrative Problem 3.5 [DSB-AM demodulation] The message signal $m(t)$ is defined as

$$m(t) = \begin{cases} 1, & 0 \leq t < \frac{t_0}{3} \\ -2, & \frac{t_0}{3} \leq t < \frac{2t_0}{3} \\ 0, & \text{otherwise} \end{cases}$$

This message DSB-AM modulates the carrier $c(t) = \cos 2\pi f_c t$, and the resulting modulated signal is denoted by $u(t)$. It is assumed that $t_0 = 0.15$ s and $f_c = 250$ Hz.

1. Obtain the expression for $u(t)$.

2. Derive the spectra of $m(t)$ and $u(t)$.

3. Demodulate the modulated signal $u(t)$ and recover $m(t)$. Plot the results in the time and frequency domains.

SOLUTION

1, 2. The first two parts of this problem are the same as the first two parts of Illustrative Problem 3.1, and we repeat only those results here:

$$u(t) = \left[\Pi \left(\frac{t - 0.025}{0.05} \right) - 2\Pi \left(\frac{t - 0.075}{0.05} \right) \right] \cos(500\pi t)$$

and

$$\mathcal{F}[m(t)] = \frac{t_0}{3} e^{-j\pi f t_0/3} \operatorname{sinc} \left(\frac{t_0 f}{3} \right) - \frac{2t_0}{3} e^{-j\pi f t_0} \operatorname{sinc} \left(\frac{t_0 f}{3} \right)$$

$$= \frac{t_0}{3} e^{-j\pi f t_0/3} \operatorname{sinc} \left(\frac{t_0 f}{3} \right) \left(1 - 2e^{-j 2\pi t_0 f/3} \right)$$

$$= 0.05 e^{-0.05 j\pi f} \operatorname{sinc}(0.05 f) \left(1 - 2e^{-0.01 j\pi f} \right)$$

Therefore,

$$U(f) = 0.025 e^{-0.05 j\pi (f-250)} \operatorname{sinc}(0.05(f - 250)) \left(1 - 2e^{-0.1 j\pi (f-250)} \right)$$

$$+ 0.025 e^{-0.05 j\pi (f+250)} \operatorname{sinc}(0.05(f + 250)) \left(1 - 2e^{-0.1 j\pi (f+250)} \right)$$

3. To demodulate, we multiply $u(t)$ by $\cos(2\pi f_c t)$ to obtain the mixer output $y(t)$:

$$y(t) = u(t) \cos(2\pi f_c t)$$

$$= \left[\Pi \left(\frac{t - 0.025}{0.05} \right) - 2\Pi \left(\frac{t - 0.075}{0.05} \right) \right] \cos^2(500\pi t)$$

$$= \frac{1}{2} \left[\Pi \left(\frac{t - 0.025}{0.05} \right) - 2\Pi \left(\frac{t - 0.075}{0.05} \right) \right]$$

$$+ \frac{1}{2} \left[\Pi \left(\frac{t - 0.025}{0.05} \right) - 2\Pi \left(\frac{t - 0.075}{0.05} \right) \right] \cos(1000\pi t)$$

whose Fourier transform is given by

$$Y(f) = 0.025 e^{-0.05 j\pi f} \operatorname{sinc}(0.05 f) \left(1 - 2e^{-0.01 j\pi f} \right)$$

$$+ 0.0125 e^{-0.05 j\pi (f-500)} \operatorname{sinc}(0.05(f - 500)) \left(1 - 2e^{-0.1 j\pi (f-500)} \right)$$

$$+ 0.0125 e^{-0.05 j\pi (f+500)} \operatorname{sinc}(0.05(f + 500)) \left(1 - 2e^{-0.1 j\pi (f+500)} \right)$$

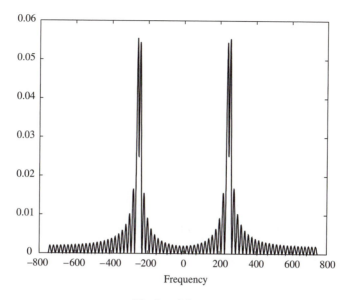

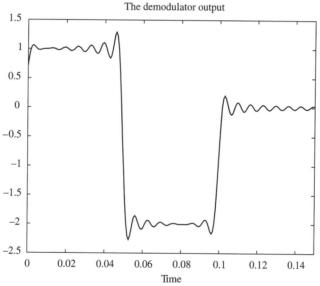

Figure 3.10   Spectra of the modulated signal and the mixer output in Illustrative
Problem 3.5

where the first term corresponds to the message signal and the last two terms correspond to the high-frequency terms at twice the carrier frequency. We see that filtering
the first term yields the original message signal (up to a proportionality constant). A
plot of the magnitudes of $U(f)$ and $Y(f)$ is shown in Figure 3.10.

As shown, the spectrum of the mixer output has a lowpass component that is quite similar to the spectrum of the message signal, except for a factor of $\frac{1}{2}$, and a bandpass component located at $\pm 2 f_c$ (in this case, 500 Hz). Using a lowpass filter, we can simply separate the lowpass component from the bandpass component. In order to recover the message signal $m(t)$, we pass $y(t)$ through a lowpass filter with a bandwidth of 150 Hz. The choice of the bandwidth of the filter is more or less arbitrary here because the message signal is not strictly bandlimited. For a strictly bandlimited message signal, the appropriate choice for the bandwidth of the lowpass filter would be W, the bandwidth of the message signal. Therefore, the ideal lowpass filter employed here has a characteristic

$$H(f) = \begin{cases} 1, & |f| \leq 150 \\ 0, & \text{otherwise} \end{cases}$$

A comparison of the spectra of $m(t)$ and the demodulator output is shown in Figure 3.11, and a comparison in the time domain is shown in Figure 3.12.

The MATLAB script for this problem follows.

M-FILE

```
% dsb_dem.m
% Matlab demonstration script for DSB-AM demodulation. The message signal
% is +1 for 0 < t < t0/3, -2 for t0/3 < t < 2t0/3, and zero otherwise.
echo on
t0=.15;                                    % signal duration
ts=1/1500;                                 % sampling interval
fc=250;                                    % carrier frequency
fs=1/ts;                                   % sampling frequency
t=[0:ts:t0];                               % time vector
df=0.3;                                    % desired frequency resolution
% message signal
m=[ones(1,t0/(3*ts)),−2*ones(1,t0/(3*ts)),zeros(1,t0/(3*ts)+1)];
c=cos(2*pi*fc.*t);                         % carrier signal
u=m.*c;                                    % modulated signal
y=u.*c;                                    % mixing
[M,m,df1]=fftseq(m,ts,df);                 % Fourier transform
M=M/fs;                                    % scaling
[U,u,df1]=fftseq(u,ts,df);                 % Fourier transform
U=U/fs;                                    % scaling
[Y,y,df1]=fftseq(y,ts,df);                 % Fourier transform
Y=Y/fs;                                    % scaling
f_cutoff=150;                              % cutoff freq. of the filter
n_cutoff=floor(150/df1);                   % design the filter
f=[0:df1:df1*(length(y)−1)]−fs/2;
H=zeros(size(f));
H(1:n_cutoff)=2*ones(1,n_cutoff);
H(length(f)−n_cutoff+1:length(f))=2*ones(1,n_cutoff);
DEM=H.*Y;                                  % spectrum of the filter output
dem=real(ifft(DEM))*fs;                    % filter output
pause % Press a key to see the effect of mixing
```

```
clf
subplot(3,1,1)
plot(f,fftshift(abs(M)))
title('Spectrum of the Message Signal')
xlabel('Frequency')
subplot(3,1,2)
plot(f,fftshift(abs(U)))
title('Spectrum of the Modulated Signal')
xlabel('Frequency')
subplot(3,1,3)
plot(f,fftshift(abs(Y)))
title('Spectrum of the Mixer Output')
xlabel('Frequency')
pause % Press a key to see the effect of filtering on the mixer output
clf
subplot(3,1,1)
plot(f,fftshift(abs(Y)))
title('Spectrum of the Mixer Output')
xlabel('Frequency')
subplot(3,1,2)
plot(f,fftshift(abs(H)))
title('Lowpass Filter Characteristics')
xlabel('Frequency')
subplot(3,1,3)
plot(f,fftshift(abs(DEM)))
title('Spectrum of the Demodulator Output')
xlabel('Frequency')
pause % Press a key to compare the spectra of the message and the received signal
clf
subplot(2,1,1)
plot(f,fftshift(abs(M)))
title('Spectrum of the Message Signal')
xlabel('Frequency')
subplot(2,1,2)
plot(f,fftshift(abs(DEM)))
title('Spectrum of the Demodulator Output')
xlabel('Frequency')
pause % Press a key to see the message and the demodulator output signals
subplot(2,1,1)
plot(t,m(1:length(t)))
title('The Message Signal')
xlabel('Time')
subplot(2,1,2)
plot(t,dem(1:length(t)))
title('The Demodulator Output')
xlabel('Time')
```

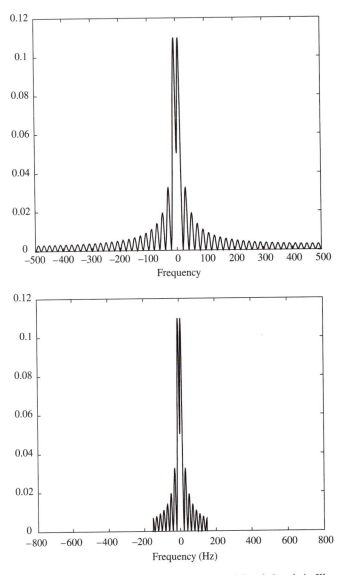

Figure 3.11   Spectra of the message and the demodulated signals in Illustrative
Problem 3.5

═══ **ILLUSTRATIVE PROBLEM** ═══

Illustrative Problem 3.6   [Effect of phase error on DSB-AM demodulation] In the demodulation of DSB-AM signals we assumed that the phase of the local oscillator is equal
to the phase of the carrier. If that is not the case —that is, if there exists a phase shift ϕ
between the local oscillator and the carrier—how would the demodulation process change?

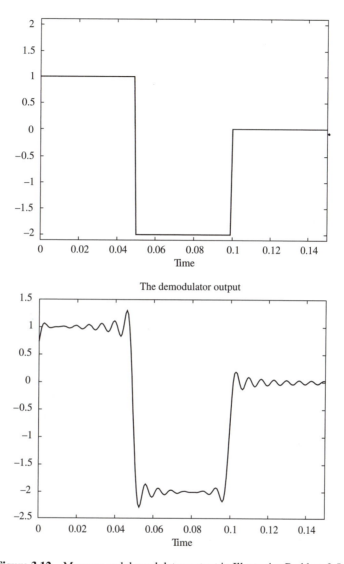

Figure 3.12   Message and demodulator output in Illustrative Problem 3.5

SOLUTION

In this case we have $u(t) = A_c m(t) \cos(2\pi f_c t)$, and the local oscillator generates a sinusoidal signal given by $\cos(2\pi f_c t + \phi)$. Mixing these two signals gives

$$y(t) = A_c m(t) \cos(2\pi f_c t) \times \cos(2\pi f_c t + \phi) \tag{3.3.3}$$

$$= \frac{A_c}{2} m(t) \cos(\phi) + \frac{A_c}{2} m(t) \cos(4\pi f_c t + \phi) \tag{3.3.4}$$

As before, there are two terms present in the mixer output. The bandpass term can be filtered out by a lowpass filter. The lowpass term, $(A_c/2)m(t)\cos(\phi)$, depends on ϕ, however. The power in the lowpass term is given by

$$P_{\text{dem}} = \frac{A_c^2}{4} P_m \cos^2\phi \tag{3.3.5}$$

where P_m denotes the power in the message signal. We can see, therefore, that in this case we can recover the message signal with essentially no distortion, but we will suffer a power loss of $\cos^2\phi$. For $\phi = \pi/4$ this power loss is 3 dB, and for $\phi = \pi/2$ nothing is recovered in the demodulation process.

3.3.2   SSB-AM Demodulation

The demodulation process of SSB-AM signals is basically the same as the demodulation process for DSB-AM signals—that is, mixing followed by lowpass filtering. In this case,

$$u(t) = \frac{A_c}{2}m(t)\cos(2\pi f_c t) \mp \frac{A_c}{2}\hat{m}(t)\sin(2\pi f_c t) \tag{3.3.6}$$

where the minus sign corresponds to the USSB and the plus sign corresponds to the LSSB. Mixing $u(t)$ with the local oscillator output, we obtain

$$\begin{aligned}
y(t) &= \frac{A_c}{2}m(t)\cos^2(2\pi f_c t) \mp \frac{A_c}{2}\hat{m}(t)\sin(2\pi f_c t)\cos(2\pi f_c t) \\
&= \frac{A_c}{4}m(t) + \frac{A_c}{4}m(t)\cos(4\pi f_c t) \mp \frac{A_c}{4}\hat{m}(t)\sin(4\pi f_c t)
\end{aligned} \tag{3.3.7}$$

which contains bandpass components at $\pm 2f_c$ and a lowpass component proportional to the message signal. The lowpass component can be filtered out using a lowpass filter to recover the message signal. This process for the USSB-AM case is depicted in Figure 3.13.

──────(**ILLUSTRATIVE PROBLEM**)──────────────────────

Illustrative Problem 3.7 [LSSB-AM example] In a USSB-AM modulation system, if the message signal is

$$m(t) = \begin{cases} 1, & 0 \le t < \frac{t_0}{3} \\ -2, & \frac{t_0}{3} \le t < \frac{2t_0}{3} \\ 0, & \text{otherwise} \end{cases}$$

with $t_0 = 0.15$ s, and the carrier has a frequency of 250 Hz, find $U(f)$ and $Y(f)$ and compare the demodulated signal with the message signal.

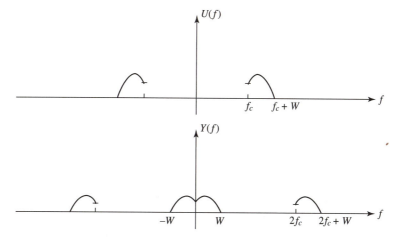

Figure 3.13   Demodulation of USSB-AM signals

SOLUTION

The modulated signal and its spectrum are given in Illustrative Problem 3.4. The expression for $U(f)$ is given by

$$U(f) = \begin{cases} 0.025e^{-0.05j\pi(f-250)}\text{sinc}(0.05(f-250))\left(1-2e^{-0.1j\pi(f-250)}\right) \\ +0.025e^{-0.05j\pi(f+250)}\text{sinc}(0.05(f+250))\left(1-2e^{-0.1j\pi(f+250)}\right), & |f| \leq f_c \\ 0, & \text{otherwise} \end{cases}$$

and

$$Y(f) = \frac{1}{2}U(f-f_c) + \frac{1}{2}U(f+f_c)$$

$$\approx \begin{cases} 0.0125e^{-0.05j\pi f}\text{sinc}(0.05f)\left(1-2e^{-0.01j\pi f}\right), & |f| \leq f_c \\ 0.0125e^{-0.05j\pi(f-500)}\text{sinc}(0.05(f-500))\left(1-2e^{-0.01j\pi(f-500)}\right), & f_c \leq f \leq 2f_c \\ 0.0125e^{-0.05j\pi(f+500)}\text{sinc}(0.05(f+500))\left(1-2e^{-0.01j\pi(f+500)}\right), & -2f_c \leq f \leq -f_c \\ 0, & \text{otherwise} \end{cases}$$

A plot of $Y(f)$ is shown in Figure 3.14. The signal $y(t)$ is filtered by a lowpass filter with a cutoff frequency of 150 Hz; the spectrum of the output is shown in Figure 3.15. In Figure 3.16 the original message signal is compared with the demodulated signal.

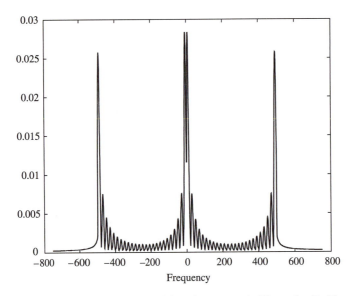

Figure 3.14   Magnitude spectrum of the mixer output in Illustrative Problem 3.7

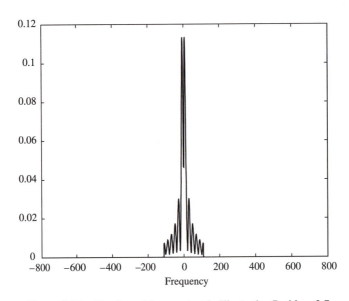

Figure 3.15   The demodulator output in Illustrative Problem 3.7

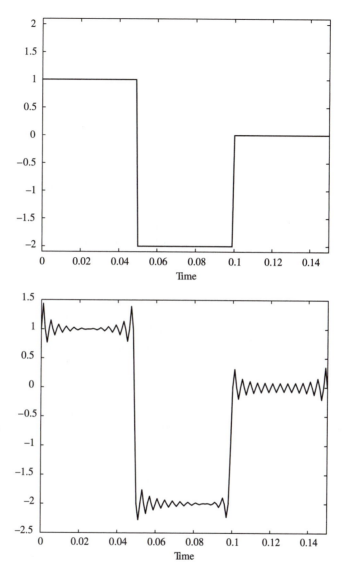

Figure 3.16   The message signal and the demodulator output in Illustrative Problem 3.7

The MATLAB script for this problem follows.

M-FILE

% lssb_dem.m
% Matlab demonstration script for LSSB-AM demodulation. The message signal
% is +1 for 0 < t < t0/3, -2 for t0/3 < t < 2t0/3, and zero otherwise.
echo on

```
t0=.15;                                        % signal duration
ts=1/1500;                                     % sampling interval
fc=250;                                        % carrier frequency
fs=1/ts;                                       % sampling frequency
df=0.25;                                       % desired freq.resolution
t=[0:ts:t0];                                   % time vector
% the message vector
m=[ones(1,t0/(3*ts)),−2*ones(1,t0/(3*ts)),zeros(1,t0/(3*ts)+1)];
c=cos(2*pi*fc.*t);                             % carrier vector
udsb=m.*c;                                     % DSB modulated signal
[UDSB,udsb,df1]=fftseq(udsb,ts,df);            % Fourier transform
UDSB=UDSB/fs;                                  % scaling
n2=ceil(fc/df1);                               % location of carrier in freq. vector
% remove the upper sideband from DSB
UDSB(n2:length(UDSB)−n2)=zeros(size(UDSB(n2:length(UDSB)−n2)));
ULSSB=UDSB;                                        % generate LSSB-AM spectrum
[M,m,df1]=fftseq(m,ts,df);                     % spectrum of the message signal
M=M/fs;                                        % scaling
f=[0:df1:df1*(length(M)−1)]−fs/2;              % frequency vector
u=real(ifft(ULSSB))*fs;                        % generate LSSB signal from spectrum
% mixing
y=u.*cos(2*pi*fc*[0:ts:ts*(length(u)−1)]);
[Y,y,df1]=fftseq(y,ts,df);                     % spectrum of the output of the mixer
Y=Y/fs;                                        % scaling
f_cutoff=150;                                  % choose the cutoff freq. of the filter
n_cutoff=floor(150/df);                        % design the filter
H=zeros(size(f));
H(1:n_cutoff)=4*ones(1,n_cutoff);
% spectrum of the filter output
H(length(f)−n_cutoff+1:length(f))=4*ones(1,n_cutoff);
DEM=H.*Y;                                      % spectrum of the filter output
dem=real(ifft(DEM))*fs;                        % filter output
pause % Press a key to see the effect of mixing
clf
subplot(3,1,1)
plot(f,fftshift(abs(M)))
title('Spectrum of the Message Signal')
xlabel('Frequency')
subplot(3,1,2)
plot(f,fftshift(abs(ULSSB)))
title('Spectrum of the Modulated Signal')
xlabel('Frequency')
subplot(3,1,3)
plot(f,fftshift(abs(Y)))
title('Spectrum of the Mixer Output')
xlabel('Frequency')
pause % Press a key to see the effect of filtering on the mixer output
clf
subplot(3,1,1)
plot(f,fftshift(abs(Y)))
title('Spectrum of the Mixer Output')
xlabel('Frequency')
subplot(3,1,2)
plot(f,fftshift(abs(H)))
```

```
title('Lowpass Filter Characteristics')
xlabel('Frequency')
subplot(3,1,3)
plot(f,fftshift(abs(DEM)))
title('Spectrum of the Demodulator Output')
xlabel('Frequency')
pause % Press a key to see the message and the demodulator output signals
subplot(2,1,1)
plot(t,m(1:length(t)))
title('The Message Signal')
xlabel('Time')
subplot(2,1,2)
plot(t,dem(1:length(t)))
title('The Demodulator Output')
xlabel('Time')
```

⬛ **ILLUSTRATIVE PROBLEM** ⬛ ─────────────────────

Illustrative Problem 3.8 [Effect of phase error on SSB-AM] What is the effect of phase error on SSB-AM?

⬛ **SOLUTION** ⬛ ─────────────────────────────────────

Assuming that the local oscillator generates a sinusoidal with a phase offset of ϕ with respect to the carrier, we have

$$y(t) = u(t)\cos(2\pi f_c t + \phi)$$

$$= \left[\frac{A_c}{2}m(t)\cos(2\pi f_c t) \mp \frac{A_c}{2}\hat{m}(t)\sin(2\pi f_c t)\right]\cos(2\pi f_c t + \phi)$$

$$= \frac{A_c}{4}m(t)\cos\phi \pm \frac{A_c}{4}\hat{m}(t)\sin\phi + \text{high-frequency terms} \qquad (3.3.8)$$

As seen, unlike the DSB-AM case, the effect of the phase offset here is not simply attenuating the demodulated signal. Here the demodulated signal is attenuated by a factor of $\cos\phi$ as well as distorted by addition of the $\pm(A_c/4)\hat{m}(t)\sin\phi$ term. In the special case of $\phi = \pi/2$, the Hilbert transform of the signal will be demodulated instead of the signal itself.

3.3.3 Conventional AM Demodulation

We have already seen that conventional AM is inferior to DSB-AM and SSB-AM when power and SNR are considered. The reason is that a usually large part of the modulated signal power is in the carrier component that does not carry information. The role of the

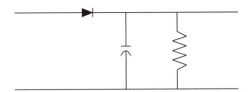

Figure 3.17   A simple envelope detector

carrier component is to make the demodulation of the conventional AM easier via enve-lope detection, as opposed to coherent demodulation required for DSB-AM and SSB-AM. Therefore, demodulation of AM signals is significantly less complex than the demodula-tion of DSB-AM and SSB-AM signals. Hence, this modulation scheme is widely used in broadcasting, where there exists a single transmitter and numerous receivers whose cost should be kept low. In envelope detection the envelope of the modulated signal is de-tected via a simple circuit consisting of a diode, a resistor, and a capacitor, as shown in Figure 3.17.

Mathematically, the envelope detector generates the envelope of the conventional AM signal, which is

$$V(t) = |1 + am_n(t)| \tag{3.3.9}$$

Because $1 + m_n(t) \geq 0$, we conclude that

$$V(t) = 1 + am_n(t) \tag{3.3.10}$$

where $m_n(t)$ is proportional to the message signal $m(t)$ and 1 corresponds to the carrier component that can be separated by a dc-block circuit. As seen in the preceding procedure, there is no need for knowledge of ϕ, the phase of the carrier signal. That is why such a demodulation scheme is called *noncoherent,* or *asynchronous, demodulation.* Recall from Chapter 1 that the envelope of a bandpass signal can be expressed as the magnitude of its lowpass equivalent signal. Thus, if $u(t)$ is the bandpass signal with central frequency f_c and the lowpass equivalent to $u(t)$ is denoted by $u_l(t)$, then the envelope of $u(t)$, denoted by $V(t)$, can be expressed as

$$V(t) = \sqrt{u_{lr}^2(t) + u_{li}^2(t)}$$
$$= \sqrt{u_c^2(t) + u_s^2(t)} \tag{3.3.11}$$

where $u_c(t)$ and $u_s(t)$ represent the in-phase and the quadrature components of the band-pass signal $u(t)$. Therefore, in order to obtain the envelope, it is enough to obtain the lowpass equivalent of the bandpass signal. The envelope is simply the magnitude of the lowpass equivalent of the bandpass signal.

─────(ILLUSTRATIVE PROBLEM)─────────────────────

Illustrative Problem 3.9 [Envelope detection] The message signal

$$
m(t) = \begin{cases} 1, & 0 \le t < \frac{t_0}{3} \\ -2, & \frac{t_0}{3} \le t < \frac{2t_0}{3} \\ 0, & \text{otherwise} \end{cases}
$$

modulates the carrier $c(t) = \cos(2\pi f_c t)$ using a conventional AM modulation scheme. It is assumed that $f_c = 250$ Hz and $t_0 = 0.15$ s, and the modulation index is $a = 0.85$.

1. Using envelope detection, demodulate the message signal.

2. If the message signal is periodic with a period equal to t_0 and if an AWGN process is added to the modulated signal such that the power in the noise process is one-hundredth the power in the modulated signal, use an envelope demodulator to demodulate the received signal. Compare this case with the case where there is no noise present.

─────(SOLUTION)─────────────────────

1. As in Illustrative Problem 3.3, we have

$$
u(t) = \left[1 + 0.85 \frac{m(t)}{2} \right] \cos(2\pi f_c t)
$$

$$
= \left[1 + 0.425 \Pi \left(\frac{t - 0.025}{0.05} \right) - 0.85 \Pi \left(\frac{t - 0.075}{0.05} \right) \right] \cos(500\pi t)
$$

If an envelope detector is used to demodulate the signal and the carrier component is removed by a dc-block, then the original message $m(t)$ is recovered. Note that a crucial point in the recovery of $m(t)$ is that for all values of t, the expression $1 + am_n(t)$ is positive; therefore, the envelope of the signal $[1 + am_n(t)] \cos(2\pi f_c t)$, which is $V(t) = |1 + am_n(t)|$, is equal to $1 + am_n(t)$, from which $m(t)$ can be recovered easily. Plots of the conventional AM modulated signal and its envelope as detected by the envelope detector are shown in Figure 3.18.

After the envelope detector separates the envelope of the modulated signal, the dc component of the signal is removed and the signal is scaled to generate the demodulator output. Plots of the original message signal and the demodulator output are shown in Figure 3.19.

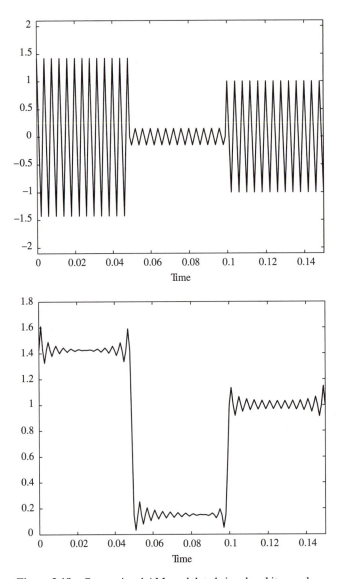

Figure 3.18   Conventional AM modulated signal and its envelope

2. When noise is present, there will also be some distortion due to the noise. In Figure 3.20 the received signal and its envelope are shown. In Figure 3.21 the message signal and the demodulated signal are compared for this case.

The MATLAB script for this problem follows.

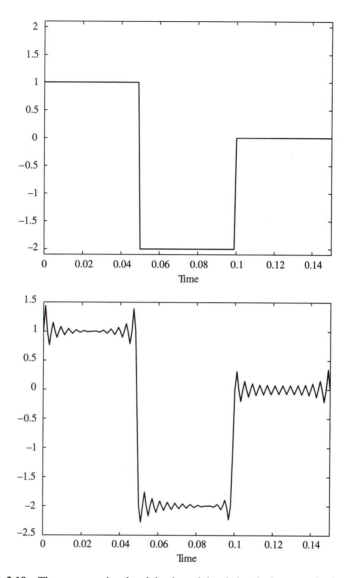

Figure 3.19   The message signal and the demodulated signal when no noise is present

M-FILE

```
% am-dem.m
% Matlab demonstration script for envelope detection. The message signal
% is +1 for 0 < t < t0/3, -2 for t0/3 < t < 2t0/3, and zero otherwise.
echo on
t0=.15;                                 % signal duration
ts=0.001;                               % sampling interval
```

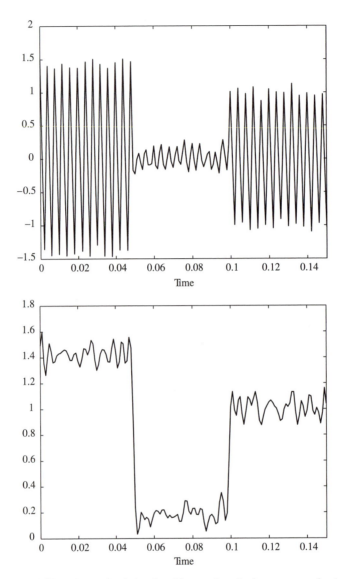

Figure 3.20   The received signal and its envelope in the presence of noise

```
fc=250;                              % carrier frequency
a=0.85;                              % modulation index
fs=1/ts;                             % sampling frequency
t=[0:ts:t0];                         % time vector
df=0.25;                             % required frequency resolution
% message signal
m=[ones(1,t0/(3*ts)),−2*ones(1,t0/(3*ts)),zeros(1,t0/(3*ts)+1)];
c=cos(2*pi*fc.*t);                   % carrier signal
```

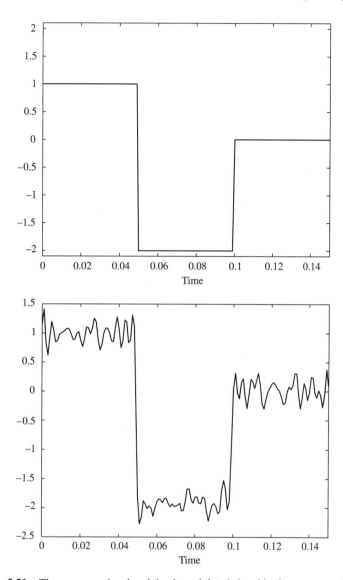

Figure 3.21   The message signal and the demodulated signal in the presence of noise

```
m_n=m/max(abs(m));                  % normalized message signal
[M,m,df1]=fftseq(m,ts,df);          % Fourier transform
f=[0:df1:df1*(length(m)−1)]−fs/2;   % frequency vector
u=(1+a*m_n).*c;                     % modulated signal
[U,u,df1]=fftseq(u,ts,df);          % Fourier transform
env=env_phas(u);                    % find the envelope
dem1=2*(env−1)/a;                   % remove dc and rescale
signal_power=spower(u(1:length(t)));  % power in modulated signal
noise_power=signal_power/100;       % noise power
```

```
noise_std=sqrt(noise_power);            % noise standard deviation
noise=noise_std*randn(1,length(u));     % generate noise
r=u+noise;                              % add noise to the modulated signal
[R,r,df1]=fftseq(r,ts,df);              % Fourier transform
env_r=env_phas(r);                      % envelope, when noise is present
dem2=2*(env_r−1)/a;                     % demodulate in the presence of noise
pause   % Press any key to see a plot of the message
subplot(2,1,1)
plot(t,m(1:length(t)))
axis([0 0.15 −2.1 2.1])
xlabel('Time')
title('The message signal')
pause   % Press any key to see a plot of the modulated signal
subplot(2,1,2)
plot(t,u(1:length(t)))
axis([0 0.15 −2.1 2.1])
xlabel('Time')
title('The modulated signal')
pause   % Press a key to see the envelope of the modulated signal
clf
subplot(2,1,1)
plot(t,u(1:length(t)))
axis([0 0.15 −2.1 2.1])
xlabel('Time')
title('The modulated signal')
subplot(2,1,2)
plot(t,env(1:length(t)))
xlabel('Time')
title('Envelope of the modulated signal')
pause   % Press a key to compare the message and the demodulated signal
clf
subplot(2,1,1)
plot(t,m(1:length(t)))
axis([0 0.15 −2.1 2.1])
xlabel('Time')
title('The message signal')
subplot(2,1,2)
plot(t,dem1(1:length(t)))
xlabel('Time')
title('The demodulated signal')
pause   % Press a key to compare in the presence of noise
clf
subplot(2,1,1)
plot(t,m(1:length(t)))
axis([0 0.15 −2.1 2.1])
xlabel('Time')
title('The message signal')
subplot(2,1,2)
plot(t,dem2(1:length(t)))
xlabel('Time')
title('The demodulated signal in the presence of noise')
```

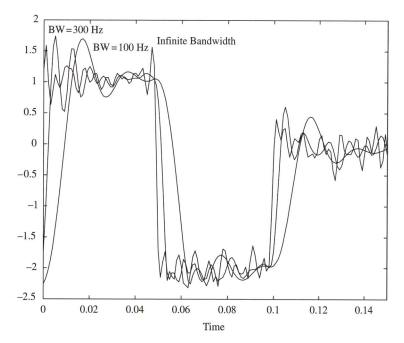

Figure 3.22   Effect of the bandwidth of the noise-limiting filter on the output of the envelope detector

---

COMMENT

In the demodulation process above, we have neglected the effect of the noise-limiting filter, which is a bandpass filter in the first stage of any receiver. In practice the received signal $r(t)$ is passed through the noise-limiting filter and then supplied to the envelope detector. In the preceding example, since the message bandwidth is not finite, passing $r(t)$ through any bandpass filter will cause distortion on the demodulated message, but it will also decrease the amount of noise in the demodulator output. In Figure 3.22 we have plotted the demodulator outputs when noise-limiting filters of different bandwidths are used. The case of infinite bandwidth is equivalent to the result shown in Figure 3.21.

3.4   Angle Modulation

Angle-modulation schemes, which include frequency modulation (FM) and phase modulation (PM), belong to the class of nonlinear modulation schemes. This family of modulation schemes is characterized by their high-bandwidth requirements and good performance in the presence of noise. These schemes can be visualized as modulation techniques that trade off bandwidth for power and, therefore, are used in situations where bandwidth is not the major concern and a high SNR is required. Frequency modulation is widely used in high-fidelity FM broadcasting, TV audio broadcasting, microwave carrier modulation, and point-to-point communication systems.

In our treatment of angle-modulation schemes, we again concentrate on their five basic properties—namely, time-domain representation, frequency-domain representation, bandwidth, power content, and, finally, SNR. Since there is a close relationship between PM and FM, we will treat them in parallel, with emphasis on FM.

The time-domain representation of angle-modulated signals, when the carrier is $c(t) = A_c \cos(2\pi f_c t)$ and the message signal is $m(t)$, is given by

$$u(t) = \begin{cases} A_c \cos\left(2\pi f_c t + k_p m(t)\right), & \text{PM} \\ A_c \cos\left(2\pi f_c t + 2\pi k_f \int_{-\infty}^{t} m(\tau)\, d\tau\right), & \text{FM} \end{cases} \quad (3.4.1)$$

where k_f and k_p represent the *deviation constants* of FM and PM, respectively. The frequency-domain representation of angle-modulated signals is, in general, very complex due to the nonlinearity of these modulation schemes. We treat only the case where the message signal $m(t)$ is a sinusoidal signal. We assume $m(t) = a\cos(2\pi f_m t)$ for PM and $m(t) = -a\sin(2\pi f_m t)$ for FM. Then the modulated signal is of the form

$$u(t) = \begin{cases} A_c \cos(2\pi f_c t + \beta_p \cos(2\pi f_m t)), & \text{PM} \\ A_c \cos(2\pi f_c t + \beta_f \cos(2\pi f_m t)), & \text{FM} \end{cases} \quad (3.4.2)$$

where

$$\begin{cases} \beta_p = k_p a \\ \beta_f = \dfrac{k_f a}{f_m} \end{cases} \quad (3.4.3)$$

and β_p and β_f are the *modulation indices* of PM and FM, respectively. In general, for a nonsinusoidal $m(t)$, the modulation indices are defined as

$$\begin{cases} \beta_p = k_p \max |m(t)| \\ \beta_f = \dfrac{k_f \max |m(t)|}{W} \end{cases} \quad (3.4.4)$$

where W is the bandwidth of the message signal $m(t)$. In the case of a sinusoidal message signal, the modulated signal can be represented by

$$u(t) = \sum_{n=-\infty}^{\infty} A_c J_n(\beta) \cos(2\pi(f_c + n f_m)t) \quad (3.4.5)$$

where $J_n(\beta)$ is the Bessel function of the first kind and of order n and β is either β_p or β_f, depending on whether we are dealing with PM or FM. In the frequency domain we have

$$U(f) = \sum_{n=-\infty}^{\infty} \left[\frac{A_c J_n(\beta)}{2} \delta(f - (f_c + n f_m)) + \frac{A_c J_n(\beta)}{2} \delta(f + (f_c + n f_m)) \right] \quad (3.4.6)$$

Obviously, the bandwidth of the modulated signal is not finite. However, we can define the *effective bandwidth* of the signal as the bandwidth containing 98% to 99% of the modulated signal power. This bandwidth is given by *Carson's rule* as

$$B_T = 2(\beta + 1)W \quad (3.4.7)$$

where β is the modulation index, W is the bandwidth of the message, and B_T is the bandwidth of the modulated signal.

The expression for the power content of the angle-modulated signals is very simple. Since the modulated signal is sinusoidal, with varying instantaneous frequency and constant amplitude, its power is constant and does not depend on the message signal. The power content for both FM and PM is given by

$$P_u = \frac{A_c^2}{2} \tag{3.4.8}$$

The SNR for angle-modulated signals, when no pre-emphasis and de-emphasis filtering is employed, is given by

$$\left(\frac{S}{N}\right)_o = \begin{cases} \dfrac{P_M \beta_p^2}{(\max|m(t)|)^2} \dfrac{P_R}{N_0 W}, & \text{PM} \\[3mm] 3 \dfrac{P_M \beta_f^2}{(\max|m(t)|)^2} \dfrac{P_R}{N_0 W}, & \text{FM} \end{cases} \tag{3.4.9}$$

Since $\max|m(t)|$ denotes the maximum magnitude of the message signal, we can interpret $P_M/(\max|m(t)|)^2$ as the power in the *normalized message signal* and denote it by P_{M_n}. When pre-emphasis and de-emphasis filters with a 3-dB cutoff frequency equal to f_0 are employed, the SNR for FM is given by

$$\left(\frac{S}{N}\right)_{\text{oPD}} = \frac{(W/f_0)^3}{3\,[W/f_0 - \arctan(W/f_0)]} \left(\frac{S}{N}\right)_o \tag{3.4.10}$$

where $(S/N)_o$ is the SNR without pre-emphasis and de-emphasis filtering given by Equation (3.4.9).

─────■ ILLUSTRATIVE PROBLEM ■─────────────────────

Illustrative Problem 3.10 [Frequency modulation] The message signal

$$m(t) = \begin{cases} 1, & 0 \le t < \frac{t_0}{3} \\[2mm] -2, & \frac{t_0}{3} \le t < \frac{2t_0}{3} \\[2mm] 0, & \text{otherwise} \end{cases}$$

modulates the carrier $c(t) = \cos(2\pi f_c t)$ using a frequency-modulation scheme. It is assumed that $f_c = 200$ Hz and $t_0 = 0.15$ s; the deviation constant is $k_f = 50$.

1. Plot the modulated signal.

2. Determine the spectra of the message and the modulated signals.

SOLUTION

1. We have

$$u(t) = A_c \cos\left(2\pi f_c t + 2\pi k_f \int_{-\infty}^{t} m(\tau)\, d\tau\right)$$

We have to find $\int_{-\infty}^{t} m(\tau)\, d\tau$. This can be done numerically or analytically, and the results are shown in Figure 3.23. Using the relation for $u(t)$ and the value of the integral of $m(t)$, as shown above, we obtain the expression for $u(t)$. A plot of $m(t)$ and $u(t)$ is shown in Figure 3.24.

2. Using MATLAB's Fourier transform routines, we obtain the expression for the spectrum of $u(t)$ shown in Figure 3.25. It is readily seen that unlike AM, in the FM case there does not exist a clear similarity between the spectrum of the message and the spectrum of the modulated signal. In this particular example the bandwidth of the message signal is not finite, and therefore to define the index of modulation, an approximate bandwidth for the message should be used in the expression

$$\beta = \frac{k_f \max |m(t)|}{W} \tag{3.4.11}$$

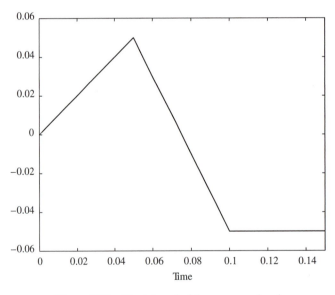

Figure 3.23   The integral of the message signal

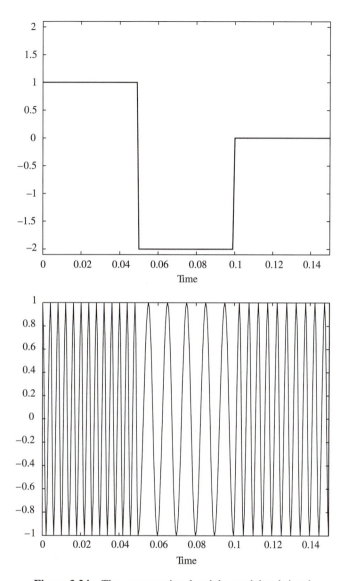

Figure 3.24   The message signal and the modulated signal

We can, for example, define the bandwidth as the width of the main lobe of the
spectrum of $m(t)$, which results in

$$W = 20 \text{ Hz}$$

and so

$$\beta = \frac{50 \times 2}{20} = 10$$

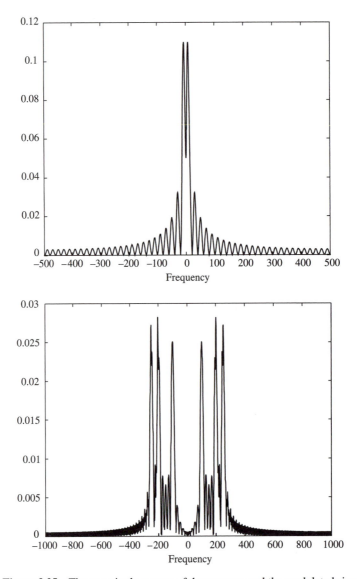

Figure 3.25   The magnitude spectra of the message and the modulated signal

The MATLAB script for this problem follows.

M-FILE

```
% fm1.m
% Matlab demonstration script for frequency modulation. The message signal
% is +1 for 0 < t < t0/3, -2 for t0/3 < t < 2t0/3, and zero otherwise.
echo on
t0=.15;                          % signal duration
ts=0.0005;                       % sampling interval
fc=200;                          % carrier frequency
kf=50;                           % modulation index
fs=1/ts;                         % sampling frequency
t=[0:ts:t0];                     % time vector
df=0.25;                         % required frequency resolution
% message signal
m=[ones(1,t0/(3*ts)),−2*ones(1,t0/(3*ts)),zeros(1,t0/(3*ts)+1)];
int_m(1)=0;
for i=1:length(t)−1              % integral of m
   int_m(i+1)=int_m(i)+m(i)*ts;
   echo off ;
end
echo on ;
[M,m,df1]=fftseq(m,ts,df);       % Fourier transform
M=M/fs;                          % scaling
f=[0:df1:df1*(length(m)−1)]−fs/2;  % frequency vector
u=cos(2*pi*fc*t+2*pi*kf*int_m);  % modulated signal
[U,u,df1]=fftseq(u,ts,df);       % Fourier transform
U=U/fs;                          % scaling
pause   % Press any key to see plot of the message and the modulated signal
subplot(2,1,1)
plot(t,m(1:length(t)))
axis([0 0.15 −2.1 2.1])
xlabel('Time')
title('The message signal')
subplot(2,1,2)
plot(t,u(1:length(t)))
axis([0 0.15 −2.1 2.1])
xlabel('Time')
title('The modulated signal')
pause    % Press any key to see plots of the magnitude of the message and the
         % modulated signal in the frequency domain.
subplot(2,1,1)
plot(f,abs(fftshift(M)))
xlabel('Frequency')
title('Magnitude spectrum of the message signal')
subplot(2,1,2)
plot(f,abs(fftshift(U)))
title('Magnitude spectrum of the modulated signal')
xlabel('Frequency')
```

ILLUSTRATIVE PROBLEM

Illustrative Problem 3.11 [Frequency modulation] Let the message signal be

$$m(t) = \begin{cases} \text{sinc}(100t), & |t| \leq t_0 \\ 0, & \text{otherwise} \end{cases}$$

where $t_0 = 0.1$. This message modulates the carrier $c(t) = \cos(2\pi f_c t)$, where $f_c = 250$ Hz. The deviation constant is $k_f = 100$.

1. Plot the modulated signal in the time and frequency domain.

2. Compare the demodulator output and the original message signal.

SOLUTION

1. We first integrate the message signal and then use the relation

$$u(t) = A_c \cos\left(2\pi f_c t + 2\pi k_f \int_{-\infty}^{t} m(\tau)\,d\tau\right)$$

to find $u(t)$. A plot of $u(t)$ together with the message signal is shown in Figure 3.26. The integral of the message signal is shown in Figure 3.27. A plot of the modulated signal in the frequency domain is shown in Figure 3.28.

2. To demodulate the FM signal, we first find the phase of the modulated signal $u(t)$. This phase is $2\pi k_f \int_{-\infty}^{t} m(\tau)\,d\tau$, which can be differentiated and divided by $2\pi k_f$ to obtain $m(t)$. Note that in order to restore the phase and undo the effect of 2π phase foldings, we employ the unwrap.m function of MATLAB. Plots of the message signal and the demodulated signal are shown in Figure 3.29. As you can see, the demodulated signal is quite similar to the message signal.

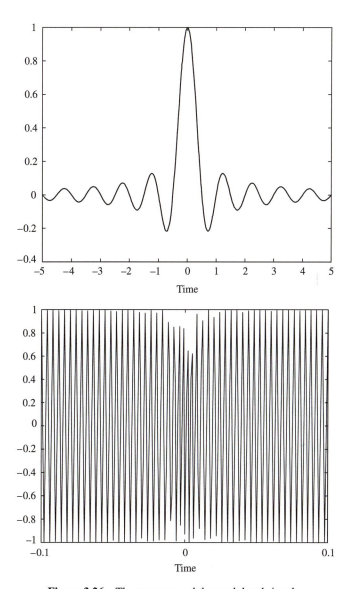

Figure 3.26   The message and the modulated signals

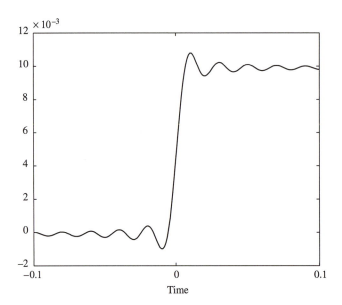

Figure 3.27   Integral of the message signal

The MATLAB script for this problem follows.

M-FILE

```
% fm2.m
% Matlab demonstration script for frequency modulation. The message signal
% is m(t)=sinc(100t).
echo on
t0=.2;                                  % signal duration
ts=0.001;                               % sampling interval
fc=250;                                 % carrier frequency
snr=20;                                 % SNR in dB (logarithmic)
fs=1/ts;                                % sampling frequency
df=0.3;                                 % required freq. resolution
t=[−t0/2:ts:t0/2];                      % time vector
kf=100;                                 % deviation constant
df=0.25;                                % required frequency resolution
m=sinc(100*t);                          % the message signal
int_m(1)=0;
for  i=1:length(t)−1                    % integral of m
  int_m(i+1)=int_m(i)+m(i)*ts;
  echo off ;
end
```

```
echo on ;
[M,m,df1]=fftseq(m,ts,df);                 % Fourier transform
M=M/fs;                                     % scaling
f=[0:df1:df1*(length(m)−1)]−fs/2;          % frequency vector
u=cos(2*pi*fc*t+2*pi*kf*int_m);            % modulated signal
[U,u,df1]=fftseq(u,ts,df);                  % Fourier transform
U=U/fs;                                     % scaling
[v,phase]=env_phas(u,ts,250);              % demodulation, find phase of u
phi=unwrap(phase);                          % restore original phase
dem=(1/(2*pi*kf))*(diff(phi)/ts);          % demodulator output, differentiate and scale phase
pause    % Press any key to see a plot of the message and the modulated signal
subplot(2,1,1)
plot(t,m(1:length(t)))
xlabel('Time')
title('The message signal')
subplot(2,1,2)
plot(t,u(1:length(t)))
xlabel('Time')
title('The modulated signal')
pause    % Press any key to see plots of the magnitude of the message and the
         % modulated signal in the frequency domain.
subplot(2,1,1)
plot(f,abs(fftshift(M)))
xlabel('Frequency')
title('Magnitude spectrum of the message signal')
subplot(2,1,2)
plot(f,abs(fftshift(U)))
title('Magnitude spectrum of the modulated signal')
xlabel('Frequency')
pause    % Press any key to see plots of the message and the demodulator output with no
         % noise
subplot(2,1,1)
plot(t,m(1:length(t)))
xlabel('Time')
title('The message signal')
subplot(2,1,2)
plot(t,dem(1:length(t)))
xlabel('Time')
title('The demodulated signal')
```

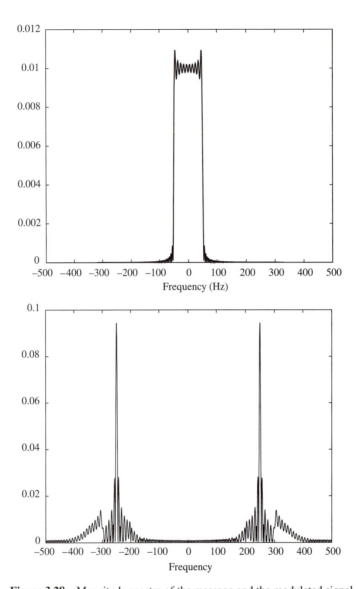

Figure 3.28   Magnitude spectra of the message and the modulated signal

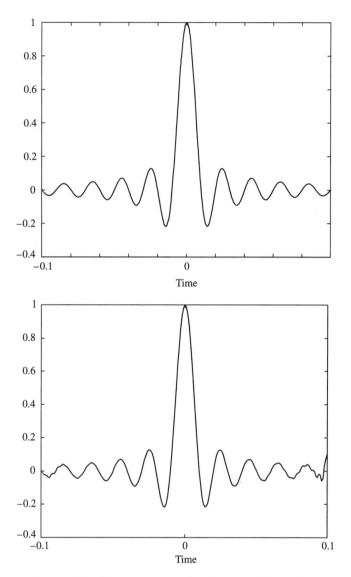

Figure 3.29   The message signal and the demodulated signal

QUESTION

A frequency-modulated signal has constant amplitude. However, in Figure 3.26, the amplitude of the signal $u(t)$ is apparently not constant. Can you explain why this happens?

Problems

3.1 Signal $m(t)$ is given by

$$m(t) = \begin{cases} t, & 0.1 \leq t < 1 \\ -t + 2, & 1 \leq t < 1.9 \\ 0.1, & \text{otherwise} \end{cases}$$

in the interval $[0, 2]$. This signal is used to DSB modulate a carrier of frequency 25 Hz and amplitude 1 to generate the modulated signal $u(t)$. Write a MATLAB m-file and use it to do the following.

    a. Plot the modulated signal.

    b. Determine the power content of the modulated signal.

    c. Determine the spectrum of the modulated signal.

    d. Determine the power-spectral density of the modulated signal and compare it with the power-spectral density of the message signal.

3.2 Repeat Problem 3.1 with

$$m(t) = \begin{cases} t, & 0 \leq t < 1 \\ -t + 2, & 1 \leq t \leq 2 \end{cases}$$

in the interval $[0, 2]$. What do you think is the difference between this problem and Problem 3.1?

3.3 Repeat Problem 3.1 with

$$m(t) = \begin{cases} \text{sinc}^2(10t), & |t| \leq 2 \\ 0, & \text{otherwise} \end{cases}$$

and a carrier with a frequency of 100 Hz.

3.4 In Problem 3.1, suppose that instead of DSB, a conventional AM scheme is used with a modulation index of $a = 0.2$.

    a. Determine and plot the spectrum of the modulated signal.

b. Change the modulation index from 0.1 to 0.9 and explain how it affects the spectrum derived in part a.

c. Plot the ratio of the power content of the sidebands to the power content of the carrier as a function of the modulation index.

3.5 In Problem 3.1, let the modulation scheme be USSB instead of DSB. First, use a DSB scheme and then remove the lower sideband.

a. Determine and plot the modulated signal.

b. Determine and plot the spectrum of the modulated signal.

c. Compare the spectrum of the modulated signal to the spectrum of the unmodulated signal.

3.6 Repeat Problem 3.5, but instead of using a filtered DSB to generate the USSB signal, use the relation

$$u(t) = \frac{A_c}{2} m(t) \cos(2\pi f_c t) - \frac{A_c}{2} \hat{m}(t) \sin(2\pi f_c t)$$

Do you observe any differences from the results obtained in Problem 3.5?

3.7 Repeat Problems 3.5 and 3.6 but substitute the message signal with

$$m(t) = \begin{cases} \text{sinc}^2(10t), & |t| \leq 2 \\ 0, & \text{otherwise} \end{cases}$$

and use a carrier frequency of 100 Hz to generate an LSSB signal.

3.8 Signal

$$m(t) = \begin{cases} t, & 0.1 \leq t < 1 \\ -t + 2, & 1 \leq t < 1.9 \\ 0.1, & \text{otherwise} \end{cases}$$

is used to modulate a carrier with a frequency of 25 Hz using a DSB scheme.

a. Determine and plot the modulated signal.

b. Assume that the local oscillator at the demodulator has a phase lag of θ with the carrier, where $0 \leq \theta \leq \pi/2$. Using MATLAB, plot the power in the demodulated signal versus θ (assume noiseless transmission).

3.9 In Problem 3.8, assume that the modulation scheme is USSB. Plot the demodulated signal for $\theta = 0°, 30°, 45°, 60°$, and $90°$.

3.10 Repeat Problem 3.8, assuming the modulation is conventional AM with a modulation index of 0.2. Use an envelope detector to demodulate the signal, in the absence of noise and plot the demodulated signal and its spectrum.

3.11 A message signal is periodic with a period of 2 s and in the time interval $[0, 2]$ is defined as

$$m(t) = \begin{cases} t, & 0.1 \leq t < 1 \\ -t + 2, & 1 \leq t < 1.9 \\ 0, & \text{otherwise} \end{cases}$$

This message signal DSB modulates a carrier of 50 Hz. Plot the output of the DSB demodulator and compare it with the message for the cases where white Gaussian noise is added to the modulated signal with a power equal to 0.001, 0.01, 0.05, 0.1, and 0.3 of the modulated signal.

3.12 Repeat Problem 3.11 with a LSSB modulation scheme. Compare your results with the results obtained in Problem 3.11.

3.13 Repeat Problem 3.11 with a conventional AM modulation and envelope demodulation.

3.14 The signal

$$m(t) = \begin{cases} t, & 0 \leq t < 1 \\ -t + 2, & 1 < t \leq 2 \\ 0.1, & \text{otherwise} \end{cases}$$

frequency-modulates a carrier with frequency 1000 Hz. The deviation constant is $k_f = 25$.

    a. Determine the range of the instantaneous frequency of the modulated signal.

    b. Determine the bandwidth of the modulated signal.

    c. Plot the spectra of the message and the modulated signal.

    d. Determine the modulation index.

3.15 Demodulate the modulated signal of Problem 3.14 using a frequency demodulation MATLAB file and compare the demodulated signal with the message signal.

3.16 Let the message signal be a periodic signal with period 2 described in the $[0, 2]$ interval by

$$m(t) = \begin{cases} t, & 0 \leq t < 1 \\ -t + 2, & 1 \leq t < 1.9 \\ 0.1 & \text{otherwise} \end{cases}$$

and let the modulation scheme be the one described in Problem 3.14. Before demodulation, additive white Gaussian noise is added to the modulated signal. Demodulate and plot the demodulated signal when the ratio of noise power to the modulated signal power is 0.001, 0.01, 0.05, 0.1, and 0.3.

3.17 In Problem 3.16, assume $m(t) = 0$. Add the noise processes to the modulated signal as described and demodulate the resulting signal. Plot the power-spectral density of the demodulator output in each case.

Chapter 4

Analog-to-Digital Conversion

4.1 Preview

The majority of information sources are analog by nature. Analog sources include speech, image, and many telemetry sources. In this chapter, we consider various methods and techniques used for converting analog sources to digital sequences in an efficient way. This is desirable because, as we will see in subsequent chapters, digital information is easier to process, to communicate, and to store. The general theme of *data compression*, of which analog-to-digital conversion is a special case, can be divided into two main branches:

1. **Quantization (or lossy data compression)**, in which the analog source is quantized into a finite number of levels. In this process, some distortion will inevitably occur, so some information will be lost. This lost information cannot be recovered. General analog-to-digital conversion techniques, such as pulse-code modulation (PCM), differential pulse-code modulation (DPCM), delta modulation (ΔM), uniform quantization, nonuniform quantization, and vector quantization, belong to this class. The fundamental limit on the performance of this class of data-compression schemes is given by the *rate-distortion bound*.

2. **Noiseless coding (or lossless data compression)**, in which the digital data (usually the result of quantization, as discussed above) are compressed with the goal of representing them with as few bits as possible, such that the original data sequence can be completely recovered from the compressed sequence. Source-coding techniques, such as Huffman coding, Lempel-Ziv coding, and arithmetic coding, belong to this class of data-compression schemes. In this class of coding schemes, no loss of information occurs. The fundamental limit on the compression achieved by this class is given by the *entropy* of the source.

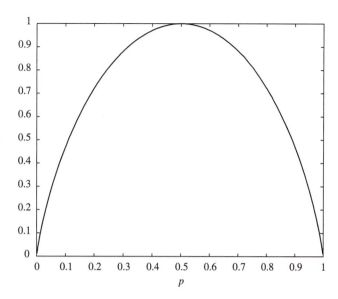

Figure 4.1   Plot of the binary entropy function

4.2   Measure of Information

The output of an information source (data, speech, video, etc.) can be modeled as a random process. For a discrete-memoryless and stationary random process, which can be thought of as independent drawings of one random variable X, the information content, or entropy, is defined as

$$H(X) = -\sum_{x \in \mathcal{X}} p(x) \log p(x) \qquad (4.2.1)$$

where $\mathcal{X}$ denotes the source alphabet and $p(x)$ is the probability of the letter x. The base of the logarithm is usually chosen to be 2, which results in the entropy being expressed in *bits*. For the binary alphabet with probabilities p and $1 - p$, the entropy is denoted by $H_b(p)$ and is given by

$$H_b(p) = -p \log p - (1 - p) \log(1 - p) \qquad (4.2.2)$$

A plot of the binary entropy function is given in Figure 4.1.

The entropy of a source provides an essential bound on the number of bits required to represent a source for full recovery. In other words, the average number of bits per source output required to encode a source for error-free recovery can be made as close to $H(X)$ as we desire but cannot be less than $H(X)$.

4.2.1   Noiseless Coding

Noiseless coding is the general term for all schemes that reduce the number of bits required for the representation of a source output for perfect recovery. The noiseless coding theorem, due to Shannon, states that for perfect reconstruction of a source, it is possible to use a code with a rate as close to the entropy of the source as we desire, but it is not possible to use a code with a rate less than the source entropy. In other words, for any $\epsilon > 0$, we can have a code with rate less than $H(X) + \epsilon$, but we cannot have a code with rate less than $H(X)$, regardless of the complexity of the encoder and the decoder. There exist various algorithms for noiseless source coding; Huffman coding and Lempel-Ziv coding are two examples. Here we discuss the Huffman coding algorithm.

Huffman Coding

In Huffman coding, we assign longer codewords to the less probable source outputs and shorter codewords to the more probable ones. To do this, we start by merging the two least probable source outputs to generate a new merged output whose probability is the sum of the corresponding probabilities. This process is repeated until only one merged output is left. In this way, we generate a *tree*. Starting from the root of the tree and assigning 0's and 1's to any two branches emerging from the same node, we generate the code. It can be shown that in this way we generate a code with minimum average length among the class of *prefix-free codes*.[1] The following example shows how to design a Huffman code.

─── **ILLUSTRATIVE PROBLEM** ───────────────────────

Illustrative Problem 4.1 [Huffman coding] Design a Huffman code for a source with alphabet $\mathcal{X} = \{x_1, x_2, \dots, x_9\}$ and corresponding probability vector

$$p = (0.2, 0.15, 0.13, 0.12, 0.1, 0.09, 0.08, 0.07, 0.06)$$

Find the average codeword length of the resulting code, and compare it with the entropy of the source.

─── **SOLUTION** ───────────────────────

We follow the algorithm outlined above to get the tree shown in Figure 4.2. The average codeword length for this code is

$$\bar{L} = 2 \times 0.2 + 3 \times (0.15 + 0.13 + 0.12 + 0.1) + 4 \times (0.09 + 0.08 + 0.07 + 0.06)$$
$$= 3.1 \quad \text{bits per source output}$$

──────────

[1] Prefix-free codes are codes in which no codeword is a prefix of another codeword.

Codewords

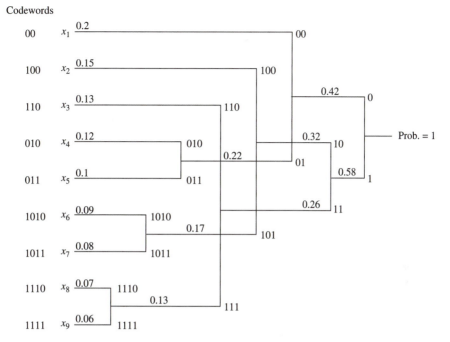

Figure 4.2   Huffman code tree

The entropy of the source is given as

$$H(X) = -\sum_{i=1}^{9} p_i \log p_i = 3.0371 \quad \text{bits per source output}$$

We observe that $\bar{L} > H(X)$, as expected.

The MATLAB function entropy.m given next calculates the entropy of a probability vector p.

M-FILE

```
function h=entropy(p)
%                 H=ENTROPY(P) returns the entropy function of
%                 the probability vector p.
if  length(find(p<0))~=0,
  error('Not a prob. vector, negative component(s)')
end
if  abs(sum(p)−1)>10e−10,
  error('Not a prob. vector, components do not add up to 1')
end
h=sum(−p.*log2(p));
```

The quantity

$$\eta = \frac{H(X)}{\bar{L}} \tag{4.2.3}$$

is called the *efficiency of the Huffman code*. Obviously, we always have $\eta \leq 1$. In general, it can be shown that the average codeword length for any Huffman code satisfies the inequalities

$$H(X) \leq \bar{L} < H(X) + 1 \tag{4.2.4}$$

If we design a Huffman code for blocks of length K instead of for single letters, we will have

$$H(X) \leq \bar{L} < H(X) + \frac{1}{K} \tag{4.2.5}$$

and therefore, by increasing K, we can get as close to $H(X)$ as we desire. Needless to say, increasing K increases the complexity considerably. It should also be noted that the Huffman coding algorithm does not result in a unique code due to the arbitrary way of assigning 0's and 1's to different tree branches. That is why we talk about *a* Huffman code rather than *the* Huffman code.

The MATLAB function huffman.m, which designs a Huffman code for a discrete-memoryless source with probability vector p and returns both the codewords and the average codeword length, is given next.

M-FILE

```
function [h,l]=huffman(p);
%HUFFMAN      Huffman code generator.
%             [h,l]=huffman(p), Huffman code generator
%             returns h the Huffman code matrix, and l the
%             average codeword length for a source with
%             probability vector p.

if  length(find(p<0))~=0,
   error('Not a prob. vector, negative component(s)')
end
if  abs(sum(p)−1)>10e−10,
   error('Not a prob. vector, components do not add up to 1')
end
n=length(p);
q=p;
m=zeros(n−1,n);
for  i=1:n−1
   [q,l]=sort(q);
   m(i,:)=[l(1:n−i+1),zeros(1,i−1)];
   q=[q(1)+q(2),q(3:n),1];
end
for  i=1:n−1
   c(i,:)=blanks(n*n);
```

```
end
c(n−1,n)='0';
c(n−1,2*n)='1';
for i=2:n−1
  c(n−i,1:n−1)=c(n−i+1,n*(find(m(n−i+1,:)==1))...
  −(n−2):n*(find(m(n−i+1,:)==1)));
  c(n−i,n)='0';
  c(n−i,n+1:2*n−1)=c(n−i,1:n−1);
  c(n−i,2*n)='1';
  for j=1:i−1
    c(n−i,(j+1)*n+1:(j+2)*n)=c(n−i+1,...
    n*(find(m(n−i+1,:)==j+1)−1)+1:n*find(m(n−i+1,:)==j+1));
  end
end
for i=1:n
  h(i,1:n)=c(1,n*(find(m(1,:)==i)−1)+1:find(m(1,:)==i)*n);
  ll(i)=length(find(abs(h(i,:))~=32));
end
l=sum(p.*ll);
```

ILLUSTRATIVE PROBLEM

Illustrative Problem 4.2 [Huffman coding] A discrete-memoryless information source with alphabet

$$\mathcal{X} = \{x_1, x_2, \ldots, x_6\}$$

and the corresponding probabilities

$$p = \{0.1, 0.3, 0.05, 0.09, 0.21, 0.25\}$$

is to be encoded using Huffman coding.

1. Determine the entropy of the source.

2. Find a Huffman code for the source and determine the efficiency of the Huffman code.

3. Now design a Huffman code for source sequences of length 2, and compare the efficiency of this code with the efficiency of the code derived in part 2.

SOLUTION

1. The entropy of the source is derived via the entropy.m function and is found to be 2.3549 bits per source symbol.

2. Using the huffman.m function, we can design a Huffman code for this source. The codewords are found to be 010, 11, 0110, 0111, 00, and 10. The average codeword length for this code is found to be 2.38 binary symbols per source output. Therefore, the efficiency of this code is

$$\eta_1 = \frac{2.3549}{2.38} = 0.9895$$

3. A new source whose outputs are letter pairs of the original source has 36 output letters of the form $\{(x_i, x_j)\}_{i,j=1}^6$. Since the source is memoryless, the probability of each pair is the product of the individual letter probabilities. Thus, in order to obtain the probability vector for the extended source, we must generate a vector with 36 components, each component being the product of two probabilities in the original probability vector p. This can be done by employing the MATLAB function kron.m in the form of kron(p, p). The Huffman codewords are given by

1110000, 01110, 10110111, 1011001, 111001, 00101, 01111, 000, 011010, 00111, 1001, 1100, 11101110, 011011, 111011110, 111011111, 1110001, 001000, 1011010, 01100, 10110110, 1011000, 101110, 111110, 111010, 1010, 1110110, 101111, 11110, 0100, 00110, 1101, 001001, 111111, 0101, 1000

The average codeword length for the extended source is 4.7420. The entropy of the extended source is found to be 4.7097, so the efficiency of this Huffman code is

$$\eta_2 = \frac{4.7097}{4.7420} = 0.9932$$

which shows an improvement compared to the efficiency of the Huffman code designed in part 2.

ILLUSTRATIVE PROBLEM

Illustrative Problem 4.3 [A Huffman code with maximum efficiency] Design a Huffman code for a source with probability vector

$$p = \left\{ \frac{1}{2}, \frac{1}{4}, \frac{1}{8}, \frac{1}{16}, \frac{1}{32}, \frac{1}{64}, \frac{1}{128}, \frac{1}{256}, \frac{1}{256} \right\}$$

SOLUTION

We use the huffman.m function to determine a Huffman code and the corresponding average codeword length. The resulting codewords are 1, 01, 001, 0001, 00001, 000001, 0000001, 00000000, and 00000001. The average codeword length is 1.9922 binary symbols per source output. If we find the entropy of the source using the entropy.m function, we see that the entropy of the source is also 1.9922 bits per source output; hence the efficiency of this code is 1.

─── **QUESTION** ──────────────────────────────────

Can you say under what conditions the efficiency of a Huffman code is equal to 1?

4.3   Quantization

In the previous section, we studied two methods for noiseless coding—that is, compression of the source output sequence such that full recovery is possible from the compressed data. In these methods, the compressed data are a deterministic function of the source output, and the source output is also a deterministic function of the compressed data. This one-to-one correspondence between the compressed data and the source output means that their entropies are equal and no information is lost in the encoding-decoding process.

In many applications, such as digital processing of the analog signals, where the source alphabet is not discrete, the number of bits required for representation of each source output is not finite. In order to process the source output digitally, the source has to be *quantized* to a finite number of levels. This process reduces the number of bits to a finite number but at the same time introduces some distortion. The information lost in the quantization process can never be recovered.

In general, quantization schemes can be classified as *scalar quantization* and *vector quantization* schemes. In scalar quantization, each source output is quantized individually, whereas in vector quantization, blocks of source output are quantized.

Scalar quantizers can be further classified as *uniform quantizers* and *nonuniform quantizers*. In uniform quantization, the quantization regions are chosen to have equal length; in nonuniform quantization, regions of various lengths are allowed. It is clear that, in general, nonuniform quantizers outperform uniform quantizers.

4.3.1   Scalar Quantization

In scalar quantization, the range of the random variable X is divided into N nonoverlapping regions $\mathcal{R}_i$, for $1 \leq i \leq N$, called *quantization intervals*, and within each region a single point called a *quantization level* is chosen. Then all values of the random variable that fall within region $\mathcal{R}_i$ are quantized to the ith quantization level, which is denoted by $\hat{x}_i$. This means that

$$x \in \mathcal{R}_i \iff Q(x) = \hat{x}_i \tag{4.3.1}$$

where

$$\hat{x}_i \in \mathcal{R}_i \tag{4.3.2}$$

Obviously, a quantization of this type introduces a mean-square error of $(x - \hat{x}_i)^2$. The mean-square quantization error is therefore given by

$$D = \sum_{i=1}^{N} \int_{\mathcal{R}_i} (x - \hat{x}_i)^2 f_X(x) \, dx \tag{4.3.3}$$

where $f_X(x)$ denotes the probability density function of the source random variable. The *signal-to-quantization-noise ratio* (SQNR) is defined as

$$\text{SQNR}_{|\text{dB}} = 10 \log_{10} \frac{E\left[X^2\right]}{D}$$

Uniform Quantization

In uniform quantization, all quantization regions except the first and the last one —that is, $\mathcal{R}_1$ and $\mathcal{R}_N$—are of equal length, which is denoted by Δ; therefore,

$$\mathcal{R}_1 = (-\infty, a]$$
$$\mathcal{R}_2 = (a, a + \Delta]$$
$$\mathcal{R}_3 = (a + \Delta, a + 2\Delta]$$
$$\vdots$$
$$\mathcal{R}_N = (a + (N - 2)\Delta, \infty)$$

The optimal quantization level in each quantization interval can be shown to be the *centroid* of that interval; that is,

$$\hat{x}_i = E\left[X | X \in \mathcal{R}_i\right] \tag{4.3.4}$$
$$= \frac{\int_{\mathcal{R}_i} x f_X(x)\, dx}{\int_{\mathcal{R}_i} f_X(x)\, dx} \qquad 1 \le i \le N$$

Therefore, the design of the uniform quantizer is equivalent to determining a and Δ. After a and Δ are determined, the values of $\hat{x}_i$'s and the resulting distortion can be determined easily using Equations (4.3.3) and (4.3.4). In some cases, it is convenient to choose the quantization levels to be simply the midpoints of the quantization regions—that is, at a distance $\Delta/2$ from the boundaries of the quantization regions.

   Plots of the quantization function $Q(x)$ for a symmetric probability density function of X and even and odd values of N are shown in Figures 4.3 and 4.4, respectively.

   For the symmetric probability density functions, the problem becomes even simpler. In such a case,

$$\mathcal{R}_i = \begin{cases} (a_{i-1}, a_i], & 1 \le i \le N - 1 \\ (a_{i-1}, a_N), & i = N \end{cases} \tag{4.3.5}$$

where

$$\begin{cases} a_0 = -\infty \\ a_i = (i - N/2)\Delta & 1 \le i \le N - 1 \\ a_N = \infty \end{cases} \tag{4.3.6}$$

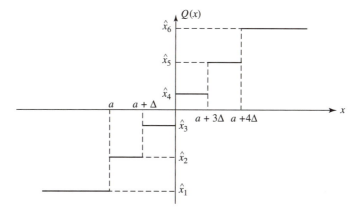

Figure 4.3   The uniform quantizer for $N = 6$. (Note that here $a + 2\Delta = 0$.)

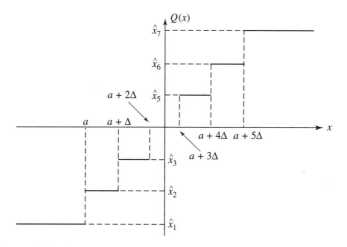

Figure 4.4   The quantization function for $N = 7$. (Note that here $\hat{x}_4 = 0$.)

We see that in this case we have only one parameter Δ, which we have to choose to achieve minimum distortion. Three m-files—centroid.m, mse_dist.m, and uq_dist.m—find the centroid of a region; the mean-square quantization error for a given distribution and given quantization region boundaries; and, finally, the distortion when a uniform quantizer is used to quantize a given source. (It is assumed that the quantization levels are set to the centroids of the quantization regions.) In order to use each of these m-files, the distribution of the source, which can depend on up to three parameters, has to be given in an m-file. These m-files are given next.

M-FILE

```
function y=centroid(funfcn,a,b,tol,p1,p2,p3)
% CENTROID    Finds the centroid of a function over a region.
%             Y=CENTROID('F',A,B,TOL,P1,P2,P3), finds the centroid of the
%             function F defined in an m-file on the [A,B] region. The
%             function can contain up to three parameters, P1, P2, P3.
%             tol=the relative error.

args=[ ];
for n=1:nargin−4
  args=[args,',p',int2str(n)];
end
args=[args,')'];
funfcn1='x_fnct';
y1=eval(['quad(funfcn1,a,b,tol,[],funfcn',args]);
y2=eval(['quad(funfcn,a,b,tol,[]',args]);
y=y1/y2;
```

M-FILE

```
function [y,dist]=mse_dist(funfcn,a,tol,p1,p2,p3)
%MSE_DIST    Returns the mean-squared quantization error.
%            [Y,DIST]=MSE_DIST(FUNFCN,A,TOL,P1,P2,P3)
%            funfcn=The distribution function given
%            in an m-file. It can depend on up to three
%            parameters, p1,p2,p3.
%            a=the vector defining the boundaries of the
%            quantization regions. (note: [a(1),a(length(a))]
%            is the support of funfcn).
%            p1,p2,p3=parameters of funfcn.
%            tol=the relative error.

args=[ ];
for n=1:nargin−3
  args=[args,',p',int2str(n)];
end
args=[args,')'];
for i=1:length(a)−1
  y(i)=eval(['centroid(funfcn,a(i),a(i+1),tol',args]);
end
dist=0;
for i=1:length(a)−1
  newfun = 'x_a2_fnct' ;
  dist=dist+eval(['quad(newfun,a(i),a(i+1),tol,[],funfcn,', num2str(y(i)), args]);
end
```

◀ M-FILE ▶

```
function [y,dist]=uq_dist(funfcn,b,c,n,delta,s,tol,p1,p2,p3)
%UQ_DIST      Returns the distortion of a uniform quantizer
%             with quantization points set to the centroids.
%             [Y,DIST]=UQ_DIST(FUNFCN,B,C,N,DELTA,S,TOL,P1,P2,P3)
%             funfcn=source density function given in an m-file
%             with at most three parameters, p1,p2,p3.
%             [b,c]=The support of the source density function.
%             n=number of levels.
%             delta=level size.
%             s=the leftmost quantization region boundary.
%             p1,p2,p3=parameters of the input function.
%             y=quantization levels.
%             dist=distortion.
%             tol=the relative error.

if (c−b<delta*(n−2))
  error('Too many levels for this range.'); return
end
if (s<b)
  error('The leftmost boundary too small.'); return
end
if (s+(n−2)*delta>c)
  error('The leftmost boundary too large.'); return
end
args=[ ];
for j=1:nargin−7
  args=[args,',p',int2str(j)];
end
args=[args,')'];
a(1)=b;
for i=2:n
  a(i)=s+(i−2)*delta;
end
a(n+1)=c;
[y,dist]=eval(['mse_dist(funfcn,a,tol',args]);
```

◀ ILLUSTRATIVE PROBLEM ▶

Illustrative Problem 4.4 [Determining the centroids] Determine the centroids of the quantization regions for a zero-mean, unit-variance Gaussian distribution, where the boundaries of the quantization regions are given by $(-5, -4, -2, 0, 1, 3, 5)$.

─── **SOLUTION** ───

The Gaussian distribution is given in the m-file normal.m. This distribution is a function of two parameters, the mean and the variance, denoted by m and s (or σ), respectively. The support of the Gaussian distribution is $(-\infty, \infty)$, but for employing the numerical routines, it is enough to use a range that is many times the standard deviation of the distribution. For example, $(m - 10\sqrt{s}, m + 10\sqrt{s})$ can be used. The following m-file determines the centroids (optimal quantization levels).

─── **M-FILE** ───

% MATLAB script for Illustrative Problem 4.4.

```
echo on ;
a=[−10,−5,−4,−2,0,1,3,5,10];
for  i=1:length(a)−1
   y(i)=centroid('normal',a(i),a(i+1),0.001,0,1);
   echo off ;
end
```

This results in the following quantization levels: $(-5.1865, -4.2168, -2.3706, 0.7228, -0.4599, 1.5101, 3.2827, 5.1865)$.

─── **ILLUSTRATIVE PROBLEM** ───

Illustrative Problem 4.5 [Mean-square error] In Illustrative Problem 4.4, determine the mean-square error.

─── **SOLUTION** ───

Letting $a = (-10, -5, -4, -2, 0, 1, 3, 5, 10)$ and using mse_dist.m, we obtain a mean-square error of 0.177.

─── **ILLUSTRATIVE PROBLEM** ───

Illustrative Problem 4.6 [Uniform quantizer distortion] Determine the mean-square error for a uniform quantizer with 12 quantization levels, each of length 1, designed for a zero-mean Gaussian source with variance of 4. It is assumed that the quantization regions are symmetric with respect to the mean of the distribution.

SOLUTION

By the symmetry assumption, the boundaries of the quantization regions are 0, ± 1, ± 2, ± 3, ± 4, and ± 5, and the quantization regions are $(-\infty, -5]$, $(-5, -4]$, $(-4, -3]$, $(-3, -2]$, $(-2, -1]$, $(-1, 0]$, $(0, 1]$, $(1, 2]$, $(2, 3]$, $(3, 4]$, $(4, 5]$, and $(5, +\infty)$. This means that in the uq_dist.m function, we can substitute $b = -20$, $c = 20$, $\Delta = 1$, $n = 12$, $s = -5$, tol $= 0.001$, $p_1 = 0$, and $p_2 = 2$. Substituting these values into uq_dist.m, we obtain a squared error distortion of 0.0851 and quantization values of ± 0.4897, ± 1.4691, ± 2.4487, ± 3.4286, ± 4.4089, and ± 5.6455.

The m-file uq_mdpnt.m determines the squared error distortion for a symmetric density function when the quantization levels are chosen to be the midpoints of the quantization intervals. In this case, the quantization levels corresponding to the first and the last quantization regions are chosen to be at distance $\Delta/2$ from the two outermost quantization boundaries. This means that if the number of quantization levels is even, then the quantization boundaries are 0, $\pm \Delta$, $\pm 2\Delta$, $\ldots$, $\pm (N/2 - 1)\Delta$, and the quantization levels are given by $\pm \Delta/2$, $\pm 3\Delta/2$, $\ldots$, $(N - 1)\Delta/2$. If the number of quantization levels is odd, then the boundaries are given by $\pm \Delta/2$, $\pm 3\Delta/2$, $\ldots$, $\pm (N/2 - 1)\Delta$, and the quantization levels are given by 0, $\pm \Delta$, $\pm 2\Delta$, $\ldots$, $(N - 1)\Delta/2$. The m-file uq_mdpnt.m is given next.

M-FILE

```
function dist=uq_mdpnt(funfcn,b,n,delta,tol,p1,p2,p3)
%UQ_MDPNT    Returns the distortion of a uniform quantizer
%            with quantization points set to the midpoints.
%            DIST=UQ_MDPNT(FUNFCN,B,N,DELTA,TOL,P1,P2,P3)
%            funfcn=source density function given in an m-file
%            with at most three parameters, p1,p2,p3. The density
%            function is assumed to be an even function.
%            [-b,b]=The support of the source density function.
%            n=number of levels.
%            delta=level size.
%            p1,p2,p3=parameters of the input function.
%            dist=distortion.
%            tol=the relative error.

if (2*b<delta*(n-1))
  error('Too many levels for this range.'); return
end
args=[ ];
for j=1:nargin-5
  args=[args,',p',int2str(j)];
end
args=[args,')'];
a(1)=-b;
a(n+1)=b;
a(2)=-(n/2-1)*delta;
y(1)=a(2)-delta/2;
for i=3:n
```

```
  a(i)=a(i−1)+delta;
  y(i−1)=a(i)−delta/2;
end
y(n)=a(n)+delta;
dist=0;
for i=1:n
  newfun = 'x_a2_fnct' ;
  dist=dist+eval(['quad(newfun,a(i),a(i+1),tol,[],funfcn,', num2str(y(i)), args]);
end
```

ILLUSTRATIVE PROBLEM

Illustrative Problem 4.7 [Uniform quantizer with levels set to the midpoints] Find the distortion when a uniform quantizer is used to quantize a zero-mean, unit-variance Gaussian random variable. The number of quantization levels is 11, and the length of each quantization region is 1.

SOLUTION

In uq_mdpnt.m, we substitute 'normal'[2] for the density function name, $p_1 = 0$ and $p_2 = 1$ for the density function parameters, $n = 11$ for the number of quantization levels, and $\Delta = 1$ for the length of the quantization levels. For the parameter b, which chooses the support set of the density function, we use the value $b = 10p_2 = 10$, and we choose the tolerance to be 0.001. The resulting distortion is 0.0833.

Nonuniform Quantization

In nonuniform quantization, the requirement that the quantization regions, except the first and the last, have equal lengths is relaxed, and each quantization region can have any length. Since in this case optimization is done under more relaxed conditions, the result is obviously superior to that of uniform quantization. The optimality conditions for this case, known as the *Lloyd-Max conditions*, can be expressed as

$$\begin{cases} \hat{x}_i = \dfrac{\int_{a_{i-1}}^{a_i} x f_X(x)\, dx}{\int_{a_{i-1}}^{a_i} f_X(x)\, dx} \\[4mm] a_i = \dfrac{(\hat{x}_{i-1} + \hat{x}_i)}{2} \end{cases} \tag{4.3.7}$$

From these equations, we conclude that the optimal quantization levels are the centroids of the quantization regions and the optimal boundaries between the quantization regions

---

[2]The name of the function should be substituted with 'normal' *including the single quotes.*

are the midpoints between the quantization levels. In order to obtain the solution to the Lloyd-Max equations, we start with a set of quantization levels $\hat{x}_i$. From this set, we can simply find the set of quantization region boundaries a_i. From this set of a_i's, a new set of quantization levels can be obtained. This process is continued until the improvement in distortion from one iteration to another is not noticeable. This algorithm is guaranteed to converge to a local minimum, but in general there is no guarantee that the global minimum can be achieved.

The procedure of designing an optimal quantizer is shown in the m-file lloydmax.m, given next.

M-FILE

```
function [a,y,dist]=lloydmax(funfcn,b,n,tol,p1,p2,p3)
%LLOYDMAX      Returns the the Lloyd-Max quantizer and the mean-squared
%              quantization error for a symmetric distribution.
%              [A,Y,DIST]=LLOYDMAX(FUNFCN,B,N,TOL,P1,P2,P3)
%              funfcn=The density function given
%              in an m-file. It can depend on up to three
%              parameters, p1,p2,p3.
%              a=The vector giving the boundaries of the
%              quantization regions.
%              [-b,b] approximates support of the density function.
%              n=The number of quantization regions.
%              y=The quantization levels.
%              p1,p2,p3=Parameters of funfcn.
%              tol=the relative error.

args=[ ];
for j=1:nargin-4
  args=[args,',p',int2str(j)];
end
args=[args,')'];
v=eval(['variance(funfcn,-b,b,tol',args]);
a(1)=-b;
d=2*b/n;
for i=2:n
  a(i)=a(i-1)+d;
end
a(n+1)=b;
dist=v;
[y,newdist]=eval(['mse_dist(funfcn,a,tol',args]);
while(newdist<0.99*dist),
  for i=2:n
    a(i)=(y(i-1)+y(i))/2;
  end
  dist=newdist;
  [y,newdist]=eval(['mse_dist(funfcn,a,tol',args]);
end
```

─(**ILLUSTRATIVE PROBLEM**)───────────────────

Illustrative Problem 4.8 [Lloyd-Max quantizer design] Design a ten-level Lloyd-Max quantizer for a zero-mean, unit-variance Gaussian source.

─(**SOLUTION**)───────────────────

Using $b = 10$, $n = 10$, tol $= 0.01$, $p_1 = 0$, and $p_2 = 1$ in lloydmax.m, we obtain the quantization boundaries and quantization-level vector $\boldsymbol{a}$ and $\boldsymbol{y}$ as

$$\boldsymbol{a} = \pm 10, \pm 2.16, \pm 1.51, \pm 0.98, \pm 0.48, 0$$
$$\boldsymbol{y} = \pm 2.52, \pm 1.78, \pm 1.22, \pm 0.72, \pm 0.24$$

and the resulting distortion is 0.02. These values are good approximations to the optimal values given in the table by Max [2].

4.4 Pulse-Code Modulation

In pulse-code modulation, an analog signal is first sampled at a rate higher than the Nyquist rate, and then the samples are quantized. It is assumed that the analog signal is distributed on an interval denoted by $[-x_{max}, x_{max}]$ and the number of quantization levels is large. The quantization levels can be equal or unequal. In the first case we are dealing with a uniform PCM, and in the second case with a nonuniform PCM.

4.4.1 Uniform PCM

In uniform PCM, the interval $[-x_{max}, x_{max}]$ of length $2x_{max}$ is divided into N equal subintervals, each of length $\Delta = 2x_{max}/N$. If N is large enough, the density function of the input in each subinterval can be assumed to be uniform, resulting in a distortion of $D = \Delta^2/12$. If N is a power of 2, or $N = 2^\nu$, then ν bits are required for representation of each level. This means that if the bandwidth of the analog signal is W, and if sampling is done at the Nyquist rate, the required bandwidth for transmission of the PCM signal is at least νW. (In practice, $1.5\nu W$ is closer to reality.) The distortion is given by

$$D = \frac{\Delta^2}{12} \tag{4.4.1}$$
$$= \frac{x_{max}^2}{3N^2}$$
$$= \frac{x_{max}^2}{3 \times 4^\nu}$$

If the power of the analog signal is denoted by $\overline{X^2}$, the signal-to-quantization-noise ratio (SQNR) is given by

$$\text{SQNR} = 3N^2 \frac{\overline{X^2}}{x_{\max}^2} \tag{4.4.2}$$

$$= 3 \times 4^\nu \frac{\overline{X^2}}{x_{\max}^2}$$

$$= 3 \times 4^\nu \overline{\check{X}^2}$$

where $\check{X}$ denotes the normalized input defined by

$$\check{X} = \frac{X}{x_{\max}}$$

The SQNR in decibels is given by

$$\text{SQNR}_{|\text{dB}} \approx 4.8 + 6\nu + \overline{\check{X}^2}_{|\text{dB}} \tag{4.4.3}$$

After quantization, the quantized levels are *encoded* using ν bits for each quantized level. The encoding scheme that is usually employed is *natural binary coding (NBC)*, meaning that the lowest level is mapped into a sequence of all 0's and the highest level is mapped into a sequence of all 1's. All the other levels are mapped in increasing order of the quantized value.

The m-file u_pcm.m given next takes as its input a sequence of sampled values and the number of desired quantization levels and finds the quantized sequence, the encoded sequence, and the resulting SQNR (in decibels).

M-FILE

```
function [sqnr,a_quan,code]=u_pcm(a,n)
%U_PCM          Uniform PCM encoding of a sequence.
%               [SQNR,A_QUAN,CODE]=U_PCM(A,N)
%               a=input sequence.
%               n=number of quantization levels (even).
%               sqnr=output SQNR (in dB).
%               a_quan=quantized output before encoding.
%               code=the encoded output.

amax=max(abs(a));
a_quan=a/amax;
b_quan=a_quan;
d=2/n;
q=d.*[0:n−1];
```

```
q=q−((n−1)/2)*d;
for i=1:n
   a_quan(find((q(i)−d/2 <= a_quan) & (a_quan <= q(i)+d/2)))=...
   q(i).*ones(1,length(find((q(i)−d/2 <= a_quan) & (a_quan <= q(i)+d/2))));
   b_quan(find( a_quan==q(i) ))=(i−1).*ones(1,length(find( a_quan==q(i) )));
end
a_quan=a_quan*amax;
nu=ceil(log2(n));
code=zeros(length(a),nu);
for i=1:length(a)
   for j=nu:−1:0
     if ( fix(b_quan(i)/(2^j)) == 1)
         code(i,(nu−j)) = 1;
         b_quan(i) = b_quan(i) − 2^j;
     end
   end
end
sqnr=20*log10(norm(a)/norm(a−a_quan));
```

ILLUSTRATIVE PROBLEM

Illustrative Problem 4.9 [Uniform PCM] Generate a sinusoidal signal with amplitude 1 and $\omega = 1$. Using a uniform PCM scheme, quantize it once to 8 levels and once to 16 levels. Plot the original signal and the quantized signals on the same axes. Compare the resulting SQNRs in the two cases.

SOLUTION

We arbitrarily choose the duration of the signal to be 10 s. Then, using the u_pcm.m m-file, we generate the quantized signals for the two cases of 8 and 16 quantization levels. The resulting SQNRs are 18.90 dB for the 8-level PCM and 25.13 dB for 16-level uniform PCM. The plots are shown in Figure 4.5.

A MATLAB script for this problem is shown next.

M-FILE

% MATLAB script for Illustrative Problem 4.9.

```
echo on
t=[0:0.01:10];
a=sin(t);
[sqnr8,aquan8,code8]=u_pcm(a,8);
[sqnr16,aquan16,code16]=u_pcm(a,16);
```

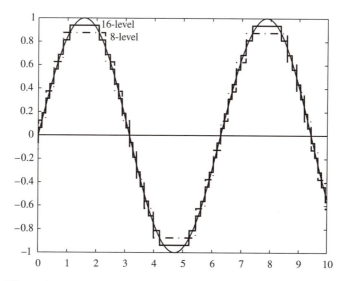

Figure 4.5   Uniform PCM for a sinusoidal signal using 8 and 16 levels

```
pause      % Press a key to see the SQNR for N = 8.
sqnr8
pause      % Press a key to see the SQNR for N = 16.
sqnr16
pause      % Press a key to see the plot of the signal and its quantized versions.
plot(t,a,' – ',t,aquan8,' – . ',t,aquan16,' – ',t,zeros(1,length(t)))
```

ILLUSTRATIVE PROBLEM

Illustrative Problem 4.10 [Uniform PCM] Generate a sequence of length 500 of zero-mean, unit-variance Gaussian random variables. Using u_pcm.m, find the resulting SQNR when the number of quantization levels is 64. Find the first five values of the sequence, the corresponding quantized values, and the corresponding codewords.

SOLUTION

The following m-file gives the solution.

◖ M-FILE ◗

% MATLAB script for Illustrative Problem 4.10.

```
echo on
a=randn(1,500);
n=64;
[sqnr,a_quan,code]=u_pcm(a,64);
pause     % Press a key to see the SQNR
sqnr
pause     % Press a key to see the first five input values.
a(1:5)
pause     % Press a key to see the first five quantized values.
a_quan(1:5)
pause     % Press a key to see the first five codewords.
code(1:5,:)
```

In a typical running of this file, the following values were observed:

$$\text{SQNR} = 31.66 \quad \text{dB}$$
$$\text{Input} = [0.1775, -0.4540, 1.0683, -2.2541, 0.5376]$$
$$\text{Quantized values} = [0.1569, -0.4708, 1.0985, -2.2494, 0.5754]$$
$$\text{Codewords} = \begin{cases} 1 & 0 & 0 & 0 & 0 & 1 \\ 0 & 1 & 1 & 0 & 1 & 1 \\ 1 & 0 & 1 & 0 & 1 & 0 \\ 0 & 0 & 1 & 0 & 1 & 0 \\ 1 & 0 & 0 & 1 & 0 & 1 \end{cases}$$

Note that different runnings of the program result in different values for the input, the quantized values, and the codewords. However, the resulting SQNRs are very close.

◖ ILLUSTRATIVE PROBLEM ◗

Illustrative Problem 4.11 [Quantization error] In Illustrative Problem 4.10, plot the quantization error, defined as the difference between the input value and the quantized value. Also, plot the quantized value as a function of the input value.

◖ SOLUTION ◗

The two desired plots are shown in Figure 4.6.

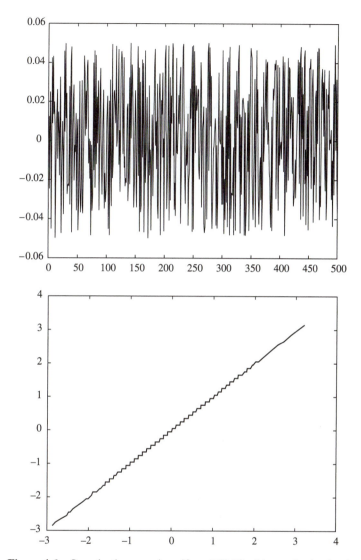

Figure 4.6   Quantization error in uniform PCM for 64 quantization levels

ILLUSTRATIVE PROBLEM

Illustrative Problem 4.12 [Quantization error] Repeat Illustrative Problem 4.11 with
the number of quantization levels set once to 16 and set once to 128. Compare the results.

—◖ SOLUTION ◗——————————————————————

The result for 16 quantization levels is shown in Figure 4.7, and the result for 128 quanti-zation levels is shown in Figure 4.8. From Figures 4.6, 4.7, and 4.8, it is obvious that the larger the number of quantization levels, the smaller the quantization error, as expected. Also note that for a large number of quantization levels, the relation between the input and the quantized values tends to a line with slope 1 passing through the origin; that is, the input and the quantized values become almost equal. For a small number of quantization levels (16 for instance), this relation is far from equality, as shown in Figure 4.7.

4.4.2 Nonuniform PCM

In nonuniform PCM, the input signal is first passed through a nonlinear element to reduce its dynamic range, and the output is applied to a uniform PCM system. At the receiving end, the output is passed through the inverse of the nonlinear element used in the transmit-ter. The overall effect is equivalent to a PCM system with nonuniform spacing between levels. In general, for transmission of speech signals, the nonlinearities that are employed are either μ-law or A-law nonlinearities.

A μ-law nonlinearity is defined by the relation

$$y = g(x) = \frac{\log(1 + \mu|x|)}{\log(1 + \mu)} \text{sgn}(x) \tag{4.4.4}$$

where x is the normalized input ($|x| \le 1$) and μ is a parameter that in standard μ-law nonlinearity is equal to 255. A plot of this nonlinearity for different values of μ is shown in Figure 4.9.

The inverse of μ-law nonlinearity is given by

$$x = \frac{(1 + \mu)^{|y|} - 1}{\mu} \text{sgn}(y) \tag{4.4.5}$$

The two m-files mulaw.m and invmulaw.m given next implement μ-law nonlinearity and its inverse.

—◖ M-FILE ◗——————————————————————

```
function [y,a]=mulaw(x,mu)
%MULAW          mu-law nonlinearity for nonuniform PCM.
%               Y=MULAW(X,MU)
%               X=input vector.

a=max(abs(x));
y=(log(1+mu*abs(x/a))./log(1+mu)).*signum(x);
```

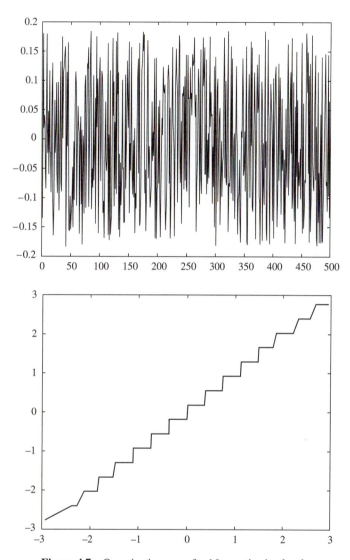

Figure 4.7   Quantization error for 16 quantization levels

M-FILE

```
function  x=invmulaw(y,mu)
%INVMULAW          The inverse of mu-law nonlinearity
%X=INVMULAW(Y,MU)    Y=Normalized output of the mu-law nonlinearity

x=(((1+mu).^(abs(y))−1)./mu).*signum(y);
```

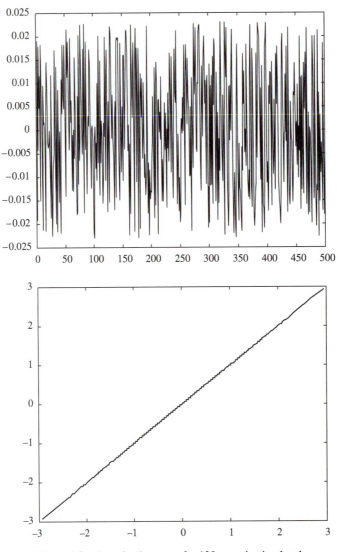

Figure 4.8   Quantization error for 128 quantization levels

The m-file mula_pcm.m is the equivalent of the m-file u_pcm.m when using a μ-law PCM scheme. This file is given next.

M-FILE

```
function [sqnr,a_quan,code]=mula_pcm(a,n,mu)
%MULA_PCM    mu-law PCM encoding of a sequence.
%                  [SQNR,A_QUAN,CODE]=MULA_PCM(A,N,MU)
```

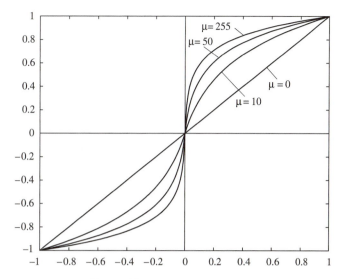

Figure 4.9   The μ-law compander

```
%               a=input sequence.
%               n=number of quantization levels (even).
%               sqnr=output SQNR (in dB).
%               a_quan=quantized output before encoding.
%               code=the encoded output.

[y,maximum]=mulaw(a,mu);
[sqnr,y_q,code]=u_pcm(y,n);
a_quan=invmulaw(y_q,mu);
a_quan=maximum*a_quan;
sqnr=20*log10(norm(a)/norm(a−a_quan));
```

───

◖ILLUSTRATIVE PROBLEM◗

Illustrative Problem 4.13 [Nonuniform PCM] Generate a sequence of random variables of length 500 according to an $\mathcal{N}(0, 1)$ distribution. Using 16, 64, and 128 quantization levels and a μ-law nonlinearity with $\mu = 255$, plot the error and the input-output relation for the quantizer in each case. Also determine the SQNR in each case.

◖SOLUTION◗

Let the vector $\boldsymbol{a}$ be the vector of length 500 generated according to $\mathcal{N}(0, 1)$; that is, let

$$a = \text{randn}(1, 500)$$

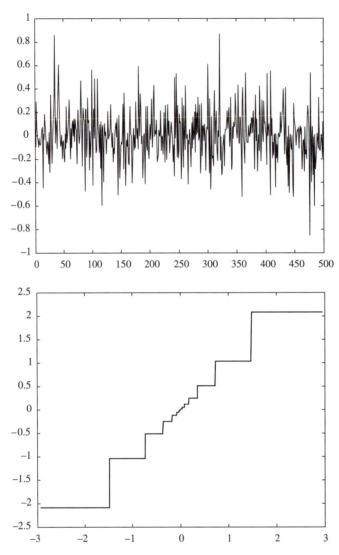

Figure 4.10   Quantization error and quantizer input-output relation for a 16-level
μ-law PCM

Then, by using
$$[\text{dist,a\_quan,code}] = \text{mula\_pcm}(a, 16, 255)$$
we can obtain the quantized sequence and the SQNR for a 16-level quantization. The
SQNR will be 13.76 dB. For the case of 64 levels we obtain SQNR = 25.89 dB, and for
128 levels we have SQNR = 31.76 dB. Comparing these results with the uniform PCM,
we observe that in all cases the performance is inferior to the uniform PCM. Plots of the
input-output relation for the quantizer and the quantization error are given in Figures 4.10,
4.11, and 4.12.

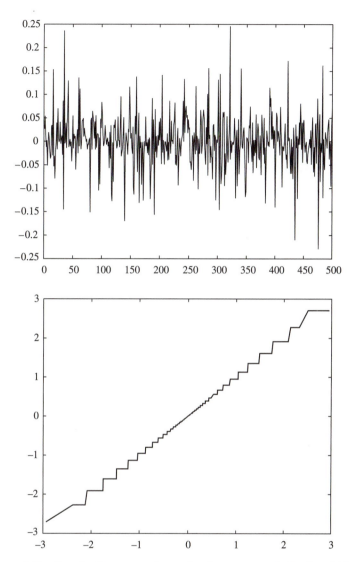

Figure 4.11   Quantization error and quantizer input-output relation for a 64-level
μ-law PCM

Comparing the input-output relation for the uniform and the nonuniform PCM shown in Figures 4.7 and 4.10 clearly shows why the former is called uniform PCM and the latter is called nonuniform PCM.

From the preceding example, we see that the performance of the nonuniform PCM, in this case, is not as good as uniform PCM. The reason is that in the example, the dynamic range of the input signal is not very large. The next example examines the case where the performance of the nonuniform PCM is superior to the performance of the uniform PCM.

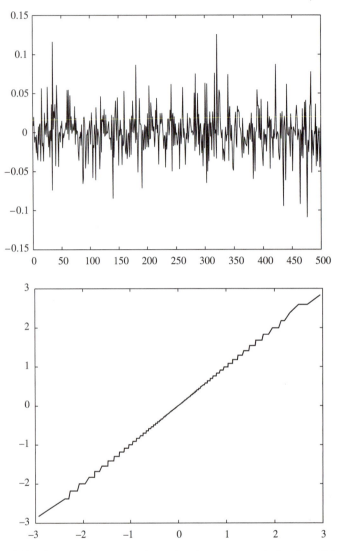

Figure 4.12   Quantization error and quantizer input-output relation for a 128-level
μ-law PCM

════ILLUSTRATIVE PROBLEM════

Illustrative Problem 4.14 [Nonuniform PCM] The nonstationary sequence a of length
500 consists of two parts. The first 20 samples are generated according to a Gaussian random variable with mean 0 and variance 400 ($\sigma = 20$), and the next 480 samples are drawn

according to a Gaussian random variable with mean 0 and variance 1. This sequence is quantized once using a uniform PCM scheme and once using a nonuniform PCM scheme. Compare the resulting SQNR in the two cases.

SOLUTION

The sequence is generated by the MATLAB command

$$a = [20 * \text{randn}(1, 20)\text{randn}(1, 480)]$$

Now we can apply the u_pcm.m and the mula_pcm.m files to determine the resulting SQNR. The resulting SQNRs are 20.49 dB and 24.95 dB, respectively. In this case, the performance of the nonuniform PCM is definitely superior to the performance of the uniform PCM.

Problems

4.1 Design a Huffman code for an information source with probabilities

$$p = \{0.1, 0.05, 0.21, 0.07, 0.02, 0.2, 0.2, 0.15\}$$

Determine the efficiency of the code by computing the average codeword length and the entropy of the source.

4.2 A discrete-memoryless information source is described by the probability vector $p = \{0.2, 0.3, 0.1, 0.4\}$.

    a. Write a MATLAB file to compute the probabilities of the Kth extension of this source for a given K.

    b. Design Huffman codes for this source and its Kth extensions for $K = 1, 2, 3, 4, 5$.

    c. Plot the average codeword length (per source output) as a function of K.

4.3 The probabilities of the letters of the alphabet occurring in printed English are given in Table 4.1.

    a. Determine the entropy of printed English.

    b. Design a Huffman code for printed English.

    c. Determine the average codeword length and the efficiency of the Huffman code.

4.4 Repeat Problem 4.2 for a discrete-memoryless source with a probability vector $p = \{0.5, 0.25, 0.125, 0.125\}$. Explain why the result in this case is different from the result obtained in Problem 4.2.

Table 4.1 Probabilities of letters in printed English

Letter	Probability
A	0.0642
B	0.0127
C	0.0218
D	0.0317
E	0.1031
F	0.0208
G	0.0152
H	0.0467
I	0.0575
J	0.0008
K	0.0049
L	0.0321
M	0.0198
N	0.0574
O	0.0632
P	0.0152
Q	0.0008
R	0.0484
S	0.0514
T	0.0796
U	0.0228
V	0.0083
W	0.0175
X	0.0013
Y	0.0164
Z	0.0005
Word space	0.1859

4.5 A continuous information source has a zero-mean, unit-variance Gaussian distribution. This source is quantized using a uniform symmetric quantizer, where the length of the quantization regions is unity. The number of quantization levels is N. For $N = 2, 3, 4, 5, 6, 7, 8$, determine the entropy of the quantized source output and plot it as a function of N. On the same graph, plot $\log_2 N$ versus N and explain why the two curves differ.

4.6 A zero-mean, unit-variance Gaussian source is quantized using a uniform quantizer. The quantizer uniformly quantizes the interval $[-10, 10]$. Assuming the quantization levels are located at midpoints of the quantization regions, determine and plot the mean-square distortion for $N = 3, 4, 5, 6, 7, 8, 9, 10$ as a function of N, the number of quantization levels.

4.7 On the same figure that you plotted in Problem 4.6, plot the mean-square distortion when the quantization levels are taken to be the centroids of the quantization regions. For what values of N are the two plots closer, and why?

4.8 For a zero-mean, unit-variance Gaussian source, design optimal nonuniform quantizers with a number of levels $N = 2, 3, 4, 5, 6, 7, 8$. For each case, determine $H(\hat{X})$, the entropy of the quantized source, and R, the average codeword length of a Huffman code designed for that source. Plot $H(\hat{X})$, R, and $\log_2 N$ as a function of N on the same figure.

4.9 A Laplacian random variable is defined by the probability density function

$$f(x) = \frac{\lambda}{2} e^{-\lambda |x|}$$

where $\lambda > 0$ is a given constant.

    a. Verify that the variance of a Laplacian random variable is equal to $2/\lambda^2$.

    b. Assuming $\lambda = 1$, design uniform quantizers with $N = 2, 3, 4, 5, 6, 7, 8$ levels for this source. As usual, take the interval of interest to be $[-10\sigma, 10\sigma]$, where σ is the standard deviation of the source.

    c. Plot the entropy of the quantized source and $\log_2 N$ as functions of N on the same figure.

4.10 Repeat Problem 4.9, substituting the uniform quantizer with the optimal nonuniform quantizer.

4.11 Design an optimal 8-level quantizer for a Laplacian source, and plot the resulting mean-square distortion as a function of λ as λ changes in the interval $[0.1, 5]$.

4.12 Design optimal nonuniform quantizers with $N = 2, 3, 4, 5, 6, 7, 8$ for the Laplacian source given in Problem 4.9 with $\lambda = \sqrt{2}$. (Note that this choice of λ results in a zero-mean, unit-variance Laplacian source.) Plot the mean-square error as a function of N for this source. Compare these results with those obtained from quantizing a zero-mean, unit-variance Gaussian source.

4.13 The periodic signal $x(t)$ has a period of 2 and in the interval $[0, 2]$ is defined as

$$x(t) = \begin{cases} t, & 0 \le t < 1 \\ -t + 2, & 1 \le t < 2 \end{cases}$$

    a. Design an 8-level uniform PCM quantizer for this signal, and plot the quantized output of this system.

    b. Plot the quantization error for this system.

    c. By calculating the power in the error signal, determine the SQNR for this system in decibels.

d. Repeat parts a, b, and c using a 16-level uniform PCM system.

4.14 Generate a Gaussian sequence with mean equal to 0 and variance equal to 1 with 1000 elements. Design 4-level, 8-level, 16-level, 32-level, and 64-level uniform PCM schemes for this sequence, and plot the resulting SQNR (in decibels) as a function of the number of bits allocated to each source output.

4.15 Generate a zero-mean, unit-variance Gaussian sequence with a length of 1000, and quantize it using a 6-bit-per-symbol uniform PCM scheme. The resulting 6000 bits are transmitted to the receiver via a noisy channel. The error probability of the channel is denoted by p. Plot the overall SQNR in decibels as a function of p for values of $p = 10^{-3}, 5 \times 10^{-3}, 10^{-2}, 5 \times 10^{-2}, 0.1, 0.2$. For simulation of the effect of noise, you can generate binary random sequences with these probabilities and add them (modulo 2) to the encoded sequence.

4.16 Repeat Problem 4.13 using a nonuniform μ-law PCM with $\mu = 255$.

4.17 Repeat Problem 4.14 using a nonuniform μ-law PCM with $\mu = 255$.

4.18 Repeat Problem 4.15 using a nonuniform μ-law PCM with $\mu = 255$.

Chapter 5

Baseband Digital Transmission

5.1 Preview

In this chapter, we consider several baseband digital modulation and demodulation techniques for transmitting digital information through an additive white Gaussian noise channel. We begin with binary pulse modulation, and then we introduce several nonbinary modulation methods. We describe the optimum receivers for these different signals and consider the evaluation of their performance in terms of the average probability of error.

5.2 Binary Signal Transmission

In a binary communication system, binary data consisting of a sequence of 0's and 1's are transmitted by means of two signal waveforms, say, $s_0(t)$ and $s_1(t)$. Suppose that the data rate is specified as R bits per second. Then each bit is mapped into a corresponding signal waveform according to the rule

$$0 \rightarrow s_0(t), \qquad 0 \leq t \leq T_b$$
$$1 \rightarrow s_1(t), \qquad 0 \leq t \leq T_b$$

where $T_b = 1/R$ is defined as the bit time interval. We assume that the data bits 0 and 1 are equally probable—that is, each occurs with probability $\frac{1}{2}$—and are mutually statistically independent.

   The channel through which the signal is transmitted is assumed to corrupt the signal by the addition of noise, denoted as $n(t)$, which is a sample function of a white Gaussian process with power spectrum $N_0/2$ watts/hertz. Such a channel is called an additive white Gaussian noise (AWGN) channel. Consequently, the received signal waveform is expressed as

$$r(t) = s_i(t) + n(t), \qquad i = 0, 1, \quad 0 \leq t \leq T_b \qquad (5.2.1)$$

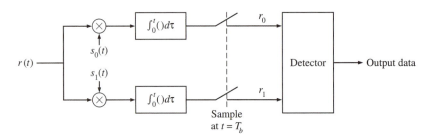

Figure 5.1   Cross-correlation of the received signal $r(t)$ with the two transmitted signals

The task of the receiver is to determine whether a 0 or a 1 was transmitted after observing the received signal $r(t)$ in the interval $0 \le t \le T_b$. The receiver is designed to minimize the probability of error. Such a receiver is called the *optimum receiver.*

5.2.1   Optimum Receiver for the AWGN Channel

In nearly all basic digital communication texts, it is shown that the optimum receiver for the AWGN channel consists of two building blocks. One is either a *signal correlator* or a *matched filter.* The other is a *detector.*

Signal Correlator

The signal correlator cross-correlates the received signal $r(t)$ with the two possible transmitted signals $s_0(t)$ and $s_1(t)$, as illustrated in Figure 5.1. That is, the signal correlator computes the two outputs

$$r_0(t) = \int_0^t r(\tau)s_0(\tau)\,d\tau$$

$$r_1(t) = \int_0^t r(\tau)s_1(\tau)\,d\tau \tag{5.2.2}$$

in the interval $0 \le t \le T_b$, samples the two outputs at $t = T_b$, and feeds the sampled outputs to the detector.

─────(**ILLUSTRATIVE PROBLEM**)─────────────────────────

Illustrative Problem 5.1 [Signal correlator] Suppose the signal waveforms $s_0(t)$ and $s_1(t)$ are the ones shown in Figure 5.2, and let $s_0(t)$ be the transmitted signal. Then the received signal is

$$r(t) = s_0(t) + n(t), \qquad 0 \le t \le T_b \tag{5.2.3}$$

Determine the correlator outputs at the sampling instants.

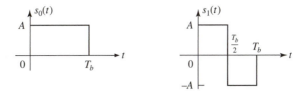

Figure 5.2   Signal waveforms $s_0(t)$ and $s_1(t)$ for a binary communication system

SOLUTION

When the signal $r(t)$ is processed by the two signal correlators shown in Figure 5.1, the outputs r_0 and r_1 at the sampling instant $t = T_b$ are

$$r_0 = \int_0^{T_b} r(t)s_0(t)\, dt$$

$$= \int_0^{T_b} s_0^2(t)\, dt + \int_0^{T_b} n(t)s_0(t)\, dt$$

$$= \mathcal{E} + n_0 \tag{5.2.4}$$

and

$$r_1 = \int_0^{T_b} r(t)s_1(t)\, dt$$

$$= \int_0^{T_b} s_0(t)s_1(t)\, dt + \int_0^{T_b} n(t)s_1(t)\, dt$$

$$= n_1 \tag{5.2.5}$$

where n_0 and n_1 are the noise components at the output of the signal correlators; that is,

$$n_0 = \int_0^{T_b} n(t)s_0(t)\, dt$$

$$n_1 = \int_0^{T_b} n(t)s_1(t)\, dt \tag{5.2.6}$$

and $\mathcal{E} = A^2 T_b$ is the energy of the signals $s_0(t)$ and $s_1(t)$. We also note that the two signal waveforms are *orthogonal*; that is,

$$\int_0^{T_b} s_0(t)s_1(t)\, dt = 0 \tag{5.2.7}$$

On the other hand, when $s_1(t)$ is the transmitted signal, the received signal is

$$r(t) = s_1(t) + n(t), \qquad 0 \le t \le T_b$$

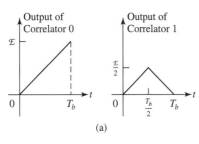
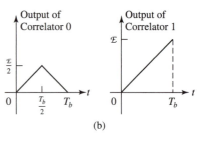

Figure 5.3   Noise-free correlator outputs. (a) $s_0(t)$ was transmitted; (b) $s_1(t)$ was transmitted.

It is easy to show that, in this case, the signal correlator outputs are

$$r_0 = n_0$$
$$r_1 = \mathcal{E} + n_1 \tag{5.2.8}$$

Figure 5.3 illustrates the two noise-free correlator outputs in the interval $0 \le t \le T_b$ for each of the two cases—that is, when $s_0(t)$ is transmitted and when $s_1(t)$ is transmitted.

Since $n(t)$ is a sample function of a white Gaussian process with power spectrum $N_0/2$, the noise components n_0 and n_1 are Gaussian with zero means—that is,

$$E(n_0) = \int_0^{T_b} s_0(t) E[n(t)] \, dt = 0$$

$$E(n_1) = \int_0^{T_b} s_1(t) E[n(t)] \, dt = 0 \tag{5.2.9}$$

and variances σ_i^2, for $i = 1, 2$, where

$$
\begin{aligned}
\sigma_i^2 &= E(n_i^2) \\
&= \int_0^{T_b} \int_0^{T_b} s_i(t) s_i(\tau) E[n(t)n(\tau)] \, dt \, d\tau \\
&= \frac{N_0}{2} \int_0^{T_b} s_i(t) s_i(\tau) \delta(t - \tau) \, dt \, d\tau \\
&= \frac{N_0}{2} \int_0^{T_b} s_i^2(t) \, dt \tag{5.2.10} \\
&= \frac{\mathcal{E} N_0}{2}, \qquad i = 0, 1 \tag{5.2.11}
\end{aligned}
$$

Therefore, when $s_0(t)$ is transmitted, the probability density functions of r_0 and r_1 are

$$p(r_0 \mid s_0(t) \text{ was transmitted}) = \frac{1}{\sqrt{2\pi}\,\sigma} e^{-(r_0 - \mathcal{E})^2/2\sigma^2}$$

$$p(r_1 \mid s_0(t) \text{ was transmitted}) = \frac{1}{\sqrt{2\pi}\,\sigma} e^{-r_1^2/2\sigma^2} \tag{5.2.12}$$

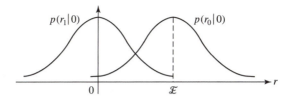

Figure 5.4   Probability density functions $p(r_0 \mid 0)$ and $p(r_1 \mid 0)$ when $s_0(t)$ is transmitted

These two probability density functions, denoted as $p(r_0 \mid 0)$ and $p(r_1 \mid 0)$, are illustrated in Figure 5.4. Similarly, when $s_1(t)$ is transmitted, r_0 is zero-mean Gaussian with variance σ^2, and r_1 is Gaussian with mean value $\mathcal{E}$ and variance σ^2.

Matched Filter

The matched filter provides an alternative to the signal correlator for demodulating the received signal $r(t)$. A filter that is matched to the signal waveform $s(t)$, where $0 \leq t \leq T_b$, has an impulse response

$$h(t) = s(T_b - t), \qquad 0 \leq t \leq T_b \tag{5.2.13}$$

Consequently, the signal waveform—say, $y(t)$—at the output of the matched filter when the input waveform is $s(t)$ is given by the convolution integral

$$y(t) = \int_0^t s(\tau)h(t - \tau)\, dt \tag{5.2.14}$$

If we substitute in (5.2.14) for $h(t - \tau)$ from (5.2.13), we obtain

$$y(t) = \int_0^t s(\tau)s(T_b - t + \tau)\, dt \tag{5.2.15}$$

and if we sample $y(t)$ at $t = T_b$, we obtain

$$y(T_b) = \int_0^{T_b} s^2(t)\, dt = \mathcal{E} \tag{5.2.16}$$

where $\mathcal{E}$ is the energy of the signal $s(t)$. Therefore, the matched filter output at the sampling instant $t = T_b$ is identical to the output of the signal correlator.

———(ILLUSTRATIVE PROBLEM)———————————————

Illustrative Problem 5.2 [Matched filter] Consider the use of matched filters for the demodulation of the two signal waveforms shown in Figure 5.2, and determine the outputs.

Figure 5.5   Impulse responses of matched filters for signals $s_0(t)$ and $s_1(t)$

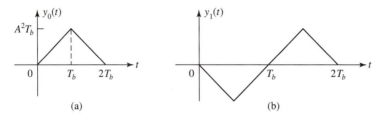

(a)   (b)

Figure 5.6   Signal outputs of matched filters when $s_0(t)$ is transmitted

SOLUTION

The impulse responses of the two matched filters are

$$h_0(t) = s_0(T_b - t)$$
$$h_1(t) = s_1(T_b - t) \tag{5.2.17}$$

as illustrated in Figure 5.5. Note that each impulse response is obtained by folding the signal $s(t)$ to obtain $s(-t)$ and then delaying the folded signal $s(-t)$ by T_b to obtain $s(T_b - t)$.

Now suppose the signal waveform $s_0(t)$ is transmitted. Then the received signal $r(t) = s_0(t) + n(t)$ is passed through the two matched filters. The response of the filter with impulse response $h_0(t)$ to the signal component $s_0(t)$ is illustrated in Figure 5.6(a). Also, the response of the filter with impulse response $h_1(t)$ to the signal component $s_0(t)$ is illustrated in Figure 5.6(b). Hence, at the sampling instant $t = T_b$, the outputs of the two matched filters with impulse responses $h_0(t)$ and $h_1(t)$ are

$$r_0 = \mathcal{E} + n_0$$
$$r_1 = n_1 \tag{5.2.18}$$

respectively. Note that these outputs are identical to the outputs obtained from sampling the signal correlator outputs at $t = T_b$.

The Detector

The detector observes the correlator or matched filter outputs r_0 and r_1 and decides whether the transmitted signal waveform is $s_0(t)$ or $s_1(t)$, which correspond to the transmission of either a 0 or a 1, respectively. The *optimum detector* is defined as the detector that minimizes the probability of error.

ILLUSTRATIVE PROBLEM

Illustrative Problem 5.3 [Binary detection] Let us consider the detector for the signals shown in Figure 5.2, which are equally probable and have equal energies. The optimum detector for these signals compares r_0 and r_1 and decides that a 0 was transmitted when $r_0 > r_1$ and that a 1 was transmitted when $r_1 > r_0$. Determine the probability of error.

SOLUTION

When $s_0(t)$ is the transmitted signal waveform, the probability of error is

$$P_e = P(r_1 > r_0) = P(n_1 > \mathcal{E} + n_0) = P(n_1 - n_0 > \mathcal{E}) \tag{5.2.19}$$

Since n_1 and n_0 are zero-mean Gaussian random variables, their difference $x \equiv n_1 - n_0$ is also zero-mean Gaussian. The variance of the random variable x is

$$E(x^2) = E[(n_1 - n_0)^2] = E(n_1^2) + E(n_0^2) - 2E(n_1 n_0) \tag{5.2.20}$$

But $E(n_1 n_0) = 0$ because the signal waveforms are orthogonal; that is,

$$\begin{aligned}
E(n_1 n_0) &= E \int_0^{T_b} \int_0^{T_b} s_0(t) s_1(\tau) n(t) n(\tau)\, dt\, d\tau \\
&= \frac{N_0}{2} \int_0^{T_b} \int_0^{T_b} s_0(t) s_1(\tau) \delta(t - \tau)\, dt\, d\tau \\
&= \frac{N_0}{2} \int_0^{T_b} s_0(t) s_1(t)\, dt \\
&= 0 \tag{5.2.21}
\end{aligned}$$

Therefore,

$$E(x^2) = 2\left(\frac{\mathcal{E} N_0}{2}\right) = \mathcal{E} N_0 \equiv \sigma_x^2 \tag{5.2.22}$$

Hence, the probability of error is

$$\begin{aligned}
P_e &= \frac{1}{\sqrt{2\pi}\, \sigma_x} \int_{\mathcal{E}}^{\infty} e^{-x^2/2\sigma_x^2}\, dx \\
&= \frac{1}{\sqrt{2\pi}} \int_{\sqrt{\mathcal{E}/N_0}}^{\infty} e^{-x^2/2}\, dx \\
&= Q\left(\sqrt{\frac{\mathcal{E}}{N_0}}\right) \tag{5.2.23}
\end{aligned}$$

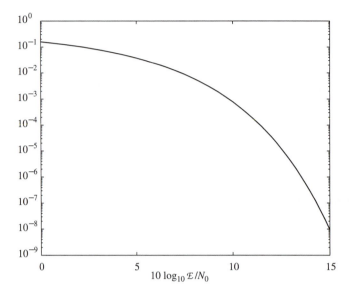

Figure 5.7   Probability of error for orthogonal signals

The ratio $\mathcal{E}/N_0$ is called the signal-to-noise ratio (SNR).

The derivation of the detector performance given in the example was based on the transmission of the signal waveform $s_0(t)$. The reader may verify that the probability of error that is obtained when $s_1(t)$ is transmitted is identical to that obtained when $s_0(t)$ is transmitted. Because the 0's and 1's in the data sequence are equally probable, the average probability of error is that given by (5.2.23). This expression for the probability of error is evaluated by using the MATLAB script given next and is plotted in Figure 5.7 as a function of the SNR, where the SNR is displayed on a logarithmic scale ($10\log_{10}\mathcal{E}/N_0$). As expected, the probability of error decreases exponentially as the SNR increases.

■ M-FILE

```
% The MATLAB script that generates the probability of error versus the signal to noise ratio
initial_snr=0;
final_snr=15;
snr_step=0.25;
snr_in_dB=initial_snr:snr_step:final_snr;
for i=1:length(snr_in_dB),
  snr=10^(snr_in_dB(i)/10);
  Pe(i)=Qfunct(sqrt(snr));
  echo off;
end;
echo on;
semilogy(snr_in_dB,Pe);
```

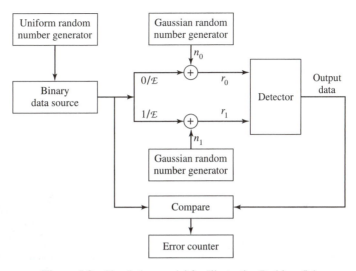

Figure 5.8   Simulation model for Illustrative Problem 5.4

5.2.2   Monte Carlo Simulation of a Binary Communication System

Monte Carlo computer simulations are usually performed in practice to estimate the probability of error of a digital communication system, especially in cases where the analysis of the detector performance is difficult to perform. We demonstrate the method for estimating the probability of error for the binary communication system described earlier.

──(**ILLUSTRATIVE PROBLEM**)──────────────────────────────

Illustrative Problem 5.4 [Monte Carlo Simulation] Use Monte Carlo simulation to estimate and plot P_e versus SNR for a binary communication system that employs correlators or matched filters. The model of the system is illustrated in Figure 5.8.

──(**SOLUTION**)──────────────────────────────────────

We simulate the generation of the random variables r_0 and r_1, which constitute the input to the detector. We begin by generating a binary sequence of 0's and 1's that occur with equal probability and are mutually statistically independent. To accomplish this task, we use a random number generator that generates a uniform random number with range $(0, 1)$. If the number generated is in the range $(0, 0.5)$, the binary source output is a 0. Otherwise, it is a 1. If a 0 is generated, then $r_0 = \mathcal{E} + n_0$, and $r_1 = n_1$. If a 1 is generated, then $r_0 = n_0$, and $r_1 = \mathcal{E} + n_1$.

The additive noise components n_0 and n_1 are generated by means of two Gaussian noise generators. Their means are zero, and their variances are $\sigma^2 = \mathcal{E} N_0/2$. For convenience, we may normalize the signal energy $\mathcal{E}$ to unity ($\mathcal{E} = 1$) and vary σ^2. Note that the

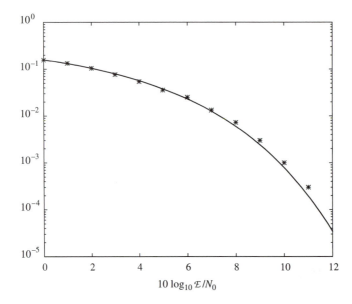

Figure 5.9   Error probability from Monte Carlo simulation compared with theoretical error probability for orthogonal signaling

SNR, which is defined as $\mathcal{E}/N_0$, is then equal to $1/2\sigma^2$. The detector output is compared with the binary transmitted sequence, and an error counter is used to count the number of bit errors.

Figure 5.9 illustrates the results of this simulation for the transmission of $N = 10{,}000$ bits at several different values of SNR. Note the agreement between the simulation results and the theoretical value of P_e given by (5.2.23). We should also note that a simulation of $N = 10{,}000$ data bits allows us to estimate the error probability reliably down to about $P_e = 10^{-3}$. In other words, with $N = 10{,}000$ data bits, we should have at least ten errors for a reliable estimate of P_e. MATLAB scripts for this problem are given next.

M-FILE

```
% MATLAB script for Illustrative Problem 5.4.
echo on
SNRindB1=0:1:12;
SNRindB2=0:0.1:12;
for i=1:length(SNRindB1),
    % simulated error rate
    smld_err_prb(i)=smldPe54(SNRindB1(i));
    echo off ;
end;
echo on ;
for i=1:length(SNRindB2),
    SNR=exp(SNRindB2(i)*log(10)/10);
    % theoretical error rate
```

```
      theo_err_prb(i)=Qfunct(sqrt(SNR));
      echo off ;
end;
echo on;
% Plotting  commands  follow
semilogy(SNRindB1,smld_err_prb,' * ');
hold
semilogy(SNRindB2,theo_err_prb);
```

M-FILE

```
function [p]=smldPe54(snr_in_dB)
% [p]=smldPe54(snr_in_dB)
%              SMLDPE54  finds the probability of error for the given
%              snr_in_dB, signal to noise ratio in dB.
E=1;
SNR=exp(snr_in_dB*log(10)/10);          % signal to noise ratio
sgma=E/sqrt(2*SNR);                     % sigma, standard deviation of noise
N=10000;
% generation of the binary data source
for i=1:N,
  temp=rand;                            % a uniform random variable over (0,1)
  if (temp<0.5),
    dsource(i)=0;                       % with probability 1/2, source output is 0
  else
    dsource(i)=1;                       % with probability 1/2, source output is 1
  end
end;
% detection, and probability of error calculation
numoferr=0;
for i=1:N,
  % matched filter outputs
  if (dsource(i)==0),
    r0=E+gngauss(sgma);
    r1=gngauss(sgma);                   % if the source output is "0"
  else
    r0=gngauss(sgma);
    r1=E+gngauss(sgma);                 % if the source output is "1"
  end;
  % detector follows
  if (r0>r1),
    decis=0;                            % decision is "0"
  else
    decis=1;                            % decision is "1"
  end;
  if (decis~=dsource(i)),               % if it is an error, increase the error counter
    numoferr=numoferr+1;
  end;
end;
p=numoferr/N;                           % probability of error estimate
```

QUESTION

In Figure 5.9, simulation and theoretical results completely agree at low signal-to-noise ratios, whereas at higher SNRs, they agree less. Can you explain why? How should we change the simulation process to result in better agreement at higher signal-to-noise ratios?

5.2.3   Other Binary Signal Transmission Methods

The binary signal transmission method described above was based on the use of orthogonal signals. Next, we describe two other methods for transmitting binary information through a communication channel. One method employs antipodal signals, and the other employs an on-off-type signal.

Antipodal Signals for Binary Signal Transmission

Two signal waveforms are said to be *antipodal* if one signal waveform is the negative of the other. For example, one pair of antipodal signals is illustrated in Figure 5.10(a), and a second pair is illustrated in Figure 5.10(b).

Suppose we use antipodal signal waveforms $s_0(t) = s(t)$ and $s_1(t) = -s(t)$ to transmit binary information, where $s(t)$ is some arbitrary waveform having energy $\mathcal{E}$. The received signal waveform from an AWGN channel may be expressed as

$$r(t) = \pm s(t) + n(t), \qquad 0 \le t \le T_b \tag{5.2.24}$$

The optimum receiver for recovering the binary information employs a single correlator or a single matched filter matched to $s(t)$, followed by a detector, as illustrated in Figure 5.11. Let us suppose that $s(t)$ was transmitted, so that the received signal is

$$r(t) = s(t) + n(t) \tag{5.2.25}$$

The output of the correlator or matched filter at the sampling instant $t = T_b$ is

$$r = \mathcal{E} + n \tag{5.2.26}$$

where $\mathcal{E}$ is the signal energy and n is the additive noise component, which is expressed as

$$n = \int_0^{T_b} n(t)s(t)\, dt \tag{5.2.27}$$

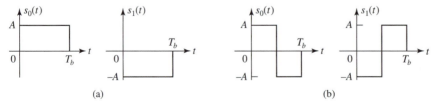

Figure 5.10   Examples of pairs of antipodal signals

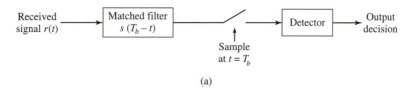

(a)

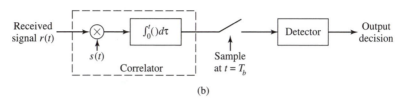

(b)

Figure 5.11   Optimum receiver for antipodal signals. (a) Matched filter demodulator;
(b) correlator demodulator

Since the additive noise process $n(t)$ is zero mean, it follows that $E(n) = 0$. The variance of the noise component n is

$$\sigma^2 = E(n^2)$$
$$= \int_0^{T_b} \int_0^{T_b} E\left[n(t)n(\tau)\right] s(t)s(\tau) \, dt \, d\tau$$
$$= \frac{N_0}{2} \int_0^{T_b} \int_0^{T_b} \delta(t - \tau)s(t)s(\tau) \, dt \, d\tau$$
$$= \frac{N_0}{2} \int_0^{T_b} s^2(t) \, dt$$
$$= \frac{N_0 \mathcal{E}}{2} \tag{5.2.28}$$

Consequently, the probability density function of r when $s(t)$ is transmitted is

$$p(r \mid s(t) \text{ was transmitted}) \equiv p(r \mid 0) = \frac{1}{\sqrt{2\pi}\,\sigma}e^{-(r-\mathcal{E})^2/2\sigma^2} \tag{5.2.29}$$

Similarly, when the signal waveform $-s(t)$ is transmitted, the input to the detector is

$$r = -\mathcal{E} + n \tag{5.2.30}$$

and the probability density function of r is

$$p(r \mid -s(t) \text{ was transmitted}) \equiv p(r \mid 1) = \frac{1}{\sqrt{2\pi}\,\sigma}e^{-(r+\mathcal{E})^2/2\sigma^2} \tag{5.2.31}$$

These two probability density functions are illustrated in Figure 5.12.

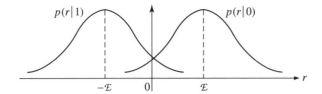

Figure 5.12   Probability density functions for the input to the detector

For equally probable signal waveforms, the optimum detector compares r with the threshold zero. If $r > 0$, the decision is made that $s(t)$ was transmitted. If $r < 0$, the decision is made that $-s(t)$ was transmitted.

The noise that corrupts the signal causes errors at the detector. The probability of a detector error is easily computed. Suppose that $s(t)$ was transmitted. Then the probability of error is equal to the probability that $r < 0$; that is,

$$
\begin{aligned}
P_e &= P(r < 0) \\
&= \frac{1}{\sqrt{2\pi}\,\sigma} \int_{-\infty}^{0} e^{-(r-\mathcal{E})^2/2\sigma^2}\, dr \\
&= \frac{1}{\sqrt{2\pi}} \int_{-\infty}^{-\mathcal{E}/\sigma} e^{-r^2/2}\, dr \\
&= Q\left(\frac{\mathcal{E}}{\sigma}\right) \\
&= Q\left(\sqrt{\frac{2\mathcal{E}}{N_0}}\right)
\end{aligned}
\tag{5.2.32}
$$

A similar result is obtained when $-s(t)$ is transmitted. Consequently, when the two signal waveforms are equally probable, the average probability of error is given by (5.2.32).

When we compare the probability of error for antipodal signals with that for orthogonal signals given by (5.2.23), we observe that, for the same transmitted signal energy $\mathcal{E}$, antipodal signals result in better performance. Alternatively, we may say that antipodal signals yield the same performance (same error probability) as orthogonal signals by using one-half of the transmitted energy of orthogonal signals. Hence, antipodal signals are 3 dB more efficient than orthogonal signals.

━━━━(ILLUSTRATIVE PROBLEM)━━━━━━━━━━━━━━━━━━━━━━━

Illustrative Problem 5.5 [Binary antipodal simulation] Use Monte Carlo simulation to estimate and plot the error probability performance of a binary antipodal communication system. The model of the system is illustrated in Figure 5.13.

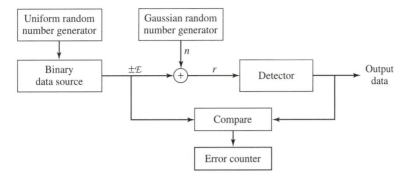

Figure 5.13   Model of the binary communication system employing antipodal signals

---

SOLUTION

As shown, we simulate the generation of the random variable r, which is the input to the detector. A uniform random number generator is used to generate the binary information sequence from the binary data source. The sequence of 0's and 1's is mapped into a sequence of $\pm\mathcal{E}$, where $\mathcal{E}$ represents the signal energy. $\mathcal{E}$ may be normalized to unity. A Gaussian noise generator is used to generate the sequence of zero-mean Gaussian random numbers with variance σ^2. The detector compares the random variable r with the threshold of 0. If $r > 0$, the decision is made that the transmitted bit is a 0. If $r < 0$, the decision is made that the transmitted bit is a 1. The output of the detector is compared with the transmitted sequence of information bits, and the bit errors are counted. Figure 5.14 illustrates the results of this simulation for the transmission of 10,000 bits at several different values of SNR. The theoretical value for P_e given by (5.2.32) is also plotted in Figure 5.14 for comparison.

The MATLAB scripts for this problem are given next.

M-FILE

```
% MATLAB script for Illustrative Problem 5.5.
echo on
SNRindB1=0:1:10;
SNRindB2=0:0.1:10;
for i=1:length(SNRindB1),
    % simulated error rate
    smld_err_prb(i)=smldPe55(SNRindB1(i));
    echo off;
end;
echo on;
for i=1:length(SNRindB2),
    SNR=exp(SNRindB2(i)*log(10)/10);
    % theoretical error rate
    theo_err_prb(i)=Qfunct(sqrt(2*SNR));
```

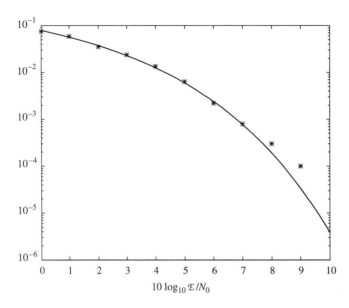

Figure 5.14 Error probability from Monte Carlo simulation compared with theoretical error probability for antipodal signals

---

```
   echo off;
end;
echo on;
% Plotting commands follow
semilogy(SNRindB1,smld_err_prb,'*');
hold
semilogy(SNRindB2,theo_err_prb);
```

M-FILE

```
function [p]=smldPe55(snr_in_dB)
% [p]=smldPe55(snr_in_dB)
%                SMLDPE55   simulates the probability of error for the particular
%                value of snr_in_dB, signal to noise ratio in dB.
E=1;
SNR=exp(snr_in_dB*log(10)/10);              % signal to noise ratio
sgma=E/sqrt(2*SNR);                         % sigma, standard deviation of noise
N=10000;
% generation of the binary data source follows
for i=1:N,
   temp=rand;                               % a uniform random variable over (0,1)
   if (temp<0.5),
```

```
        dsource(i)=0;                    % with probability 1/2, source output is 0
    else
        dsource(i)=1;                    % with probability 1/2, source output is 1
    end
end;
% the detection, and probability of error calculation follows
numoferr=0;
for i=1:N,
    % The matched filter outputs
    if (dsource(i)==0),
        r=−E+gngauss(sgma);              % if the source output is "0"
    else
        r=E+gngauss(sgma);               % if the source output is "1"
    end;
    % detector follows
    if (r<0),
        decis=0;                         % decision is "0"
    else
        decis=1;                         % decision is "1"
    end;
    if (decis˜=dsource(i)),              % if it is an error, increase the error counter
        numoferr=numoferr+1;
    end;
end;
p=numoferr/N;                            % probability of error estimate
```

On-Off Signals for Binary Signal Transmission

A binary information sequence may also be transmitted by use of on-off signals. To transmit a 0, no signal is transmitted in the time interval of duration T_b. To transmit a 1, a signal waveform $s(t)$ is transmitted. Consequently, the received signal waveform may be represented as

$$r(t) = \begin{cases} n(t), & \text{if a 0 is transmitted} \\ s(t) + n(t), & \text{if a 1 is transmitted} \end{cases} \tag{5.2.33}$$

where $n(t)$ represents the additive white Gaussian noise.

As in the case of antipodal signals, the optimum receiver consists of a correlator or a matched filter matched to $s(t)$, whose output is sampled at $t = T_b$, and followed by a detector that compares the sampled output to the threshold, denoted as α. If $r > \alpha$, a 1 is declared to have been transmitted; otherwise, a 0 is declared to have been transmitted.

The input to the detector may be expressed as

$$r = \begin{cases} n, & \text{if a 0 is transmitted} \\ \mathcal{E} + n, & \text{if a 1 is transmitted} \end{cases} \tag{5.2.34}$$

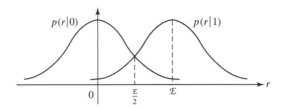

Figure 5.15   The probability density functions for the received signal at the output of the correlator for on-off signals

where n is a zero-mean Gaussian random variable with variance $\sigma^2 = \mathcal{E} N_0/2$. Therefore, the conditional probability density functions of the random variable r are

$$p(r \mid 0) = \frac{1}{\sqrt{2\pi}\,\sigma} e^{-r^2/2\sigma^2}, \qquad \text{if a 0 is transmitted}$$

$$p(r \mid 1) = \frac{1}{\sqrt{2\pi}\,\sigma} e^{-(r-\mathcal{E})^2/2\sigma^2}, \quad \text{if a 1 is transmitted}$$

These probability density functions are illustrated in Figure 5.15.

When a 0 is transmitted, the probability of error is

$$P_{e0}(\alpha) = P(r > \alpha) = \frac{1}{\sqrt{2\pi}\,\sigma} \int_\alpha^\infty e^{-r^2/2\sigma^2}\, dr \tag{5.2.35}$$

where α is the threshold. On the other hand, when a 1 is transmitted, the probability of error is

$$P_{e1}(\alpha) = P(r < \alpha) = \frac{1}{\sqrt{2\pi}\,\sigma} \int_{-\infty}^\alpha e^{-(r-\mathcal{E})^2/2\sigma^2}\, dr \tag{5.2.36}$$

Assuming that the binary information bits are equally probable, we have for the average probability of error:

$$P_e(\alpha) = \frac{1}{2} P_{e0}(\alpha) + \frac{1}{2} P_{e1}(\alpha) \tag{5.2.37}$$

The value of the threshold α that minimizes the average probability of error is found by differentiating $P_e(\alpha)$ and solving for the optimum threshold. It is easily shown that the optimum threshold is

$$\alpha_{\text{opt}} = \frac{\mathcal{E}}{2} \tag{5.2.38}$$

Substitution of this optimum value into (5.2.35), (5.2.36), and (5.2.37) yields the probability of error

$$P_e(\alpha_{\text{opt}}) = Q\left(\sqrt{\frac{\mathcal{E}}{2N_0}}\right) \tag{5.2.39}$$

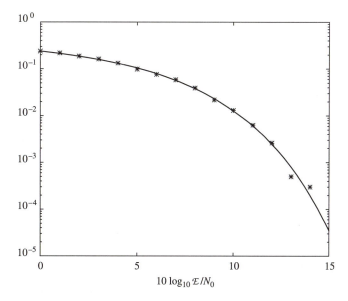

$$10 \log_{10} \mathcal{E}/N_0$$

Figure 5.16   Error probability from Monte Carlo simulation compared with theoretical error probability for on-off signals

We observe that error-rate performance with on-off signals is not as good as with antipodal signals. It appears to be 6 dB worse than with antipodal signals and 3 dB worse than with orthogonal signals. However, the average transmitted energy for the on-off signals is 3 dB less than that for antipodal and orthogonal signals. Consequently, this difference should be factored in when making comparisons of the performance with other signal types.

━━━━ ◖ILLUSTRATIVE PROBLEM◗ ━━━━

Illustrative Problem 5.6 [On-off signaling simulation] Use Monte Carlo simulation to estimate and plot the performance of a binary communication system employing on-off signaling.

━━━━ ◖SOLUTION◗ ━━━━

The model for the system to be simulated is similar to that shown in Figure 5.13, except that one of the signals is 0. Thus, we generate a sequence of random variables $\{r_i\}$ as given by (5.2.34). The detector compares the random variables $\{r_i\}$ to the optimum threshold $\mathcal{E}/2$ and makes the appropriate decisions. Figure 5.16 illustrates the estimated error probability based on 10,000 binary digits. The theoretical error rate given by (5.2.39) is also illustrated in this figure.

The MATLAB scripts for this problem are given next.

M-FILE

```
% MATLAB script for Illustrative Problem 5.6.
echo on
SNRindB1=0:1:15;
SNRindB2=0:0.1:15;
for i=1:length(SNRindB1),
    smld_err_prb(i)=smldPe56(SNRindB1(i));    % simulated error rate
    echo off;
end;
echo on;
for i=1:length(SNRindB2),
    SNR=exp(SNRindB2(i)*log(10)/10);          % signal to noise ratio
    theo_err_prb(i)=Qfunct(sqrt(SNR/2));      % theoretical error rate
    echo off;
end;
echo on;
% Plotting commands follow
semilogy(SNRindB1,smld_err_prb,'*');
hold
semilogy(SNRindB2,theo_err_prb);
```

M-FILE

```
function [p]=smldPe56(snr_in_dB)
% [p]=smldPe56(snr_in_dB)
%              SMLDPE56   simulates the probability of error for a given
%              snr_in_dB, signal to noise ratio in dB.

E=1;
alpha_opt=1/2;
SNR=exp(snr_in_dB*log(10)/10);            % signal to noise ratio
sgma=E/sqrt(2*SNR);                       % sigma, standard deviation of noise
N=10000;
% generation of the binary data source follows
for i=1:N,
    temp=rand;                            % a uniform random variable over (0,1)
    if (temp<0.5),
        dsource(i)=0;                     % with probability 1/2, source output is 0
    else
        dsource(i)=1;                     % with probability 1/2, source output is 1
    end
end;
% detection, and probability of error calculation
numoferr=0;
for i=1:N,
    % The matched filter outputs
    if (dsource(i)==0),
        r=gngauss(sgma);                  % if the source output is "0"
```

```
    else
        r=E+gngauss(sgma);              % if the source output is "1"
    end;
    % detector follows
    if (r<alpha_opt),
        decis=0;                        % decision is "0"
    else
        decis=1;                        % decision is "1"
    end;
    if (decis~=dsource(i)),             % if it is an error, increase the error counter
        numoferr=numoferr+1;
    end;
end;
p=numoferr/N;                           % probability of error estimate
```

5.2.4   Signal Constellation Diagrams for Binary Signals

The three types of binary signals—namely, antipodal, on-off, and orthogonal—may be characterized geometrically as points in "signal space." In the case of antipodal signals, where the signals are $s(t)$ and $-s(t)$, each having energy $\mathcal{E}$, the two signal points fall on the real line at $\pm\sqrt{\mathcal{E}}$, as shown in Figure 5.17(a). The one-dimensional geometric representation of antipodal signals follows from the fact that only one signal waveform or basis function—namely, $s(t)$—suffices to represent the antipodal signals in the signal space.

On-off signals are also one-dimensional. Hence, the two signal points also fall on the real line at 0 and $\sqrt{\mathcal{E}}$, as shown in Figure 5.17(b).

On the other hand, binary orthogonal signals require a two-dimensional geometric representation, since there are two linearly independent functions $s_0(t)$ and $s_1(t)$ that constitute the two signal waveforms. Consequently, the signal points corresponding to these points are $(\sqrt{\mathcal{E}}, 0)$ and $(0, \sqrt{\mathcal{E}})$, as shown in Figure 5.17(c).

The geometric representations of the binary signals shown in Figure 5.17 are called *signal constellations*.

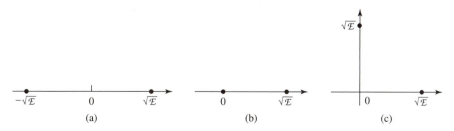

Figure 5.17   Signal constellations for binary signals. (a) Antipodal signals; (b) On-off signals; (c) Orthogonal signals

─────(ILLUSTRATIVE PROBLEM)──────────────────────────────

Illustrative Problem 5.7 [Noise effect and the constellation] The effect of noise on the performance of a binary communication system can be observed from the received signal plus noise at the input to the detector. For example, let us consider binary orthogonal signals, for which the input to the detector consists of the pair of random variables (r_0, r_1), where either

$$(r_0, r_1) = (\sqrt{\mathcal{E}} + n_0, n_1)$$

or

$$(r_0, r_1) = (n_0, \sqrt{\mathcal{E}} + n_1)$$

The noise random variables n_0 and n_1 are zero-mean, independent Gaussian random variables with variance σ^2. As in Illustrative Problem 5.4, use Monte Carlo simulation to generate 100 samples of (r_0, r_1) for each value of $\sigma = 0.1$, $\sigma = 0.3$, and $\sigma = 0.5$, and plot these 100 samples for each σ on different two-dimensional plots. The energy $\mathcal{E}$ of the signal may be normalized to unity.

─────(SOLUTION)──

The results of the Monte Carlo simulation are shown in Figure 5.18. Note that at a low noise power level (σ small), the effect of the noise on performance (error rate) of the communication system is small. As the noise power level increases, the noise components increase in size and cause more errors.

The MATLAB script for this problem for $\sigma = 0.5$ is given next.

─────(M-FILE)──

```
% MATLAB script for Illustrative Problem 5.7.
echo on
n0=.5*randn(100,1);
n1=.5*randn(100,1);
n2=.5*randn(100,1);
n3=.5*randn(100,1);
x1=1.+n0;
y1=n1;
x2=n2;
y2=1.+n3;
plot(x1,y1,'o',x2,y2,'*')
axis('square')
```

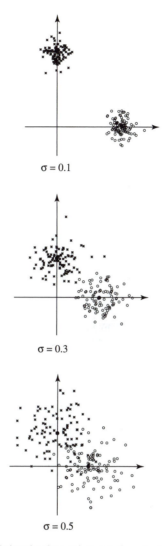

$\sigma = 0.1$

$\sigma = 0.3$

$\sigma = 0.5$

Figure 5.18   Received signal points at input to the selector for orthogonal signals (Monte Carlo simulation)

5.3   Multiamplitude Signal Transmission

In the preceding section, we treated the transmission of digital information by use of binary signal waveforms. Thus, each signal waveform conveyed one bit of information. In this section, we use signal waveforms that take on multiple amplitude levels. Thus, we can transmit multiple bits per signal waveform.

5.3.1   Signal Waveforms with Four Amplitude Levels

Let us consider a set of signal waveforms of the form

$$s_m(t) = A_m g(t), \qquad 0 \leq t \leq T \tag{5.3.1}$$

where A_m is the amplitude of the mth waveform and $g(t)$ is a rectangular pulse defined as

$$g(t) = \begin{cases} \sqrt{1/T}, & 0 \leq t \leq T \\ 0, & \text{otherwise} \end{cases} \tag{5.3.2}$$

where the energy in the pulse $g(t)$ is normalized to unity. In particular, we consider the case in which the signal amplitude takes on one of four possible equally spaced values— namely, $\{A_m\} = \{-3d, -d, d, 3d\}$ or, equivalently,

$$A_m = (2m - 3)d, \qquad m = 0, 1, 2, 3 \tag{5.3.3}$$

where $2d$ is the Euclidean distance between two adjacent amplitude levels. The four signal waveforms are illustrated in Figure 5.19. We call this set of waveforms *pulse-amplitude-modulated* (PAM) signals.

The four PAM signal waveforms shown in Figure 5.19 can be used to transmit two bits of information per waveform. Thus, we assign the following pairs of information bits to the four signal waveforms:

$$00 \rightarrow s_0(t)$$
$$01 \rightarrow s_1(t)$$
$$11 \rightarrow s_2(t)$$
$$10 \rightarrow s_3(t)$$

Each pair of information bits $\{00, 01, 10, 11\}$ is called a *symbol*, and the time duration T is called the *symbol interval*. Note that if the bit rate is $R = 1/T_b$, the symbol interval is $T = 2T_b$. Since all the signal waveforms are scaled versions of the signal basis function $g(t)$, these signal waveforms may be represented geometrically as points on the real line.

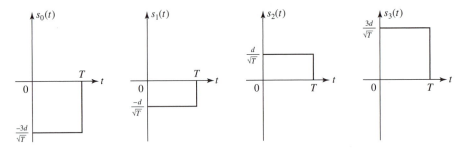

Figure 5.19   Multiamplitude signal waveforms

Figure 5.20   Signal constellation for the four PAM signal waveforms

Therefore, the geometric representation of the four PAM signals is the *signal constellation diagram* shown in Figure 5.20.

As in the case of binary signals, we assume that the PAM signal waveforms are transmitted through an AWGN channel. Consequently, the received signal is represented as

$$r(t) = s_i(t) + n(t), \qquad i = 0, 1, 2, 3, \quad 0 \le t \le T \tag{5.3.4}$$

where $n(t)$ is a sample function of a white Gaussian noise process with power spectrum $N_0/2$ watts/hertz. The task of the receiver is to determine which of the four signal waveforms was transmitted after observing the received signal $r(t)$ in the interval $0 \le t \le T$. The optimum receiver is designed to minimize the probability of a symbol error.

5.3.2   Optimum Receiver for the AWGN Channel

The receiver that minimizes the probability of error is implemented by passing the signal through a signal correlator or matched filter followed by an amplitude detector. Since the signal correlator and the matched filter yield the same output at the sampling instant, we consider only the signal correlator in our treatment.

Signal Correlator

The signal correlator cross-correlates the received signal $r(t)$ with the signal pulse $g(t)$, and its output is sampled at $t = T$. Thus, the signal correlator output is

$$
\begin{aligned}
r &= \int_0^T r(t)g(t)\,dt \\
&= \int_0^T A_i g^2(t)\,dt + \int_0^T g(t)n(t)\,dt \\
&= A_i + n
\end{aligned}
\tag{5.3.5}
$$

where n represents the noise component, defined as

$$n = \int_0^T g(t)n(t)\,dt \tag{5.3.6}$$

We note that n is a Gaussian random variable with mean

$$E(n) = \int_0^T g(t)E[n(t)]\,dt = 0 \tag{5.3.7}$$

and variance

$$\sigma^2 = E(n^2)$$

$$= \int_0^T \int_0^T g(t)g(\tau)E\left[n(t)n(\tau)\right] dt\, d\tau$$

$$= \frac{N_0}{2} \int_0^T \int_0^T g(t)g(\tau)\delta(t-\tau)\, dt\, d\tau$$

$$= \frac{N_0}{2} \int_0^T g^2(t)\, dt$$

$$= \frac{N_0}{2} \qquad\qquad (5.3.8)$$

Therefore, the probability density function for the output r of the signal correlator is

$$p(r \mid s_i(t) \text{was transmitted}) = \frac{1}{\sqrt{2\pi}\sigma} e^{-(r-A_i)^2/2\sigma^2} \qquad (5.3.9)$$

where A_i is one of the four possible amplitude values.

The Detector

The detector observes the correlator output r and decides which of the four PAM signals was transmitted in the signal interval. In the following development of the performance of the optimum detector, we assume that the four possible amplitude levels are equally probable.

Since the received signal amplitude A_i can take the values $\pm d, \pm 3d$, as illustrated in the signal constellation in Figure 5.20, the optimum amplitude detector compares the correlator output r with the four possible transmitted amplitude levels and selects the amplitude level that is closest in Euclidean distance to r. Thus, the optimum amplitude detector computes the distances

$$D_i = |r - A_i|, \qquad i = 0, 1, 2, 3 \qquad (5.3.10)$$

and selects the amplitude corresponding to the smallest distance.

We note that a decision error occurs when the noise variable n exceeds in magnitude one-half of the distance between amplitude levels—that is, when $|n| > d$. However, when the amplitude level $+3d$ or $-3d$ is transmitted, an error can occur in one direction only. Since the four amplitude levels are equally probable, the average probability of a symbol error is

$$P_4 = \frac{3}{4} P(|r - A_m| > d)$$

$$= \frac{3}{2} \int_d^\infty \frac{1}{\sqrt{2\pi}\,\sigma} e^{-x^2/2\sigma^2}\, dx$$

$$= \frac{3}{2} \int_{d/\sigma}^\infty \frac{1}{\sqrt{2\pi}} e^{-x^2/2}\, dx$$

$$= \frac{3}{2} Q\left(\sqrt{\frac{d^2}{\sigma^2}}\right)$$

$$= \frac{3}{2} Q\left(\sqrt{\frac{2d^2}{N_0}}\right) \tag{5.3.11}$$

We observe that the squared distance between successive amplitude levels is $(2d)^2 \equiv \delta^2$. Therefore, the average probability of error may be expressed as

$$P_4 = \frac{3}{2} Q\left(\sqrt{\frac{\delta^2}{4N_0}}\right) \tag{5.3.12}$$

Alternatively, the average probability of error may be expressed in terms of the signal energy. Since all four amplitude levels are equally probable, the average transmitted signal energy per symbol is

$$\mathcal{E}_{\text{av}} = \frac{1}{4} \sum_{k=1}^{4} \int_0^T s_k^2(t)\, dt = 5d^2 \tag{5.3.13}$$

Consequently, $d^2 = \mathcal{E}_{\text{av}}/5$, and hence,

$$P_4 = \frac{3}{2} Q\left(\sqrt{\frac{2\mathcal{E}_{\text{av}}}{5N_0}}\right) \tag{5.3.14}$$

Since each transmitted symbol consists of two information bits, the transmitted average energy per bit is $\mathcal{E}_{\text{av}}/2 \equiv \mathcal{E}_{\text{av}b}$.

The average probability of error P_4 is plotted in Figure 5.21 as a function of the SNR defined as $10 \log_{10}(\mathcal{E}_{\text{av}b}/N_0)$.

─────────(ILLUSTRATIVE PROBLEM)─────────────────────

Illustrative Problem 5.8 [Multiamplitude signal simulation] Perform a Monte Carlo simulation of the four-level (quaternary) PAM communication system that employs a signal correlator, as described earlier, followed by an amplitude detector. The model for the system to be simulated is shown in Figure 5.22.

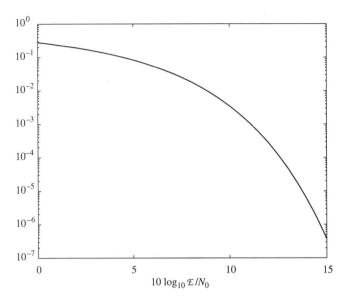

Figure 5.21   Probability of symbol error for four-level PAM

SOLUTION

As shown, we simulate the generation of the random variable r, which is the output of the signal correlator and the input to the detector. We begin by generating a sequence of quaternary symbols that are mapped into corresponding amplitude levels $\{A_m\}$. To accomplish this task, we use a random number generator that generates a uniform random number in the range $(0, 1)$. This range is subdivided into four equal intervals, $(0, 0.25)$, $(0.25, 0.5)$, $(0.5, 0.75)$, $(0.75, 1.0)$, where the subintervals correspond to the symbols (pairs of information bits) 00, 01, 11, 10, respectively. Thus, the output of the uniform random number generator is mapped into the corresponding signal amplitude levels $(-3d, -d, d, 3d)$, respectively.

The additive noise component having mean 0 and variance σ^2 is generated by means of a Gaussian random number generator. For convenience, we may normalize the distance parameter $d = 1$ and vary σ^2. The detector observes $r = A_m + n$ and computes the distance between r and the four possible transmitted signal amplitudes. Its output $\hat{A}_m$ is the signal amplitude level corresponding to the smallest distance. $\hat{A}_m$ is compared with the actual transmitted signal amplitude, and an error counter is used to count the errors made by the detector.

Figure 5.23 illustrates the results of the simulation for the transmissions of $N = 10,000$ symbols at different values of the average bit SNR, which is defined as

$$\frac{\mathscr{E}_{avb}}{N_0} = \frac{5}{4}\left(\frac{d^2}{\sigma^2}\right) \qquad (5.3.15)$$

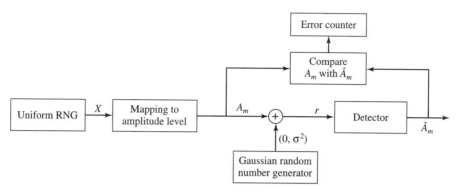

Figure 5.22 Block diagram of four-level PAM for Monte Carlo simulation

Note the agreement between the simulation results and the theoretical values of P_4 computed from (5.3.14).

The MATLAB scripts for this problem are given next.

M-FILE

```
% MATLAB script for Illustrative Problem 5.8.

echo on
SNRindB1=0:1:12;
SNRindB2=0:0.1:12;
for i=1:length(SNRindB1),
   % simulated error rate
   smld_err_prb(i)=smldPe58(SNRindB1(i));
   echo off;
end;
echo on;
for i=1:length(SNRindB2),
   % signal to noise ratio
   SNR_per_bit=exp(SNRindB2(i)*log(10)/10);
   % theoretical error rate
   theo_err_prb(i)=(3/2)*Qfunct(sqrt((4/5)*SNR_per_bit));
   echo off;
end;
echo on;
% Plotting commands follow
semilogy(SNRindB1,smld_err_prb,' * ');
hold
semilogy(SNRindB2,theo_err_prb);
```

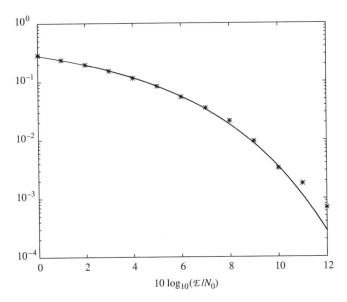

$$10 \log_{10}(\mathcal{E}/N_0)$$

Figure 5.23   Error probability for Monte Carlo simulation compared with theoretical error probability for $M = 4$ PAM

M-FILE

```
function [p]=smldPe58(snr_in_dB)
% [p]=smldPe58(snr_in_dB)
%                 SMLDPE58   simulates the probability of error for the given
%                 snr_in_dB, signal to noise ratio in dB.
d=1;
SNR=exp(snr_in_dB*log(10)/10);      % signal to noise ratio per bit
sgma=sqrt((5*d^2)/(4*SNR));         % sigma, standard deviation of noise
N=10000;                            % number of symbols being simulated
% generation of the quarternary data source follows
for i=1:N,
    temp=rand;                      % a uniform random variable over (0,1)
    if (temp<0.25),
        dsource(i)=0;               % with probability 1/4, source output is "00"
    elseif (temp<0.5),
        dsource(i)=1;               % with probability 1/4, source output is "01"
    elseif (temp<0.75),
        dsource(i)=2;               % with probability 1/4, source output is "10"
    else
        dsource(i)=3;               % with probability 1/4, source output is "11"
    end
end;
% detection, and probability of error calculation
numoferr=0;
for i=1:N,
```

```
% The matched filter outputs
if (dsource(i)==0),
    r=−3*d+gngauss(sgma);              % if the source output is "00"
elseif (dsource(i)==1),
    r=−d+gngauss(sgma);                % if the source output is "01"
elseif (dsource(i)==2)
    r=d+gngauss(sgma);                 % if the source output is "10"
else
    r=3*d+gngauss(sgma);               % if the source output is "11"
end;
% detector follows
if (r<−2*d),
    decis=0;                           % decision is "00"
elseif (r<0),
    decis=1;                           % decision is "01"
elseif (r<2*d),
    decis=2;                           % decision is "10"
else
    decis=3;                           % decision is "11"
end;
if (decis˜=dsource(i)),                % if it is an error, increase the error counter
    numoferr=numoferr+1;
end;
end;
p=numoferr/N;                          % probability of error estimate
```

5.3.3   Signal Waveforms with Multiple Amplitude Levels

It is relatively straightforward to construct multiamplitude signals with more than four levels. In general, a set of $M = 2^k$ multiamplitude signal waveforms is represented as

$$s_m(t) = A_m g(t), \qquad 0 \leq t \leq T, \quad m = 0, 1, 2, \ldots, M - 1$$

where the M amplitude values are equally spaced and given as

$$A_m = (2m - M + 1)d, \qquad m = 0, 1, \ldots, M - 1 \qquad (5.3.16)$$

and $g(t)$ is a rectangular pulse, which has been defined in (5.3.2). Each signal waveform conveys $k = \log_2 M$ bits of information. When the bit rate is $R = 1/T_b$, the corresponding symbol rate is $1/T = 1/kT_b$. As in the case of four-level PAM, the optimum receiver consists of a signal correlator (or matched filter) followed by an amplitude detector that computes the Euclidean distances given by (5.3.10) for $m = 0, 1, \ldots, M - 1$. For equally probable amplitude levels, the decision is made in favor of the amplitude level that corresponds to the smallest distance.

The probability of error for the optimum detector in an M-level PAM system is easily shown to be

$$P_M = \frac{2(M - 1)}{M} Q \left(\sqrt{\frac{6(\log_2 M)\mathcal{E}_{\text{avb}}}{(M^2 - 1)N_0}} \right) \qquad (5.3.17)$$

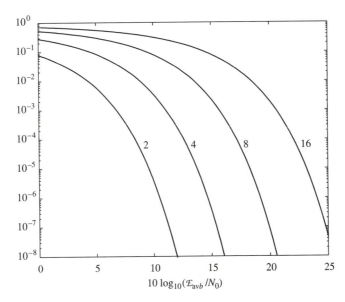

Figure 5.24   Symbol-error probability for M-level PAM for $M = 2, 4, 8, 16$

where $\mathcal{E}_{avb}$ is the average energy for an information bit. Figure 5.24 illustrates the probability of a symbol error for $M = 2, 4, 8, 16$.

ILLUSTRATIVE PROBLEM

Illustrative Problem 5.9 [PAM simulation] Perform a Monte Carlo simulation of a 16-level PAM digital communication system, and measure its error-rate performance.

SOLUTION

The basic block diagram shown in Figure 5.22 applies in general. A uniform random number generator is used to generate the sequence of information symbols, which are viewed as blocks of four information bits. The 16-ary symbols may be generated directly by subdividing the interval $(0, 1)$ into 16 equal-width subintervals and mapping the 16-ary symbols into the 16-ary signal amplitudes. A white Gaussian noise sequence is added to the 16-ary information symbol sequence, and the resulting signal plus noise is fed to the detector. The detector computes the distance metrics given by (5.3.10) and selects the amplitude corresponding to the smallest metric. The output of the detector is compared with the transmitted information symbol sequence and errors are counted. Figure 5.25 illustrates the measured symbol-error rate for 10,000 transmitted symbols and the theoretical symbol-error rate given by (5.3.17) with $M = 16$.

The MATLAB scripts for this problem are given next.

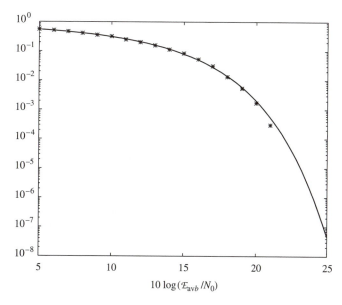

Figure 5.25   Error rate from Monte Carlo simulation compared with the theoretical error probability for $M = 16$ PAM

M-FILE

```
% MATLAB script for Illustrative Problem 5.9.
echo on
SNRindB1=5:1:25;
SNRindB2=5:0.1:25;
M=16;
for i=1:length(SNRindB1),
  % simulated error rate
  smld_err_prb(i)=smldPe59(SNRindB1(i));
  echo off;
end;
echo on ;
for i=1:length(SNRindB2),
  SNR_per_bit=exp(SNRindB2(i)*log(10)/10);
  % theoretical error rate
  theo_err_prb(i)=(2*(M−1)/M)*Qfunct(sqrt((6*log2(M)/(M^2−1))*SNR_per_bit));
  echo off;
end;
echo on;
% Plotting commands follow
semilogy(SNRindB1,smld_err_prb,' * ');
hold
semilogy(SNRindB2,theo_err_prb);
```

M-FILE

```
function [p]=smldPe59(snr_in_dB)
% [p]=smldPe59(snr_in_dB)
%               SMLDPE59   simulates the error probability for the given
%               snr_in_dB, signal to noise ratio in dB.
M=16;                              % 16-ary PAM
d=1;
SNR=exp(snr_in_dB*log(10)/10);     % signal to noise ratio per bit
sgma=sqrt((85*d^2)/(8*SNR));       % sigma, standard deviation of noise
N=10000;                           % number of symbols being simulated
% generation of the data source
for i=1:N,
   temp=rand;                      % a uniform random variable over (0,1)
   index=floor(M*temp);            % the index is an integer from 0 to M-1, where
                                   % all the possible values are equally likely

   dsource(i)=index;
end;
% detection, and probability of error calculation
numoferr=0;
for i=1:N,
   % matched filter outputs
   % (2*dsource(i)-M+1)*d is the mapping to the 16-ary constellation
   r=(2*dsource(i)-M+1)*d+gngauss(sgma);
   % the detector
   if (r>(M-2)*d),
     decis=15;
   elseif (r>(M-4)*d),
     decis=14;
   elseif (r>(M-6)*d),
     decis=13;
   elseif (r>(M-8)*d),
     decis=12;
   elseif (r>(M-10)*d),
     decis=11;
   elseif (r>(M-12)*d),
     decis=10;
   elseif (r>(M-14)*d),
     decis=9;
   elseif (r>(M-16)*d),
     decis=8;
   elseif (r>(M-18)*d),
     decis=7;
   elseif (r>(M-20)*d),
     decis=6;
   elseif (r>(M-22)*d),
     decis=5;
   elseif (r>(M-24)*d),
     decis=4;
   elseif (r>(M-26)*d),
     decis=3;
   elseif (r>(M-28)*d),
     decis=2;
```

```
    elseif (r>(M−30)*d),
      decis=1;
    else
      decis=0;
    end;
    if (decis~=dsource(i)),    % if it is an error, increase the error counter
      numoferr=numoferr+1;
    end;
  end;
  p=numoferr/N;                    % probability of error  estimate
```

5.4   Multidimensional Signals

In the preceding section, we constructed multiamplitude signal waveforms, which allowed us to transmit multiple bits per signal waveform. Thus, with signal waveforms having $M = 2^k$ amplitude levels, we are able to transmit $k = \log_2 M$ bits of information per signal waveform. We also observed that the multiamplitude signals can be represented geometrically as signal points on the real line (see Figure 5.20). Such signal waveforms are called *one-dimensional signals*.

In this section, we consider the construction of a class of $M = 2^k$ signal waveforms that have a multidimensional representation. That is, the set of signal waveforms can be represented geometrically as points in N-dimensional space. We have already observed that binary orthogonal signals are represented geometrically as points in two-dimensional space.

5.4.1   Multidimensional Orthogonal Signals

There are many ways to construct multidimensional signal waveforms with various properties. In this section, we consider the construction of a set of $M = 2^k$ waveforms $s_i(t)$, for $i = 0, 1, \ldots, M - 1$, which have the properties of (a) mutual orthogonality and (b) equal energy. These two properties may be succinctly expressed as

$$\int_0^T s_i(t)s_k(t)\,dt = \mathcal{E}\delta_{ik}, \qquad i, k = 0, 1, \ldots, M - 1 \tag{5.4.1}$$

where $\mathcal{E}$ is the energy of each signal waveform, and δ_{ik} is called the Kronecker delta, which is defined as

$$\delta_{ik} = \begin{cases} 1, & i = k \\ 0, & i \neq k \end{cases} \tag{5.4.2}$$

As in our previous discussion, we assume that an information source is providing a sequence of information bits, which are to be transmitted through a communication channel. The information bits occur at a uniform rate of R bits per second. The reciprocal of R is the bit interval, T_b. The modulator takes k bits at a time and maps them into one of $M = 2^k$

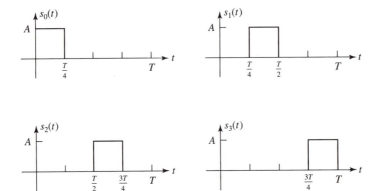

Figure 5.26   An example of four orthogonal, equal-energy signal waveforms

signal waveforms. Each block of k bits is called a *symbol*. The time interval available to transmit each symbol is $T = kT_b$. Hence, T is the symbol interval.

The simplest way to construct a set of $M = 2^k$ equal-energy orthogonal waveforms in the interval $(0, T)$ is to subdivide the interval into M equal subintervals of duration T/M and to assign a signal waveform for each subinterval. Figure 5.26 illustrates such a construction for $M = 4$ signals. All signal waveforms constructed in this manner have identical energy, given as

$$\mathcal{E} = \int_0^T s_i^2(t)\, dt, \qquad i = 0, 1, 2, \ldots, M - 1$$

$$= \frac{A^2 T}{M} \tag{5.4.3}$$

Such a set of orthogonal waveforms can be represented as a set of M-dimensional orthogonal vectors—that is,

$$s_0 = (\sqrt{\mathcal{E}}, 0, 0, \ldots, 0)$$
$$s_1 = (0, \sqrt{\mathcal{E}}, 0, \ldots, 0)$$
$$\vdots$$
$$s_M = (0, 0, \ldots, 0, \sqrt{\mathcal{E}}) \tag{5.4.4}$$

Figure 5.27 illustrates the signal points (signal constellations) corresponding to $M = 2$ and $M = 3$ orthogonal signals.

Let us assume that these orthogonal signal waveforms are used to transmit information through an AWGN channel. Consequently, if the transmitted waveform is $s_i(t)$, the received waveform is

$$r(t) = s_i(t) + n(t), \qquad 0 \le t \le T, \quad i = 0, 1, \ldots, M - 1 \tag{5.4.5}$$

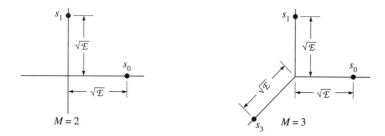

Figure 5.27   Signal constellations for $M = 2$ and $M = 3$ orthogonal signals

where $n(t)$ is a sample function of a white Gaussian noise process with power spectrum $N_0/2$ watts/hertz. The receiver observes the signal $r(t)$ and decides which of the M signal waveforms was transmitted.

Optimum Receiver for the AWGN Channel

The receiver that minimizes the probability of error first passes the signal $r(t)$ through a parallel bank of M matched filters or M correlators. Since the signal correlators and matched filters yield the same output at the sampling instant, let us consider the case in which signal correlators are used, as shown in Figure 5.28.

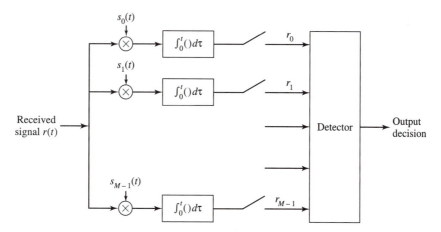

Figure 5.28   Optimum receiver for multidimensional orthogonal signals

Signal Correlators

The received signal $r(t)$ is cross-correlated with each of the M signal waveforms and the correlator outputs are sampled at $t = T$. Thus, the M correlator outputs are

$$r_i = \int_0^T r(t)s_i(t)\, dt, \qquad i = 0, 1, \ldots, M-1 \tag{5.4.6}$$

which may be represented in vector form as $r = \left[r_0, r_1, \ldots, r_{M-1} \right]^t$. Suppose that signal waveform $s_0(t)$ is transmitted. Then

$$r_0 = \int_0^T s_0^2(t)\, dt + \int_0^T n(t)s_0(t)\, dt$$

$$= \mathcal{E} + n_0 \tag{5.4.7}$$

and

$$r_i = \int_0^T s_0(t)s_i(t)\, dt + \int_0^T n(t)s_i(t)\, dt$$

$$= \int_0^T n(t)s_i(t)\, dt = n_i, \qquad i = 1, 2, 3, \ldots, M-1 \tag{5.4.8}$$

where

$$n_i = \int_0^T n(t)s_i(t)\, dt \tag{5.4.9}$$

Therefore, the output r_0 consists of a signal component $\mathcal{E}$ and a noise component n_0. The other $M - 1$ outputs consist of noise only. Each of the noise components is Gaussian, with mean zero and variance

$$\sigma^2 = E(n_i^2)$$

$$= \int_0^T \int_0^T s_i(t)s_i(\tau)E\left[n(t)n(\tau)\right]\, dt\, d\tau$$

$$= \frac{N_0}{2} \int_0^T \int_0^T s_i(t)s_i(\tau)\delta(t - \tau)\, dt\, d\tau$$

$$= \frac{N_0}{2} \int_0^T s_i^2(t)\, dt$$

$$= \frac{N_0\mathcal{E}}{2} \tag{5.4.10}$$

The reader is encouraged to show that $E(n_i n_j) = 0$, where $i \neq j$. Consequently, the probability density functions for the correlator outputs are

$$p(r_0 \mid s_0(t) \text{ was transmitted}) = \frac{1}{\sqrt{2\pi}\,\sigma} e^{-(r_0 - \mathcal{E})^2/2\sigma^2}$$

$$p(r_i \mid s_0(t) \text{ was transmitted}) = \frac{1}{\sqrt{2\pi}\,\sigma} e^{-r_i/2\sigma^2}, \qquad i = 1, 2, \ldots, M-1$$

The Detector

The optimum detector observes the M correlator outputs r_i, where $i = 0, 1, \ldots, M - 1$, and selects the signal corresponding to the largest correlator output. In the case where $s_0(t)$ was transmitted, the probability of a correct decision is simply the probability that $r_0 > r_i$ for $i = 1, 2, \ldots, M - 1$, or

$$P_c = P(r_0 > r_1, r_0 > r_2, \ldots, r_0 > r_{M-1}) \tag{5.4.11}$$

and the probability of a symbol error is simply

$$
\begin{aligned}
P_M &= 1 - P_c \\
&= 1 - P(r_0 > r_1, r_0 > r_2, \ldots, r_0 > r_{M-1})
\end{aligned}
\tag{5.4.12}
$$

It can be shown that P_M can be expressed in integral form as

$$P_M = \frac{1}{\sqrt{2\pi}} \int_{-\infty}^{\infty} \left\{ 1 - [1 - Q(y)]^{M-1} \right\} e^{-(y - \sqrt{2\mathcal{E}/N_0})^2/2} \, dy \tag{5.4.13}$$

For the special case $M = 2$, the expression in (5.4.13) reduces to

$$P_2 = Q\left(\sqrt{\frac{\mathcal{E}_b}{N_0}} \right)$$

which is the result we obtained in Section 5.2 for binary orthogonal signals.

The same expression for the probability of error is obtained when any one of the other $M - 1$ signals is transmitted. Since all the M signals are equally likely, the expression for P_M given by (5.4.13) is the average probability of a symbol error. This integral can be evaluated numerically.

Sometimes, it is desirable to convert the probability of a symbol error into an equivalent probability of a binary digit error. For equiprobable orthogonal signals, all symbol errors are equiprobable and occur with probability

$$\frac{P_M}{M - 1} = \frac{P_M}{2^k - 1} \tag{5.4.14}$$

Furthermore, there are $\binom{k}{n}$ ways in which n bits out of k may be in error. Hence, the average number of bit errors per k-bit symbol is

$$\sum_{n=1}^{k} n \binom{k}{n} \frac{P_M}{2^k - 1} = k \frac{2^{k-1}}{2^k - 1} P_M \tag{5.4.15}$$

and the average bit-error probability is just the result in (5.4.15) divided by k, the number of bits per symbol. Thus,

$$P_b = \frac{2^{k-1}}{2^k - 1} P_M \tag{5.4.16}$$

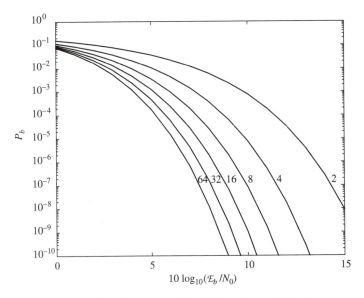

P_b

$10 \log_{10}(\mathcal{E}_b/N_0)$

Figure 5.29   Bit-error probability for orthogonal signals

The graphs of the probability of a binary digit error as a function of the SNR per bit, $\mathcal{E}_b/N_0$, are shown in Figure 5.29 for $M = 2, 4, 8, 16, 32, 64$, where $\mathcal{E}_b = \mathcal{E}/k$ is the energy per bit. This figure illustrates that by increasing the number M of waveforms, one can reduce the SNR per bit required to achieve a given probability of a bit error.

The MATLAB script for the computation of the error probability in (5.4.13) is given next.

M-FILE

```
% MATLAB script that generates the probability of error versus the signal to noise ratio
initial_snr=0;
final_snr=15;
snr_step=1;
tolerance=1e−7;                     % Tolerance used for the integration
minus_inf=−20;                      % This is practically -infinity
plus_inf=20;                        % This is practically infinity
snr_in_dB=initial_snr:snr_step:final_snr;
for i=1:length(snr_in_dB),
    snr=10^(snr_in_dB(i)/10);
    Pe_2(i)=Qfunct(sqrt(snr));
    Pe_4(i)=(2/3)*quad8('bdt_int',minus_inf,plus_inf,tolerance,[ ],snr,4);
    Pe_8(i)=(4/7)*quad8('bdt_int',minus_inf,plus_inf,tolerance,[ ],snr,8);
    Pe_16(i)=(8/15)*quad8('bdt_int',minus_inf,plus_inf,tolerance,[ ],snr,16);
    Pe_32(i)=(16/31)*quad8('bdt_int',minus_inf,plus_inf,tolerance,[ ],snr,32);
    Pe_64(i)=(32/63)*quad8('bdt_int',minus_inf,plus_inf,tolerance,[ ],snr,64);
end;
% Plotting commands follow
```

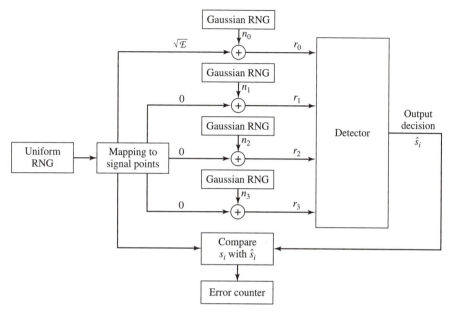

Figure 5.30   Block diagram of system with $M = 4$ orthogonal signals for Monte Carlo simulation

Illustrative Problem 5.10 [Orthogonal signaling simulation] Perform a Monte Carlo simulation of a digital communication system that employs $M = 4$ orthogonal signals. The model of the system to be simulated is illustrated in Figure 5.30.

As shown, we simulate the generation of the random variables r_0, r_1, r_2, r_3, which constitute the input to the detector. We may first generate a binary sequence of 0's and 1's that occur with equal probability and are mutually statistically independent, as in Illustrative Problem 5.4. The binary sequence is grouped into pairs of bits, which are mapped into the corresponding signal components. An alternative to generating the individual bits is to generate the pairs of bits, as in Illustrative Problem 5.8. In any case, we have the mapping of the four symbols into the signal points:

$$00 \rightarrow s_0 = (\sqrt{\mathcal{E}}, 0, 0, 0)$$
$$01 \rightarrow s_1 = (0, \sqrt{\mathcal{E}}, 0, 0)$$
$$10 \rightarrow s_2 = (0, 0, \sqrt{\mathcal{E}}, 0)$$
$$11 \rightarrow s_3 = (0, 0, 0, \sqrt{\mathcal{E}}) \tag{5.4.17}$$

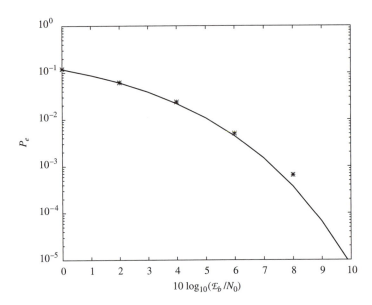

Figure 5.31   Bit-error probability for $M = 4$ orthogonal signals from a Monte Carlo simulation compared with theoretical error probability

The additive noise components n_0, n_1, n_2, n_3 are generated by means of four Gaussian noise generators, each having mean zero and variance $\sigma^2 = N_0 \mathcal{E}/2$. For convenience, we may normalize the symbol energy to $\mathcal{E} = 1$ and vary σ^2. Since $\mathcal{E} = 2\mathcal{E}_b$, it follows that $\mathcal{E}_b = \frac{1}{2}$. The detector output is compared with the transmitted sequence of bits, and an error counter is used to count the number of bit errors.

Figure 5.31 illustrates the results of this simulation for the transmission of 20,000 bits at several different values of the SNR $\mathcal{E}_b/N_0$. Note the agreement between the simulation results and the theoretical value of P_b given by (5.4.16).

The MATLAB scripts for this problem are given next.

M-FILE

```
% MATLAB script for Illustrative Problem 5.10.
echo on
SNRindB=0:2:10;
for i=1:length(SNRindB),
   % simulated error rate
   smld_err_prb(i)=smldP510(SNRindB(i));
   echo off;
end;
echo on;
% Plotting commands follow
semilogy(SNRindB,smld_err_prb,'*');
```

M-FILE

```
function [p]=smldP510(snr_in_dB)
% [p]=smldP510(snr_in_dB)
%              SMLDP510  simulates the probability of error for the given
%              snr_in_dB, signal to noise ratio in dB.
M=4;                               % quarternary orthogonal signalling
E=1;
SNR=exp(snr_in_dB*log(10)/10);     % signal to noise ratio per bit
sgma=sqrt(E^2/(4*SNR));            % sigma, standard deviation of noise
N=10000;                           % number of symbols being simulated
% generation of the quarternary data source
for i=1:N,
  temp=rand;                       % a uniform random variable over (0,1)
  if (temp<0.25),
    dsource1(i)=0;
    dsource2(i)=0;
  elseif (temp<0.5),
    dsource1(i)=0;
    dsource2(i)=1;
  elseif (temp<0.75),
    dsource1(i)=1;
    dsource2(i)=0;
  else
    dsource1(i)=1;
    dsource2(i)=1;
  end
end;
% detection, and probability of error calculation
numoferr=0;
for i=1:N,
  % matched filter outputs
  if ((dsource1(i)==0) & (dsource2(i)==0)),
    r0=sqrt(E)+gngauss(sgma);
    r1=gngauss(sgma);
    r2=gngauss(sgma);
    r3=gngauss(sgma);
  elseif ((dsource1(i)==0) & (dsource2(i)==1)),
    r0=gngauss(sgma);
    r1=sqrt(E)+gngauss(sgma);
    r2=gngauss(sgma);
    r3=gngauss(sgma);
  elseif ((dsource1(i)==1) & (dsource2(i)==0)),
    r0=gngauss(sgma);
    r1=gngauss(sgma);
    r2=sqrt(E)+gngauss(sgma);
    r3=gngauss(sgma);
  else
    r0=gngauss(sgma);
    r1=gngauss(sgma);
    r2=gngauss(sgma);
    r3=sqrt(E)+gngauss(sgma);
  end;
```

```
% the detector
max_r=max([r0 r1 r2 r3]);
if (r0==max_r),
    decis1=0;
    decis2=0;
elseif (r1==max_r),
    decis1=0;
    decis2=1;
elseif (r2==max_r),
    decis1=1;
    decis2=0;
else
    decis1=1;
    decis2=1;
end;
% count the number of bit errors made in this decision
if (decis1~=dsource1(i)),          % if it is an error, increase the error counter
    numoferr=numoferr+1;
end;
if (decis2~=dsource2(i)),          % if it is an error, increase the error counter
    numoferr=numoferr+1;
end;
end;
p=numoferr/(2*N);                  % bit error probability estimate
```

5.4.2   Biorthogonal Signals

As we observed in the preceding section, a set of $M = 2^k$ equal-energy orthogonal wave-forms can be constructed by subdividing the symbol interval T into M equal subintervals of duration T/M and assigning a rectangular signal pulse to each subinterval. A similar method can be used to construct another set of $M = 2^k$ multidimensional signals that have the property of being biorthogonal. In such a signal set, one-half the waveforms are orthogonal, and the other half are the negative of these orthogonal waveforms; that is, $s_0(t), s_1(t), \ldots, s_{M/2-1}(t)$ are orthogonal signal waveforms. The other $M/2$ waveforms are simply $s_{i+M/2}(t) = -s_i(t)$, for $i = 0, 1, \ldots, (M/2) - 1$. Thus, we obtain M signals, each having $M/2$ dimensions.

The $M/2$ orthogonal waveforms can be easily constructed by subdividing the symbol interval $T = kT_b$ into $M/2$ nonoverlapping subintervals, each of duration $2T/M$, and assigning a rectangular pulse to each subinterval. Figure 5.32 illustrates a set of $M = 4$ biorthogonal waveforms constructed in this manner. The geometric representation of a set of M signals constructed in this manner is given by the following $(M/2)$-dimensional signal points:

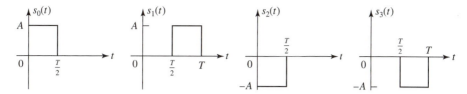

Figure 5.32   A set of $M = 4$ biorthogonal signal waveforms

$$s_0 = (\sqrt{\mathcal{E}}, 0, 0, \ldots, 0)$$
$$s_1 = (0, \sqrt{\mathcal{E}}, 0, \ldots, 0)$$
$$\vdots$$
$$s_{M/2-1} = (0, 0, 0, \ldots, \sqrt{\mathcal{E}})$$
$$s_{M/2} = (-\sqrt{\mathcal{E}}, 0, 0, \ldots, 0)$$
$$\vdots$$
$$s_{M-1} = (0, 0, \ldots, -\sqrt{\mathcal{E}}) \tag{5.4.18}$$

As in the case of orthogonal signals, let us assume that the biorthogonal signal waveforms are used to transmit information through an AWGN channel. Then the received signal waveform may be expressed as

$$r(t) = s_i(t) + n(t), \qquad 0 \le t \le T \tag{5.4.19}$$

where $s_i(t)$ is the transmitted waveform and $n(t)$ is a sample function of a white Gaussian noise process with power spectrum $N_0/2$ watts/hertz.

Optimum Receiver

The optimum receiver may be implemented by cross-correlating the received signal $r(t)$ with each of the $M/2$ orthogonal signal waveforms, sampling the correlator outputs at $t = T$, and passing the $M/2$ correlator outputs to the detector. Thus, we have

$$r_i = \int_0^T r(t)s_i(t)\,dt, \qquad i = 0, 1, \ldots, \frac{M}{2} - 1 \tag{5.4.20}$$

Suppose that the transmitted signal waveform is $s_0(t)$. Then

$$r_i = \int_0^T s_i(t)\,dt, \qquad i = 0, 1, \ldots, \frac{M}{2} - 1$$
$$= \begin{cases} \mathcal{E} + n_0, & i = 0 \\ n_i, & i \neq 0 \end{cases} \tag{5.4.21}$$

where

$$n_i = \int_0^T n(t) s_i(t)\, dt, \qquad i = 0, 1, \ldots, \frac{M}{2} - 1 \tag{5.4.22}$$

and $\mathcal{E}$ is the symbol energy for each signal waveform. The noise components are zero-mean Gaussian and have a variance of $\sigma^2 = \mathcal{E} N_0 / 2$.

The Detector

The detector observes the $M/2$ correlator outputs $\{ r_i,\ 0 \le i \le (M/2) - 1 \}$ and selects the correlator output whose magnitude $|r_i|$ is largest. Suppose

$$|r_j| = \max_i \{ |r_i| \} \tag{5.4.23}$$

Then the detector selects the signal $s_j(t)$ if $r_j > 0$ and $-s_j(t)$ if $r_j < 0$.

To determine the probability of error, suppose that $s_0(t)$ was transmitted. Then the probability of a correct decision is equal to the probability that $r_0 = \mathcal{E} + n_0 > 0$ and $|r_0| > |r_i|$ for $i = 1, 2, \ldots, (M/2) - 1$. Thus,

$$P_c = \int_0^\infty \left[\frac{1}{\sqrt{2\pi}} \int_{-r_0/\sqrt{\mathcal{E} N_0/2}}^{r_0/\sqrt{\mathcal{E} N_0/2}} e^{-x^2/2}\, dx \right]^{M-1} p(r_0)\, dr_0 \tag{5.4.24}$$

where

$$p(r_0) = \frac{1}{\sqrt{2\pi}\, \sigma} e^{-(r_0 - \mathcal{E})^2 / 2\sigma^2} \tag{5.4.25}$$

Finally, the probability of a symbol error is simply

$$P_M = 1 - P_c \tag{5.4.26}$$

P_c and P_M may be evaluated numerically for different values of M from (5.4.24) and (5.4.25). The graph shown in Figure 5.33 illustrates P_M as a function of the signal-to-noise ratio $\mathcal{E}_b / N_0$, where $\mathcal{E} = k \mathcal{E}_b$ for $M = 2, 4, 8, 16$, and 32. We observe that this graph is similar to that for orthogonal signals. However, for biorthogonal signals, we note that $P_4 > P_2$. This is due to the fact that we have plotted the symbol-error probability P_M in Figure 5.33. If we plot the equivalent bit-error probability, we would find that the graphs for $M = 2$ and $M = 4$ coincide.

The MATLAB script for the computation of the error probability in (5.4.24) and (5.4.26) is given next.

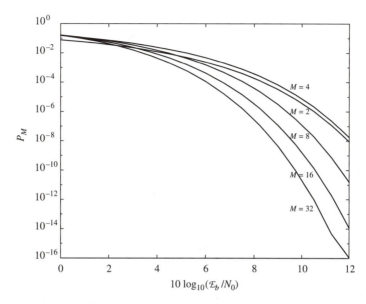

Figure 5.33   Symbol-error probability for biorthogonal signals

---

M-FILE

```
% MATLAB script that generates the probability of error versus the signal to noise ratio.
initial_snr=0;
final_snr=12;
snr_step=0.75;
tolerance=eps;                          % Tolerance used for the integration
plus_inf=20;                            % This is practically infinity
snr_in_dB=initial_snr:snr_step:final_snr;
for  i=1:length(snr_in_dB),
    snr=10^(snr_in_dB(i)/10);
    Pe_2(i)=1−quad8('bdt_int2',0,plus_inf,tolerance,[ ],snr,2);
    Pe_4(i)=1−quad8('bdt_int2',0,plus_inf,tolerance,[ ],snr,4);
    Pe_8(i)=1−quad8('bdt_int2',0,plus_inf,tolerance,[ ],snr,8);
    Pe_16(i)=1−quad8('bdt_int2',0,plus_inf,tolerance,[ ],snr,16);
    Pe_32(i)=1−quad8('bdt_int2',0,plus_inf,tolerance,[ ],snr,32);
end;
% Plotting commands follow
```

ILLUSTRATIVE PROBLEM

Illustrative Problem 5.11 [Biorthogonal signaling simulation] Perform a Monte Carlo
simulation for a digital communication system that employs $M = 4$ biorthogonal signals.
The model of the system to be simulated is illustrated in Figure 5.34.

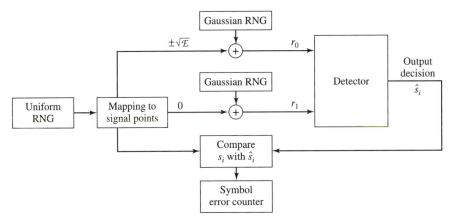

Figure 5.34 Block diagram of the system with $M = 4$ biorthogonal signals for Monte Carlo simulation

SOLUTION

As shown, we simulate the generation of the random variables r_0 and r_1, which constitute the input to the detector. We begin by generating a binary sequence of 0's and 1's that occur with equal probability and are mutually statistically independent, as in Illustrative Problem 5.4. The binary sequence is grouped into pairs of bits, which are mapped into the corresponding signal components as follows:

$$00 \rightarrow s_0 = (\sqrt{\mathcal{E}}, 0)$$
$$01 \rightarrow s_1 = (0, \sqrt{\mathcal{E}})$$
$$10 \rightarrow s_2 = (0, -\sqrt{\mathcal{E}})$$
$$11 \rightarrow s_3 = (-\sqrt{\mathcal{E}}, 0)$$

Alternatively, we may use the method in Illustrative Problem 5.8 to generate the 2-bit symbols directly.

Since $s_2 = -s_1$ and $s_3 = -s_0$, the demodulation requires only two correlators or matched filters, whose outputs are r_0 and r_1. The additive noise components n_0 and n_1 are generated by means of two Gaussian noise generators, each having mean zero and variance $\sigma^2 = N_0\mathcal{E}/2$. For convenience, we may normalize the symbol energy to $\mathcal{E} = 1$ and vary σ^2. Since $\mathcal{E} = 2\mathcal{E}_b$, it follows that $\mathcal{E}_b = \frac{1}{2}$. The detector output is compared with the transmitted sequence of bits, and an error counter is used to count the number of symbol errors and the number of bit errors.

Figure 5.35 illustrates the results of this simulation for the transmission of 20,000 bits at several different values of the SNR $\mathcal{E}_b/N_0$. Note the agreement between the simulation results and the theoretical value of P_4 given by (5.4.26) and (5.4.24).

The MATLAB scripts for this problem are given next.

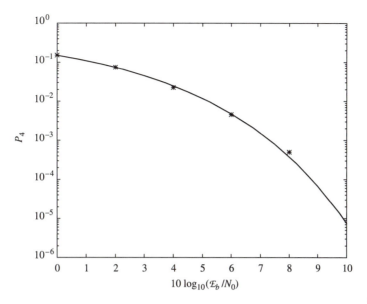

Figure 5.35   Symbol-error probability for $M = 4$ biorthogonal signals from Monte Carlo simulation compared with theoretical error probability

---

M-FILE

```
% MATLAB script for Illustrative Problem 5.11.
echo on
SNRindB=0:2:10;
for  i=1:length(SNRindB),
    % simulated error rate
    smld_err_prb(i)=smldP511(SNRindB(i));
    echo off;
end;
echo on ;
% Plotting commands follow
```

M-FILE

```
function [p]=smldP511(snr_in_dB)
% [p]=smldP511(snr_in_dB)
%              SMLDP511  simulates the probability of error for the given
%              snr_in_dB, signal to noise ratio in dB, for the system
%              described in illustrated problem 11, Chapter 5.
M=4;                                        % quarternary orthogonal signalling
E=1;
```

```
SNR=exp(snr_in_dB*log(10)/10);        % signal to noise ratio per bit
sgma=sqrt(E^2/(4*SNR));               % sigma, standard deviation of noise
N=10000;                              % number of symbols being simulated
% generation of the quarternary data source
for i=1:N,
    temp=rand;                        % uniform random variable over (0,1)
    if (temp<0.25),
        dsource(i)=0;
    elseif (temp<0.5),
        dsource(i)=1;
    elseif (temp<0.75),
        dsource(i)=2;
    else
        dsource(i)=3;
    end
end;
% detection, and error probability computation
numoferr=0;
for i=1:N,
    % The matched filter outputs
    if (dsource(i)==0)
        r0=sqrt(E)+gngauss(sgma);
        r1=gngauss(sgma);
    elseif (dsource(i)==1)
        r0=gngauss(sgma);
        r1=sqrt(E)+gngauss(sgma);
    elseif (dsource(i)==2)
        r0=-sqrt(E)+gngauss(sgma);
        r1=gngauss(sgma);
    else
        r0=gngauss(sgma);
        r1=-sqrt(E)+gngauss(sgma);
    end;
    % detector follows
    if (r0>abs(r1)),
        decis=0;
    elseif (r1>abs(r0)),
        decis=1;
    elseif (r0<-abs(r1)),
        decis=2;
    else
        decis=3;
    end;
    if (decis~=dsource(i)),            % if it is an error, increase to the error counter
        numoferr=numoferr+1;
    end;
end;
p=numoferr/N;                         % bit error probability estimate
```

Problems

5.1 Suppose the two orthogonal signals shown in Figure 5.2 are used to transmit binary information through an AWGN channel. The received signal in each bit interval of duration T_b is given by (5.2.1). Suppose that the received signal waveform is sampled at a rate of $10/T_b$—that is, at ten samples per bit interval. Hence, in discrete time, the signal waveform $s_0(t)$ with amplitude A is represented by the ten samples $(A, A, \ldots, A)$ and the signal waveform $s_1(t)$ is represented by the ten samples $(A, A, A, A, A, -A, -A, -A, -A, -A)$. Consequently, the sampled version of the received sequence when $s_0(t)$ is transmitted is

$$r_k = A + n_k, \qquad k = 1, 2, \ldots, 10$$

and when $s_1(t)$ is transmitted is

$$r_k = \begin{cases} A + n_k, & 1 \le k \le 5 \\ -A + n_k, & 6 \le k \le 10 \end{cases}$$

where the sequence $\{n_k\}$ is i.i.d., zero mean, Gaussian with each random variable having the variance σ^2. Write a MATLAB routine that generates the sequence $\{r_k\}$ for each of the two possible received signals, and perform a discrete-time correlation of the sequence $\{r_k\}$ with each of the two possible signals $s_0(t)$ and $s_1(t)$ represented by their sampled versions for different values of the additive Gaussian noise variance $\sigma^2 = 0$, $\sigma^2 = 0.1$, $\sigma^2 = 1.0$, and $\sigma^2 = 2.0$. The signal amplitude may be normalized to $A = 1$. Plot the correlator outputs at time instants $k = 1, 2, 3, \ldots, 10$.

5.2 Repeat Problem 5.1 for the two signal waveforms $s_0(t)$ and $s_1(t)$ illustrated in Figure P5.2. Describe the similarities and differences between this set of two signals and those illustrated in Figure 5.2. Is one set better than the other from the viewpoint of transmitting a sequence of binary information signals?

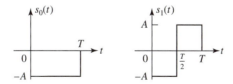

Figure P5.2

5.3 In this problem, the objective is to substitute two matched filters in place of the two correlators in Problem 5.1. The condition for generating signals is identical to Problem 5.1.

Write a MATLAB routine that generates the sequence $\{r_k\}$ for each of the two possible received signals, and perform the discrete-time matched filtering of the sequence $\{r_k\}$ with each of the two possible signals $s_0(t)$ and $s_1(t)$, represented by their sampled versions, for different values of the additive Gaussian noise variance $\sigma^2 = 0$, $\sigma^2 = 0.1$, $\sigma^2 = 1.0$, and $\sigma^2 = 2.0$. The signal amplitude may be normalized to $A = 1$. Plot the correlator outputs at time instants corresponding to $k = 1, 2, \ldots, 10$.

5.4 Repeat Problem 5.3 for the signal waveforms shown in Figure P5.2.

5.5 Run the MATLAB program that performs a Monte Carlo simulation of the binary communication system shown in Figure 5.8, based on orthogonal signals. Perform the simulation for 10,000 bits, and measure the error probability for $\sigma^2 = 0$, $\sigma^2 = 0.1$, $\sigma^2 = 1.0$, and $\sigma^2 = 2.0$. Plot the theoretical error rate and the error rate measured from the Monte Carlo simulation, and compare the two results. Also plot 1000 received signal-plus-noise samples at the input to the detector for each value of σ^2.

5.6 Repeat Problem 5.5 for the binary communication system shown in Figure 5.13, based on antipodal signals.

5.7 Repeat Problem 5.5 for the binary communication system, based on on-off signals.

5.8 Run the MATLAB program that performs a Monte Carlo simulation of a quaternary PAM communication system. Perform the simulation for 10,000 symbols (20,000 bits), and measure the symbol-error probability for $\sigma^2 = 0$, $\sigma^2 = 0.1$, $\sigma^2 = 1.0$, and $\sigma^2 = 2.0$. Plot the theoretical error rate and the error measured from the Monte Carlo simulation, and compare these results. Also plot 1000 received signal-plus-noise samples at the input to the detector for each value of σ^2.

5.9 Modify the MATLAB program in Problem 5.8 to simulate $M = 8$ PAM signals, and perform the Monte Carlo simulations as specified in Problem 5.8.

5.10 Run the MATLAB program that performs the Monte Carlo simulation of a digital communication system that employs $M = 4$ orthogonal signals, as described in Illustrative Problem 5.10. Perform the simulation for 10,000 symbols (20,000 bits), and measure the bit-error probability for $\sigma^2 = 0.1$, $\sigma^2 = 1.0$, and $\sigma^2 = 2.0$. Plot the theoretical error probability and the error rate measured from the Monte Carlo simulation, and compare these results.

5.11 Consider the four signal waveforms shown in Figure P5.11. Show that these four signal waveforms are mutually orthogonal. Will the results of the Monte Carlo simulation of Problem 5.10 apply to these signals? Why?

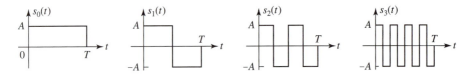

Figure P5.11

5.12 Run the MATLAB program that performs the Monte Carlo simulation of a digital communication system that employs $M = 4$ biorthogonal signals, as described in Illustrative Problem 5.11. Perform the simulations for 10,000 symbols (20,000 bits), and measure the symbol-error probability for $\sigma^2 = 0.1$, $\sigma^2 = 1.0$, and $\sigma^2 = 2.0$. Plot the theoretical symbol-error probability and the error rate measured from the Monte Carlo simulation, and compare the results. Also plot 1000 received signal-plus-noise samples at the input to the detector for each value of σ^2.

5.13 Consider the four signal waveforms shown in Figure P5.13. Show that they are biorthogonal. Will the results of the Monte Carlo simulation in Problem 5.12 apply to this set of four signal waveforms? Why?

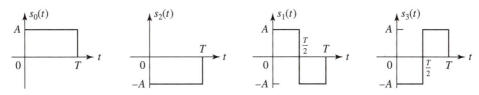

Figure P5.13

Chapter 6

Digital Transmission Through Bandlimited Channels

6.1   Preview

In this chapter, we treat several aspects of digital transmission through bandwidth-limited channels. We begin by describing the spectral characteristics of PAM signals. Second, we consider the characterization of bandlimited channels and the problem of designing signal waveforms for such channels. Then we treat the problem of designing channel equalizers that compensate for distortion caused by bandlimited channels. We show that channel distortion results in intersymbol interference (ISI), which causes errors in signal demodulation. A channel equalizer is a device that reduces the intersymbol interference and thus reduces the error rate in the demodulated data sequence.

6.2   The Power Spectrum of a Digital PAM Signal

In the preceding chapter, we considered the transmission of digital information by pulse amplitude modulation (PAM). In this section we study the spectral characteristics of such signals.

A digital PAM signal at the input to a communication channel is generally represented as

$$v(t) = \sum_{n=-\infty}^{\infty} a_n g(t - nT) \tag{6.2.1}$$

where $\{a_n\}$ is the sequence of amplitudes corresponding to the information symbols from the source, $g(t)$ is a pulse waveform, and T is the reciprocal of the symbol rate. T is also called the *symbol interval*. Each element of the sequence $\{a_n\}$ is selected from one of the possible amplitude values, which are

$$A_m = (2m - M + 1)d, \qquad m = 0, 1, \ldots, M - 1 \tag{6.2.2}$$

where d is a scale factor that determines the Euclidean distance between any pair of signal amplitudes. ($2d$ is the Euclidean distance between any adjacent signal amplitude levels.)

Since the information sequence is a random sequence, the sequence $\{a_n\}$ of amplitudes corresponding to the information symbols from the source is also random. Consequently, the PAM signal $v(t)$ is a sample function of a random process $V(t)$. To determine the spectral characteristics of the random process $V(t)$, we must evaluate the power spectrum.

First, we note that the mean value of $V(t)$ is

$$E[V(t)] = \sum_{n=-\infty}^{\infty} E(a_n)g(t - nT) \tag{6.2.3}$$

By selecting the signal amplitudes to be symmetric about zero, as given in (6.2.2), and equally probable, we have $E(a_n) = 0$, and hence, $E[V(t)] = 0$.

The autocorrelation function of $V(t)$ is

$$R_v(t + \tau; t) = E[V(t)V(t + \tau)] \tag{6.2.4}$$

It is shown in many standard texts on digital communications that the autocorrelation function is a periodic function in the variable t with period T. Random processes that have a periodic mean value and a periodic autocorrelation function are called *periodically stationary*, or *cyclostationary*. The time variable t can be eliminated by averaging $R_v(t + \tau; t)$ over a single period; that is,

$$\bar{R}_v(\tau) = \frac{1}{T} \int_{-T/2}^{T/2} R_v(t + \tau; t)\, dt \tag{6.2.5}$$

This average autocorrelation function for the PAM signal can be expressed as

$$\bar{R}_v(\tau) = \frac{1}{T} \sum_{m=-\infty}^{\infty} R_a(m) R_g(\tau - mT) \tag{6.2.6}$$

where $R_a(m) = E(a_n a_{n+m})$ is the autocorrelation of the sequence $\{a_n\}$, and $R_g(\tau)$ is defined as

$$R_g(\tau) = \int_{-\infty}^{\infty} g(t)g(t + \tau)\, dt \tag{6.2.7}$$

The power spectrum of $V(t)$ is simply the Fourier transform of the average autocorrelation function $\bar{R}_v(\tau)$; that is,

$$S_v(f) = \int_{-\infty}^{\infty} \bar{R}_v(\tau)e^{-j2\pi f\tau}\, dt$$
$$= \frac{1}{T} S_a(f)|G(f)|^2 \tag{6.2.8}$$

where $S_a(f)$ is the power spectrum of amplitude sequence $\{a_n\}$, and $G(f)$ is the Fourier transform of the pulse $g(t)$. $S_a(f)$ is defined as

$$S_a(f) = \sum_{m=-\infty}^{\infty} R_a(m)e^{-j2\pi fmT} \tag{6.2.9}$$

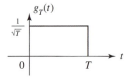

Figure 6.1   Transmitter pulse

From (6.2.8), we observe that the power spectrum of the PAM signal is a function of the power spectrum of the information symbols $\{a_n\}$ and the spectrum of the pulse $g(t)$. In the special case where the sequence $\{a_n\}$ is uncorrelated—that is,

$$R_a(m) = \begin{cases} \sigma_a^2, & m = 0 \\ 0, & m \neq 0 \end{cases} \tag{6.2.10}$$

where $\sigma_a^2 = E(a_n^2)$, it follows that $S_a(f) = \sigma_a^2$ for all f and

$$S_v(f) = \frac{\sigma_a^2}{T} |G(f)|^2 \tag{6.2.11}$$

In this case, the power spectrum of $V(t)$ is dependent entirely on the spectral characteristics of the pulse $g(t)$.

ILLUSTRATIVE PROBLEM

Illustrative Problem 6.1 [PAM power spectrum] Determine the power spectrum of $V(t)$ when $\{a_n\}$ is an uncorrelated sequence and $g(t)$ is the rectangular pulse shown in Figure 6.1.

SOLUTION

The Fourier transform of $g(t)$ is

$$G(f) = \int_{-\infty}^{\infty} g(t) e^{-j2\pi ft} \, dt$$
$$= \sqrt{T} \frac{\sin \pi fT}{\pi fT} e^{-j\pi fT} \tag{6.2.12}$$

and

$$S_v(f) = \sigma_a^2 \left(\frac{\sin \pi fT}{\pi fT} \right)^2 \tag{6.2.13}$$

This power spectrum is illustrated in Figure 6.2.

The MATLAB script for this computation is given next.

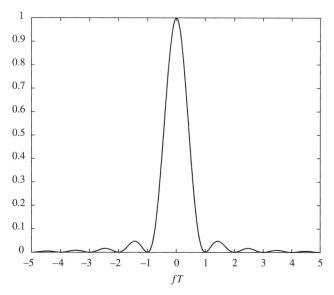

Figure 6.2   Power spectrum of the transmitted signal in Illustrative Problem 6.1
(for $\sigma_a^2 = 1$)

```
% MATLAB script for Illustrative Problem 6.1.
echo on
T=1;
delta_f=1/(100*T);
f=−5/T:delta_f:5/T;
sgma_a=1;
Sv=sgma_a^2*sinc(f*T).^2;
% plotting command follows
plot(f,Sv);
```

◖ILLUSTRATIVE PROBLEM◗

Illustrative Problem 6.2 [PAM power spectrum] Suppose the autocorrelation function
of the sequence $\{a_n\}$ is

$$R_a(m) = \begin{cases} 1, & m = 0 \\ \frac{1}{2}, & m = 1, -1 \\ 0, & \text{otherwise} \end{cases} \tag{6.2.14}$$

and $g(t)$ is the rectangular pulse shown in Figure 6.1. Evaluate $\mathcal{S}_v(f)$ in this case.

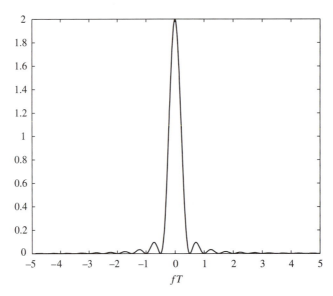

Figure 6.3   The power spectrum of the transmitted signal in Illustrative Problem 6.2
(for $\sigma_a^2 = 1$)

SOLUTION

The power spectrum of the PAM signal $V(t)$ is given by (6.2.8). The power spectrum of
the sequence $\{a_n\}$ is, from (6.2.9) and (6.2.14),

$$S_a(f) = 1 + \cos 2\pi f T$$
$$= 2\cos^2 \pi f T \tag{6.2.15}$$

Consequently,

$$S_v(f) = 2\cos^2 \pi f T \left(\frac{\sin \pi f T}{\pi f T}\right)^2 \tag{6.2.16}$$

The graph of this power spectrum is shown in Figure 6.3.

The MATLAB script for performing this computation is given next. In this case, the
overall power spectrum of the transmitted signal $V(t)$ is significantly narrower than the
spectrum in Figure 6.2.

M-FILE

```
% MATLAB script for Illustrative Problem 6.2.
echo on
T=1;
delta_f=1/(100*T);
```

```
f=−5/T:delta_f:5/T;
Sv=2*(cos(pi*f*T).*sinc(f*T)).^2;
% plotting command follows
plot(f,Sv);
```

6.3   Characterization of Bandlimited Channels and Channel Distortion

Many communication channels, including telephone channels and some radio channels, may be generally characterized as bandlimited linear filters. Consequently, such channels are described by their frequency response $C(f)$, which may be expressed as

$$C(f) = A(f)e^{j\theta(f)} \qquad (6.3.1)$$

where $A(f)$ is called the *amplitude response* and $\theta(f)$ is called the *phase response*. Another characteristic that is sometimes used in place of the phase response is the *envelope delay*, or *group delay*, which is defined as

$$\tau(f) = -\frac{1}{2\pi}\frac{d\theta(f)}{df} \qquad (6.3.2)$$

A channel is said to be *nondistorting* or *ideal* if, within the bandwidth W occupied by the transmitted signal, $A(f) = $ constant and $\theta(f)$ is a linear function of frequency [or the envelope delay $\tau(f) = $ constant]. On the other hand, if $A(f)$ and $\tau(f)$ are not constant within the bandwidth occupied by the transmitted signal, the channel distorts the signal. If $A(f)$ is not constant, the distortion is called *amplitude distortion*, and if $\tau(f)$ is not constant, the distortion on the transmitted signal is called *delay distortion*.

As a result of the amplitude and delay distortion caused by the nonideal channel frequency response characteristic $C(f)$, a succession of pulses transmitted through the channel at rates comparable to the bandwidth W are smeared to the point that they are no longer distinguishable as well-defined pulses at the receiving terminal. Instead, they overlap, so we have intersymbol interference. As an example of the effect of delay distortion on a transmitted pulse, Figure 6.4(a) illustrates a bandlimited pulse having zeros periodically spaced in time at points labeled $\pm T$, $\pm 2T$, and so on. When the information is conveyed by the pulse amplitude, as in PAM, one can transmit a sequence of pulses, each of which has a peak at the periodic zeros of the other pulses. However, transmission of the pulse through a channel modeled as having a linear envelope delay characteristic $\tau(f)$ [quadratic phase $\theta(f)$] results in the received pulse shown in Figure 6.4(b) having zero crossings that are no longer periodically spaced. Consequently, a sequence of successive pulses would be smeared into one another, and the peaks of the pulses would no longer be distinguishable. Thus, the channel delay distortion results in intersymbol interference. As will be discussed in this chapter, it is possible to compensate for the nonideal frequency response characteristic of the channel by use of a filter or equalizer at the demodulator. Figure 6.4(c) illustrates the output of a linear equalizer that compensates for the linear distortion in the channel.

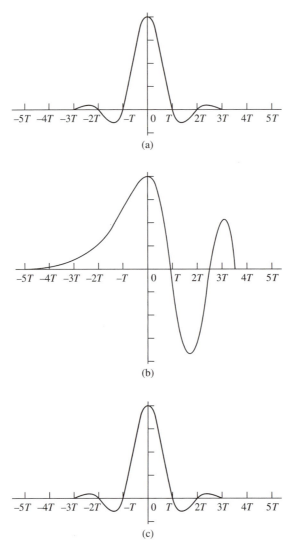

Figure 6.4 Effect of channel distortion: (a) channel input; (b) channel output; (c) equalizer output

As an example, let us consider the intersymbol interference on a telephone channel. Figure 6.5 illustrates the measured average amplitude and delay as a function of frequency for a telephone channel of the switched telecommunications network. We observe that the usable band of the channel extends from about 300 Hz to about 3200 Hz. The corresponding impulse response of the average channel is shown in Figure 6.6. Its duration is about 10 ms. In comparison, the transmitted symbol rates on such a channel may be of the order of 2500 pulses or symbols per second. Hence, intersymbol interference might extend over 20 to 30 symbols.

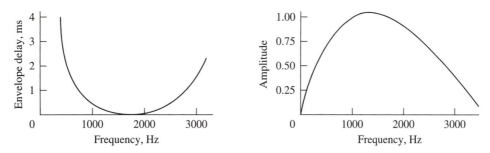

Figure 6.5   Average amplitude and delay characteristics of a medium-range telephone channel

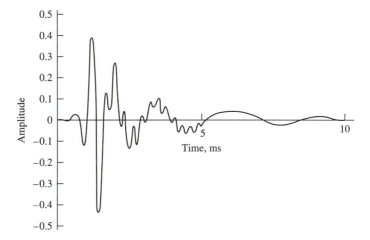

Figure 6.6   Impulse response of the average channel with amplitude and delay shown in Figure 6.5

Besides telephone channels, there are other physical channels that exhibit some form of time dispersion and thus introduce intersymbol interference. Radio channels, such as short-wave ionospheric propagation (HF), tropospheric scatter, and mobile cellular radio, are three examples of time-dispersive wireless channels. In these channels, time dispersion—and, hence, intersymbol interference—is the result of multiple propagation paths with different path delays. The number of paths and the relative time delays among the paths vary with time; for this reason, these radio channels are usually called *time-variant multipath channels*. The time-variant multipath conditions give rise to a wide variety of frequency response characteristics. Consequently, the frequency response characterization that is used for telephone channels is inappropriate for time-variant multipath channels. Instead, these radio channels are characterized statistically in terms of the scattering function, which, in brief, is a two-dimensional representation of the average received signal power as a function of relative time delay and Doppler frequency spread.

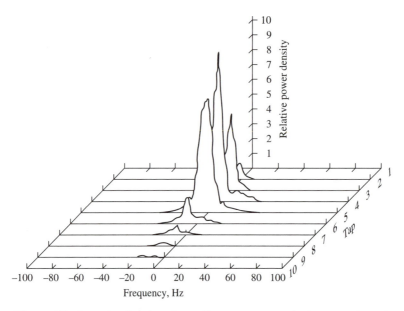

Figure 6.7   Scattering function of a medium-range tropospheric scatter channel

For illustrative purposes, a scattering function measured on a medium-range (150-mi) tropospheric scatter channel is shown in Figure 6.7. The total time duration (multipath spread) of the channel response is approximately $0.7 \mu s$ on the average, and the spread between *half-power points* in Doppler frequency is a little less than 1 Hz on the strongest path and somewhat larger on the other paths. Typically, if transmission occurs at a rate of 10^7 symbols/second over such a channel, the multipath spread of $0.7 \mu s$ will result in intersymbol interference that spans about seven symbols.

ILLUSTRATIVE PROBLEM

Illustrative Problem 6.3 [Channel as a filter] As indicated earlier, a bandlimited communication channel can be modeled as a linear filter whose frequency response characteristics match the frequency response characteristics of the channel. MATLAB may be used to design digital finite-duration impulse respone (FIR) or infinite-duration impulse response (IIR) filters that approximate the frequency response characteristics of analog communication channels. Suppose that we wish to model an ideal channel having an amplitude response $A(f) = 1$ for $|f| \leq 2000$ Hz and $A(f) = 0$ for $|f| > 2000$ Hz and constant delay (linear phase) for all f. The sampling rate for the digital filter is selected as $F_s = 10{,}000$ Hz. Since the desired phase response is linear, only an FIR filter could satisfy this condition. However, it is not possible to achieve a zero response in the stopband. Instead, we select the stopband response to be -40 dB and the stopband frequency to be 2500 Hz. In addition, we allow for a small amount, 0.5 dB, of ripple in the passband. Design a filter with these characteristics.

The impulse response and the frequency response of a length $N = 41$ FIR filter that meets these specifications are illustrated in Figure 6.8. Since N is odd, the delay through the filter is $(N + 1)/2$ taps, which corresponds to a time delay of $(N + 1)/20$ ms at the sampling rate of $F_s = 10$ KHz. In this example, the FIR filter was designed in MATLAB using the Chebyshev approximation method (Remez algorithm).

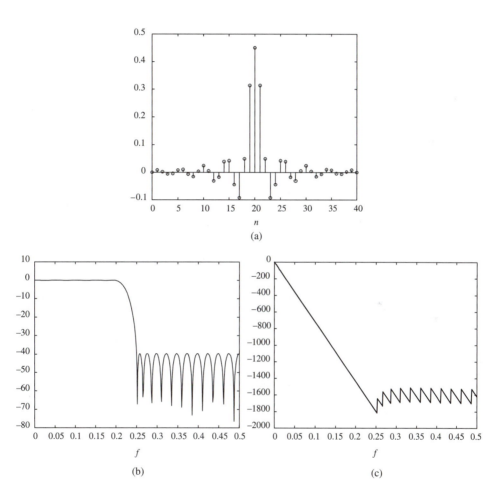

Figure 6.8 (a) Impulse response; (b, c) frequency response of linear phase FIR filter in Illustrative Problem 6.3

M-FILE

```
% MATLAB script for Illustrative Problem 6.3.
echo on
f_cutoff=2000;                          % the desired cut-off frequency
f_stopband=2500;                        % the actual stopband frequency
fs=10000;                               % the sampling frequency
f1=2*f_cutoff/fs;                       % the normalized passband frequency
f2=2*f_stopband/fs;                     % the normalized stopband frequency
N=41;                                   % this number is found by experiment
F=[0 f1 f2 1];
M=[1 1 0 0];                            % describes the low-pass filter
B=remez(N−1,F,M);                       % returns the FIR tap coefficients
% plotting command follows
figure(1);
[H,W]=freqz(B);
H_in_dB=20*log10(abs(H));
plot(W/(2*pi),H_in_dB);
figure(2);
plot(W/(2*pi),(180/pi)*unwrap(angle(H)));
% plot of the impulse response follows
figure(3);
plot(zeros(size([0:N−1])));
hold;
stem([0:N−1],B);
```

ILLUSTRATIVE PROBLEM

Illustrative Problem 6.4 [Channel as a filter] An alternative method for designing an FIR filter that approximates the desired channel characteristics is based on the window method. To be specific, if the desired channel response is $C(f)$ for $|f| \leq W$ and $C(f) = 0$ for $|f| > W$, the impulse response of the channel is

$$h(t) = \int_{-W}^{W} C(f)e^{j2\pi ft}\, df \tag{6.3.3}$$

For example, if the channel is ideal, then $C(f) = 1$, $|f| \leq W$, and hence,

$$h(t) = \frac{\sin 2\pi Wt}{\pi t} \tag{6.3.4}$$

An equivalent digital filter may be implemented by sampling $h(t)$ at $t = nT_s$, where T_s is the sampling interval and $n = 0, \pm1, \pm2, \ldots$. Design an FIR filter with $W = 2000$ Hz and $F_s = 1/T_s = 10$ KHz.

SOLUTION

The sampled version of $h(t)$—$h_n \equiv h(nT_s)$—is illustrated in Figure 6.9. Since $\{h_n\}$ has infinite length, we may truncate it at some length N. This truncation is equivalent to multiplying $\{h_n\}$ by a rectangular window sequence $w_n = 1$ for $|n| \leq (N-1)/2$ and $w_n = 0$ for $|n| \geq (N+1)/2$. The impulse response $\{h_n^1 = w_n h_n\}$ and the corresponding frequency response of the FIR (truncated) filter are illustrated in Figure 6.10 for $N = 51$. Note that the truncated filter has large sidelobes in the stopband. Hence, this FIR filter is a poor approximation to the desired channel characteristics. The size of the sidelobes can be significantly reduced by employing a smoother window function, such as a Hanning or a Hamming window, to truncate the ideal channel response. Figure 6.11 illustrates the impulse response and frequency response of $\{h_n^1 = w_n h_n\}$ when the window function is a Hanning window of length $N = 51$. MATLAB provides routines to implement several different types of window functions.

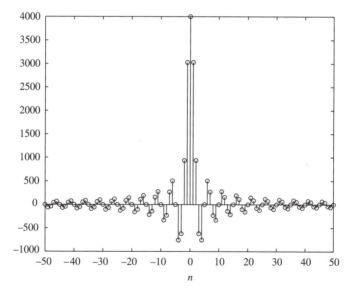

Figure 6.9   Samples of $h(n)$ in Illustrative Problem 6.4

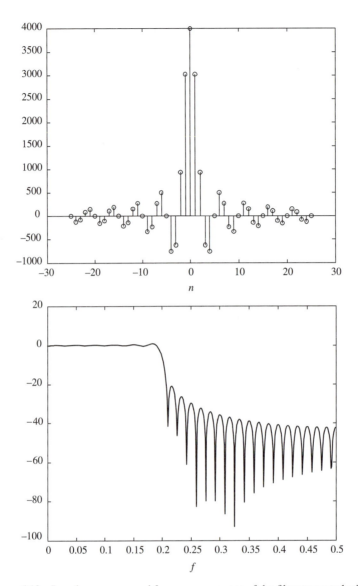

Figure 6.10   Impulse response and frequency response of the filter truncated with a
rectangular window in Illustrative Problem 6.4

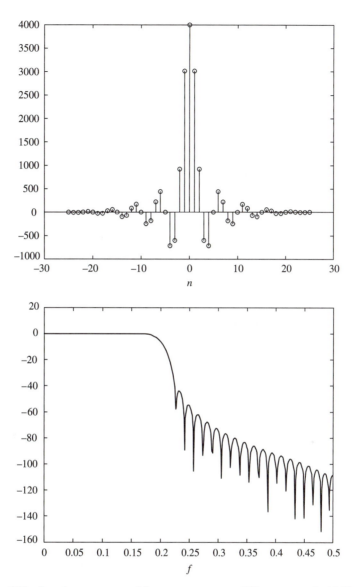

Figure 6.11   Impulse response and frequency response of filter truncated with Hanning
window in Illustrative Problem 6.4

M-FILE

```
% MATLAB script for Illustrative Problem 6.4.
echo on
Length=101;
Fs=10000;
```

```
W=2000;
Ts=1/Fs;
n=−(Length−1)/2:(Length−1)/2;
t=Ts*n;
h=2*W*sinc(2*W*t);
% The rectangular windowed version follows...
N=61;
rec_windowed_h=h((Length−N)/2+1:(Length+N)/2);
% Frequency response of rec_windowed_h follows
[rec_windowed_H,W1]=freqz(rec_windowed_h,1);
% to normalize the magnitude
rec_windowed_H_in_dB=20*log10(abs(rec_windowed_H)/abs(rec_windowed_H(1)));
% The Hanning windowed version follows...
hanning_window=hanning(N);
hanning_windowed_h=h((Length−N)/2+1:(Length+N)/2).*hanning_window.' ;
[hanning_windowed_H,W2]=freqz(hanning_windowed_h,1);
hanning_windowed_H_in_dB=20*log10(abs(hanning_windowed_H)/abs(hanning_windowed_H(1)));
% the plotting commands follow
```

ILLUSTRATIVE PROBLEM

Illustrative Problem 6.5 [Multipath channel simulation] A two-path (multipath) radio channel can be modeled in the time domain, as illustrated in Figure 6.12. Its impulse response may be expressed as

$$c(t, \tau) = b_1(t)\delta(\tau) + b_2(t)\delta(\tau - \tau_d) \tag{6.3.5}$$

where $b_1(t)$ and $b_2(t)$ are random processes that represent the time-varying propagation behavior of the channel and τ_d is the delay between the two multipath components. Simulate such a channel on the computer.

SOLUTION

We model $b_1(t)$ and $b_2(t)$ as Gaussian random processes generated by passing white Gaussian noise processes through lowpass filters. In discrete time, we may use relatively simple

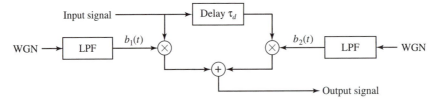

Figure 6.12   Two-path radio channel model

digital IIR filters excited by white Gaussian noise (WGN) sequences. For example, a simple lowpass digital filter having two identical poles is described by the z-transform

$$H(z) = \frac{(1-p)^2}{(1-pz^{-1})^2} = \frac{(1-p)^2}{1-2pz^{-1}+p^2z^{-2}} \tag{6.3.6}$$

or the corresponding difference equation

$$b_n = 2pb_{n-1} - p^2 b_{n-2} + (1-p)^2 w_n \tag{6.3.7}$$

where $\{w_n\}$ is the input WGN sequence, $\{b_n\}$ is the output sequence, and $0 < p < 1$ is the pole position. The position of the pole controls the bandwidth of the filter and, hence, the rate of variation of $\{b_n\}$. When p is close to unity (close to the unit circle), the filter bandwidth is narrow, whereas when p is close to zero, the bandwidth is wide. Hence, when p is close to the unit circle in the z-plane, the filter output sequence changes more slowly compared to the case when the pole is close to the origin.

Figure 6.13 illustrates the output sequences $\{b_{1n}\}$ and $\{b_{2n}\}$ generated by passing statistically independent WGN sequences through a filter having $p = 0.99$. The discrete-time channel impulse response

$$c_n = b_{1,n} + b_{2,n-d} \tag{6.3.8}$$

is also shown, with $d = 5$ samples of delay. Figure 6.14 illustrates the sequences $\{b_{1,n}\}$, $\{b_{2,n}\}$, and $\{c_n\}$ when $p = 0.9$.

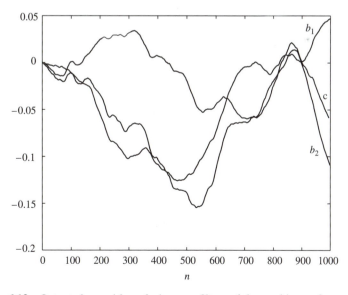

Figure 6.13   Outputs b_{1n} and b_{2n} of a lowpass filter and the resulting c_n for $p = 0.99$

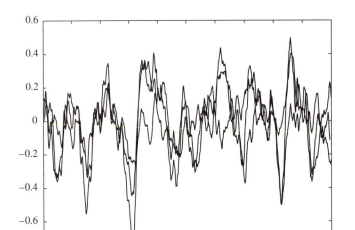

Figure 6.14   Outputs b_{1n}, b_{2n}, and c_n with pole at $p = 0.9$ (from top to bottom at $n = 1000$, b_{1n}, c_n, and b_{2n})

6.4   Characterization of Intersymbol Interference

In a digital communication system, channel distortion causes intersymbol interference (ISI). In this section, we present a model that characterizes ISI. For simplicity, we assume that the transmitted signal is a baseband PAM signal. However, this treatment is easily extended to carrier (linearly) modulated signals discussed in the next chapter.

The transmitted PAM signal is expressed as

$$s(t) = \sum_{n=0}^{\infty} a_n g(t - nT) \tag{6.4.1}$$

where $g(t)$ is the basic pulse shape that is selected to control the spectral characteristics of the transmitted signal, $\{a_n\}$ is the sequence of transmitted information symbols selected from a signal constellation consisting of M points, and T is the signal interval ($1/T$ is the symbol rate).

The signal $s(t)$ is transmitted over a baseband channel, which may be characterized by a frequency response $C(f)$. Consequently, the received signal can be represented as

$$r(t) = \sum_{n=0}^{\infty} a_n h(t - nT) + w(t) \tag{6.4.2}$$

where $h(t) = g(t) * c(t)$, $c(t)$ is the impulse response of the channel, $*$ denotes convolution, and $w(t)$ represents the additive noise in the channel. To characterize ISI, suppose that the received signal is passed through a receiving filter and then sampled at the rate $1/T$ samples/second. In general, the optimum filter at the receiver is matched to the received signal pulse $h(t)$. Hence, the frequency response of this filter is $H^*(f)$. We denote its output as

$$y(t) = \sum_{n=0}^{\infty} a_n x(t - nT) + v(t) \tag{6.4.3}$$

where $x(t)$ is the signal pulse response of the receiving filter—that is, $X(f) = H(f)H^*(f)$ $= |H(f)|^2$—and $v(t)$ is the response of the receiving filter to the noise $w(t)$. Now, if $y(t)$ is sampled at times $t = kT$, $k = 0, 1, 2, \ldots$, we have

$$y(kT) = \sum_{n=0}^{\infty} a_n x(kT - nT) + v(kT)$$

$$y_k = \sum_{n=0}^{\infty} a_n x_{k-n} + v_k, \qquad k = 0, 1, \ldots \tag{6.4.4}$$

The sample values $\{y_k\}$ can be expressed as

$$y_k = x_0 \left(a_k + \frac{1}{x_0} \sum_{\substack{n=0 \\ n \neq k}}^{\infty} a_n x_{k-n} \right) + v_k, \qquad k = 0, 1, \ldots \tag{6.4.5}$$

The term x_0 is an arbitrary scale factor, which we set equal to unity for convenience. Then

$$y_k = a_k + \sum_{\substack{n=0 \\ n \neq k}}^{\infty} a_n x_{k-n} + v_k \tag{6.4.6}$$

The term a_k represents the desired information symbol at the kth sampling instant, the term

$$\sum_{\substack{n=0 \\ n \neq k}}^{\infty} a_n x_{k-n} \tag{6.4.7}$$

represents the ISI, and v_k is the additive noise at the kth sampling instant.

The amount of ISI and noise in a digital communications system can be viewed on an oscilloscope. For PAM signals, we can display the received signal $y(t)$ on the vertical input with the horizontal sweep rate set at $1/T$. The resulting oscilloscope display is called an *eye pattern* because of its resemblance to the human eye. For example, Figure 6.15 illustrates the eye patterns for binary and four-level PAM modulation. The effect of ISI is to cause the eye to close, thereby reducing the margin for additive noise to cause errors. Figure 6.16 graphically illustrates the effect of ISI in reducing the opening of a binary eye. Note that intersymbol interference distorts the position of the zero-crossings and causes a reduction in the eye opening. Thus, it causes the system to be more sensitive to a synchronization error.

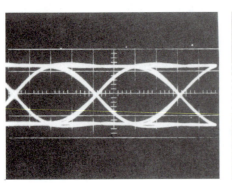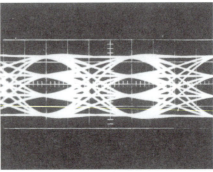

Binary   Quaternary

Figure 6.15 Examples of eye patterns for binary and quaternary
amplitude-shift keying (or PAM)

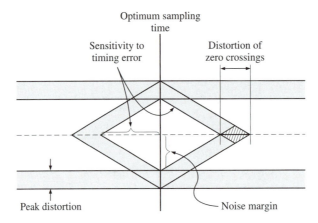

Figure 6.16 Effect of intersymbol interference on eye opening

━━━(ILLUSTRATIVE PROBLEM)━━━━━━━━━━━━━━━━━━━━━━━━━━━

Illustrative Problem 6.6 [Intersymbol interference] In this problem, we consider the effect of intersymbol interference (ISI) on the received signal sequence $\{y_k\}$ for two channels that result in the sequences $\{x_k\}$ as follows:

Channel 1

$$
x_n = \begin{cases}
1, & n = 0 \\
-0.25, & n = \pm 1 \\
0.1, & n = \pm 2 \\
0, & \text{otherwise}
\end{cases}
$$

Channel 2

$$x_n = \begin{cases} 1, & n = 0 \\ 0.5, & n = \pm 1 \\ -0.2, & n = \pm 2 \\ 0, & \text{otherwise} \end{cases}$$

Note that in these channels, the ISI is limited to two symbols on either side of the desired transmitted signal. Hence, the cascade of the transmitter and receiver filters and the channel at the sampling instants are represented by the equivalent *discrete-time FIR channel filter* shown in Figure 6.17. Now suppose that the transmitted signal sequence is binary—that is, $\{a_n = \pm 1\}$. Then, for channel 1, the received signal sequence $\{y_n\}$ in the absence of noise is shown in Figure 6.18(a), and with additive white Gaussian noise having a variance of $\sigma^2 = 0.1$, the received signal sequence is shown in Figure 6.18(b). We note that in the absence of noise, the ISI alone does not cause errors at the detector that compares the received signal sequence $\{y_n\}$ with the threshold set to zero. Hence, the eye diagram is open in the absence of noise. However, when the additive noise is sufficiently large, errors will occur.

In the case of channel 2, the noise-free and noisy ($\sigma^2 = 0.1$) sequences $\{y_n\}$ are illustrated in Figure 6.19. Now, we observe that the ISI can cause errors at the detector that compares the received sequence $\{y_n\}$ with the threshold set at zero, even in the absence of noise. Thus, for this channel characteristic, the eye is completely closed.

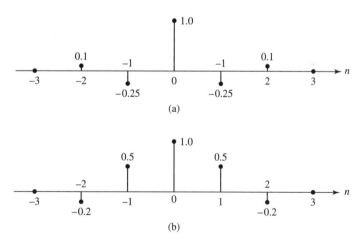

Figure 6.17   FIR channel models with ISI: (a) channel 1; (b) channel 2

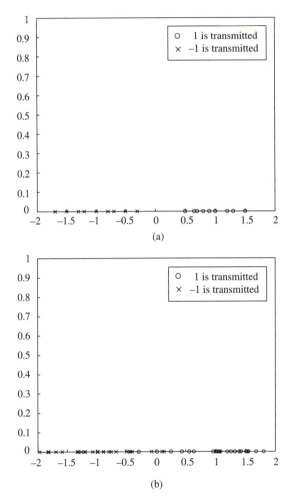

Figure 6.18   Output of channel model 1 without and with AWGN: (a) no noise;
(b) additive white Gaussian noise with $\sigma^2 = 0.1$

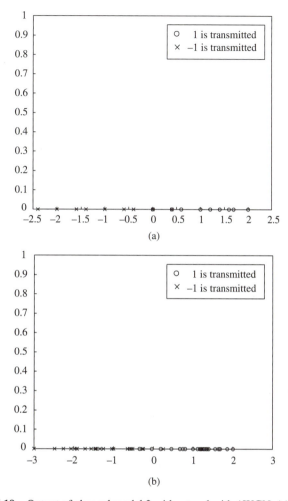

Figure 6.19   Output of channel model 2 without and with AWGN: (a) no noise;
(b) additive white Gaussian noise with variance $\sigma^2 = 0.1$

6.5   Communication System Design for Bandlimited Channels

In this section, we consider the design of the transmitter and receiver filters that are suitable for a baseband bandlimited channel. Two cases are considered. In the first case, the design is based on transmitter and receiver filters that result in zero ISI. In the second case, the design is based on transmitter and receiver filters that have a specified (predetermined) amount of ISI. Thus, the second design approach leads to a controlled amount of ISI. The corresponding transmitted signals are called *partial response signals*. In both cases, we assume that the channel is ideal; that is, $A(f)$ and $\tau(f)$ are constant within the channel bandwidth W. For simplicity, we assume that $A(f) = 1$ and $\tau(f) = 0$.

6.5.1   Signal Design for Zero ISI

The design of bandlimited signals with zero ISI was a problem considered by Nyquist about 70 years ago. He demonstrated that a necessary and sufficient condition for a signal $x(t)$ to have zero ISI—that is,

$$x(nT) = \begin{cases} 1, & n = 0 \\ 0, & n \neq 0 \end{cases} \tag{6.5.1}$$

is that its Fourier transform $X(f)$ satisfy

$$\sum_{m=-\infty}^{\infty} X\left(f + \frac{m}{T}\right) = T \tag{6.5.2}$$

where $1/T$ is the symbol rate.

In general, many signals can be designed to have this property. One of the most commonly used signals in practice has a raised-cosine frequency response characteristic, which is defined as

$$X_{\text{rc}}(f) = \begin{cases} T, & 0 \leq |f| \leq \dfrac{1-\alpha}{2T} \\ \dfrac{T}{2}\left[1 + \cos\dfrac{\pi T}{\alpha}\left(|f| - \dfrac{1-\alpha}{2T}\right)\right], & \dfrac{1-\alpha}{2T} < |f| \leq \dfrac{1+\alpha}{2T} \\ 0, & |f| > \dfrac{1+\alpha}{2T} \end{cases} \tag{6.5.3}$$

where α is called the *roll-off* factor, which takes values in the range $0 \leq \alpha \leq 1$, and $1/T$ is the symbol rate. The frequency response $X_{\text{rc}}(f)$ is illustrated in Figure 6.20(a) for $\alpha = 0$, $\alpha = \frac{1}{2}$, and $\alpha = 1$. Note that when $\alpha = 0$, $X_{\text{rc}}(f)$ reduces to an ideal "brick wall" physically nonrealizable frequency response with bandwidth occupancy $1/2T$. The frequency $1/2T$ is called the *Nyquist frequency*. For $\alpha > 0$, the bandwidth occupied by the desired signal $X_{\text{rc}}(f)$ beyond the Nyquist frequency $1/2T$ is called the *excess bandwidth*, usually expressed as a percentage of the Nyquist frequency. For example, when $\alpha = \frac{1}{2}$, the excess bandwidth is 50%, and when $\alpha = 1$, the excess bandwidth is 100%. The signal pulse $x_{\text{rc}}(t)$ having the raised-cosine spectrum is

$$x_{\text{rc}}(t) = \frac{\sin \pi t/T}{\pi t/T} \frac{\cos(\pi \alpha t/T)}{1 - 4\alpha^2 t^2/T^2} \tag{6.5.4}$$

Figure 6.20(b) illustrates $x_{\text{rc}}(t)$ for $\alpha = 0, \frac{1}{2}, 1$. Since $X_{\text{rc}}(f)$ satisfies (6.5.2), we note that $x_{\text{rc}}(t) = 1$ at $t = 0$ and $x_{\text{rc}}(t) = 0$ at $t = kT, k = \pm 1, \pm 2, \ldots$. Consequently, at the sampling instants $t = kT, k \neq 0$, there is no ISI from adjacent symbols when there is no channel distortion. However, in the presence of channel distortion, the ISI given by

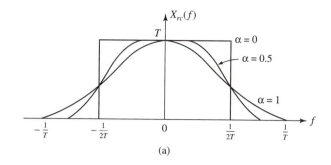

(a)

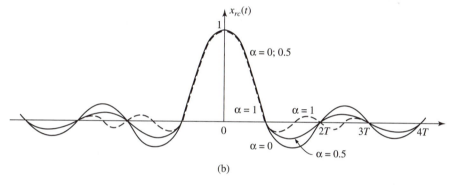

(b)

Figure 6.20   (a) Raised-cosine frequency response; (b) pulse shapes for raised-cosine
frequency response

(6.4.7) is no longer zero, and a channel equalizer is needed to minimize its effect on system
performance. Channel equalizers are considered in Section 6.6.

   In an ideal channel, the transmitter and receiver filters are jointly designed for zero
ISI at the desired sampling instants $t = nT$. Thus, if $G_T(f)$ is the frequency response
of the transmitter filter and $G_R(f)$ is the frequency response of the receiver filter, then
the product (cascade of the two filters) $G_T(f)G_R(f)$ is designed to yield zero ISI. For
example, if the product $G_T(f)G_R(f)$ is selected as

$$G_T(f)G_R(f) = X_{\text{rc}}(f) \tag{6.5.5}$$

where $X_{\text{rc}}(f)$ is the raised-cosine frequency response characteristic, then the ISI at the
sampling times $t = nT$ is zero.

═══ ILLUSTRATIVE PROBLEM ═══

Illustrative Problem 6.7 [Optimum transmitter and receiver filters] Design a digital
implementation of the transmitter and receiver filters $G_T(f)$ and $G_R(f)$ such that their
product satisfies (6.5.5) and $G_R(f)$ is the matched filter to $G_T(f)$.

SOLUTION

The simplest way to design and implement the transmitter and receiver filters in digital form is to employ FIR filters with linear phase (symmetric impulse response). The desired magnitude response is

$$|G_T(f)| = |G_R(f)| = \sqrt{X_{\rm rc}(f)} \tag{6.5.6}$$

where $X_{\rm rc}(f)$ is given by (6.5.3). The frequency response is related to the impulse response of the digital filter by the equation

$$G_T(f) = \sum_{n=-(N-1)/2}^{(N-1)/2} g_T(n)e^{-j2\pi f n T_s} \tag{6.5.7}$$

where T_s is the sampling interval and N is the length of the filter. Note that N is odd. Since $G_T(f)$ is bandlimited, we may select the sampling frequency F_s to be at least $2/T$. Our choice is

$$F_s = \frac{1}{T_s} = \frac{4}{T}$$

or, equivalently, $T_s = T/4$. Hence, the folding frequency is $F_s/2 = 2/T$. Since $G_T(f) = \sqrt{X_{\rm rc}(f)}$, we may sample $X_{\rm rc}(f)$ at equally spaced points in frequency, with frequency separation $\Delta f = F_s/N$. Thus, we have

$$\sqrt{X_{\rm rc}(m\Delta f)} = \sqrt{X_{\rm rc}\left(\frac{mF_s}{N}\right)} = \sum_{n=-(N-1)/2}^{(N-1)/2} g_T(n)e^{-j2\pi mn/N} \tag{6.5.8}$$

The inverse transform relation is

$$g_T(n) = \sum_{m=-(N-1)/2}^{(N-1)/2} \sqrt{X_{\rm rc}\left(\frac{4m}{NT}\right)} e^{j2\pi mn/N}, \qquad n = 0, \pm 1, \ldots, \pm\frac{N-1}{2} \tag{6.5.9}$$

Since $g_T(n)$ is symmetric, the impulse response of the desired linear phase transmitter filter is obtained by delaying $g_T(n)$ by $(N-1)/2$ samples.

The MATLAB scripts for this computation are given next.

M-FILE

```
% MATLAB script for Illustrative Problem 6.7.
echo on
N=31;
T=1;
alpha=1/4;
n=-(N-1)/2:(N-1)/2;              % The indices for g_T
```

```
% The expression for g_T is obtained next
for i=1:length(n),
    g_T(i)=0;
    for m=-(N-1)/2:(N-1)/2,
        g_T(i)=g_T(i)+sqrt(xrc(4*m/(N*T),alpha,T))*exp(j*2*pi*m*n(i)/N);
        echo off ;
    end;
end;
echo on ;
g_T=real(g_T) ; % The imaginary part is due to the finite machine precision
% derive g_T(n-(N-1)/2)
n2=0:N-1;
% get the frequency response characteristics
[G_T,W]=freqz(g_T,1);
% normalized magnitude response
magG_T_in_dB=20*log10(abs(G_T)/max(abs(G_T)));
% Impulse response of the cascade of the transmitter and the receiver filters.
g_R=g_T;
imp_resp_of_cascade=conv(g_R,g_T);
% plotting commands follow
```

M-FILE

```
function [y] = xrc(f,alpha,T);
% [y]=xrc(f,alpha,T)
%                 evaluates the expression Xrc(f). The parameters alpha and T
%                 must also be given as inputs to the function.
if (abs(f) > ((1+alpha)/(2*T))),
    y=0;
elseif (abs(f) > ((1-alpha)/(2*T))),
    y=(T/2)*(1+cos((pi*T/alpha)*(abs(f)-(1-alpha)/(2*T))));
else
    y=T;
end;
```

Figure 6.21(a) illustrates $g_T(n - \frac{N-1}{2})$, where $n = 0, 1, \ldots, N - 1$, for $\alpha = \frac{1}{4}$ and $N = 31$. The corresponding frequency response characteristics are shown in Figure 6.21(b). Note that the frequency response is no longer zero for $|f| \geq (1 + \alpha)/T$ because the digital filter has finite duration. However, the sidelobes in the spectrum are relatively small. Further reduction in the sidelobes may be achieved by increasing N.

Finally, in Figure 6.22, we show the impulse response of the cascade of the transmitter and receiver FIR filters. This may be compared with the ideal impulse response obtained by sampling $x_{rc}(t)$ at a rate $F_s = 4/T$.

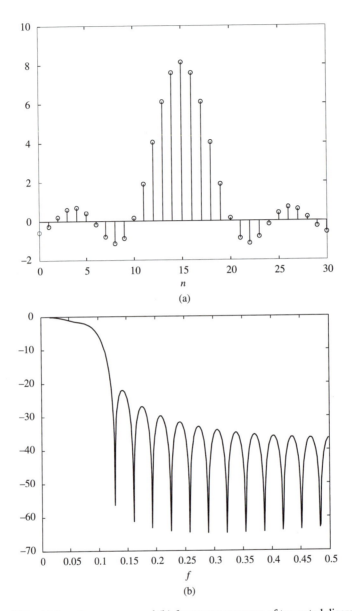

Figure 6.21   (a) Impulse response and (b) frequency response of truncated discrete-time
FIR filter at transmitter

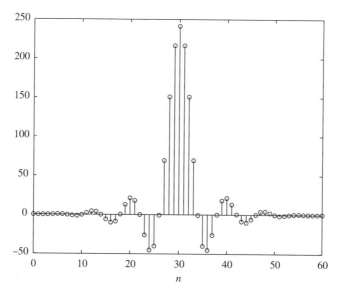

Figure 6.22   Impulse response of the cascade of the transmitter filter with the matched filter at the receiver

6.5.2   Signal Design for Controlled ISI

As we have observed from our discussion of signal design for zero ISI, a transmit filter with excess bandwidth may be employed to realize practical transmitting and receiving filters for bandlimited channels. On the other hand, suppose we choose to relax the condition of zero ISI and thus achieve a symbol transmission in a bandwidth $W = 1/2T$—that is, with no excess bandwidth. By allowing for a controlled amount of ISI, we can achieve the rate of $2W$ symbols/second.

We have already seen that the condition of zero ISI is $x(nT) = 0$ for $n \neq 0$. However, suppose that we design the bandlimited signal to have controlled ISI at one time instant. This means that we allow one additional nonzero value in the samples $\{x(nT)\}$. The ISI that we introduce is deterministic, or "controlled"; hence, it can be taken into account at the receiver, as discussed later.

In general, a signal $x(t)$ that is bandlimited to W hertz—that is

$$X(f) = 0, \qquad |f| > W \tag{6.5.10}$$

can be represented as

$$x(t) = \sum_{n=-\infty}^{\infty} x\left(\frac{n}{2W}\right) \frac{\sin 2\pi W(t - n/2W)}{2\pi W(t - n/2W)} \tag{6.5.11}$$

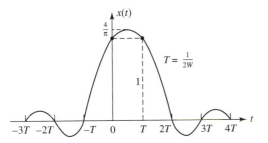
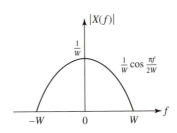

Figure 6.23   Duobinary signal pulse and its spectrum

This representation follows from the sampling theorem for bandlimited signals. The spectrum of the bandlimited signal is

$$X(f) = \int_{-\infty}^{\infty} x(t)e^{-j2\pi ft}\,dt$$

$$= \begin{cases} \dfrac{1}{2W} \displaystyle\sum_{n=-\infty}^{\infty} x\left(\dfrac{n}{2W}\right) e^{-jn\pi f/W}, & |f| \leq W \\ 0, & |f| > W \end{cases} \qquad (6.5.12)$$

One special case that leads to physically realizable transmitting and receiving filters is specified by the samples

$$x\left(\dfrac{n}{2W}\right) \equiv x(nT) = \begin{cases} 1, & n = 0, 1 \\ 0, & \text{otherwise} \end{cases} \qquad (6.5.13)$$

The corresponding signal spectrum is

$$X(f) = \begin{cases} \dfrac{1}{2W}\left[1 + e^{-j\pi f/W}\right], & |f| < W \\ 0, & \text{otherwise} \end{cases}$$

$$= \begin{cases} \dfrac{1}{W} e^{-j2\pi f/W} \cos\left(\dfrac{\pi f}{2W}\right), & |f| < W \\ 0, & \text{otherwise} \end{cases} \qquad (6.5.14)$$

Therefore, $x(t)$ is given by

$$x(t) = \text{sinc}(2Wt) + \text{sinc}(2Wt - 1) \qquad (6.5.15)$$

where $\text{sinc}(t) = \sin \pi t / \pi t$. This pulse is called a *duobinary signal pulse*. It is illustrated along with its magnitude spectrum in Figure 6.23. We note that the spectrum decays to zero smoothly, which means that physically realizable filters can be designed that approximate this spectrum very closely. Thus, a symbol rate of $2W$ is achieved.

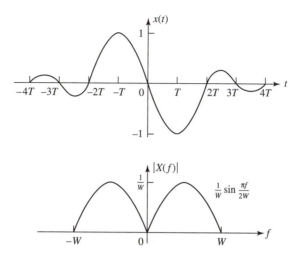

Figure 6.24   Modified duobinary signal pulse and its spectrum

Another special case that leads to physically realizable transmitting and receiving filters is specified by the samples

$$x\left(\frac{n}{2W}\right) = x(nT) = \begin{cases} 1, & n = 1 \\ -1, & n = -1 \\ 0, & \text{otherwise} \end{cases} \tag{6.5.16}$$

The corresponding pulse $x(t)$ is given as

$$x(t) = \text{sinc}(2Wt + 1) - \text{sinc}(2Wt - 1) \tag{6.5.17}$$

and its spectrum is

$$X(f) = \begin{cases} \dfrac{1}{2W}(e^{j\pi f/W} - e^{-j\pi f/W}) = \dfrac{j}{W}\sin\dfrac{\pi f}{W}, & |f| \leq W \\ 0, & |f| > W \end{cases} \tag{6.5.18}$$

This pulse and its magnitude spectrum are illustrated in Figure 6.24. It is called a *modified duobinary signal pulse*. It is interesting to note that the spectrum of this signal has a zero at $f = 0$, making it suitable for transmission over a channel that does not pass dc.

One can obtain other interesting and physically realizable filter characteristics by se-lecting different values for the samples $\{x(n/2W)\}$ and more than two nonzero samples. However, as we select more nonzero samples, the problem of unraveling the controlled ISI becomes more cumbersome and impractical.

The signals obtained when controlled ISI is purposely introduced by selecting two or more nonzero samples from the set $\{x(n/2W)\}$ are called *partial response signals*. The resulting signal pulses allow us to transmit information symbols at the Nyquist rate of $2W$

symbols/second. Thus, greater bandwidth efficiency is obtained compared to raised-cosine signal pulses.

___(ILLUSTRATIVE PROBLEM)_____

Illustrative Problem 6.8 [Duobinary signaling] Design a digital implementation of the transmitter and receiver filters $G_T(f)$ and $G_R(f)$ such that their product is equal to the spectrum of a duobinary pulse and $G_R(f)$ is the matched filter to $G_T(f)$.

___(SOLUTION)_____

To satisfy the frequency-domain specification, we have

$$|G_T(f)| \, |G_R(f)| = \begin{cases} \dfrac{1}{W} \cos\left(\dfrac{\pi f}{2W}\right), & |f| \le W \\ 0, & |f| > W \end{cases} \tag{6.5.19}$$

and, hence,

$$|G_T(f)| = \begin{cases} \sqrt{\dfrac{1}{W} \cos\left(\dfrac{\pi f}{2W}\right)}, & |f| \le W \\ 0, & |f| > W \end{cases} \tag{6.5.20}$$

Now, we follow the same approach as in Illustrative Problem 6.7 to obtain the impulse responses for an FIR implementation of the transmitter and receiver filters. Hence, with $W = 1/2T$ and $F_s = 4/T$, we have

$$g_T(n) = \sum_{m=-(N-1)/2}^{(N-1)/2} \left| G_T\left(\frac{4m}{NT}\right) \right| e^{j2\pi mn/N}, \qquad n = 0, \pm 1, \pm 2, \ldots, \pm \frac{N-1}{2} \tag{6.5.21}$$

and $g_R(n) = g_T(n)$.

The MATLAB script for this computation is given next.

___(M-FILE)_____

```
% MATLAB script for Illustrative Problem 6.8.
echo on
N=31;
T=1;
W=1/(2*T);
n=-(N-1)/2:(N-1)/2;              % The indices for g_T
```

```
% The expression for g_T is obtained next
for i=1:length(n),
  g_T(i)=0;
  for m=-(N-1)/2:(N-1)/2,
    if ( abs((4*m)/(N*T)) <= W ),
      g_T(i)=g_T(i)+sqrt((1/W)*cos((2*pi*m)/(N*T*W)))*exp(j*2*pi*m*n(i)/N);
    end;
    echo off ;
  end;
end;
echo on ;
g_T=real(g_T) ; % The imaginary part is due to the finite machine precision
% obtain g_T(n-(N-1)/2)
n2=0:N-1;
% obtain the frequency response characteristics
[G_T,W]=freqz(g_T,1);
% normalized magnitude response
magG_T_in_dB=20*log10(abs(G_T)/max(abs(G_T)));
% Impulse response of the cascade of the transmitter and the receiver filters.
g_R=g_T;
imp_resp_of_cascade=conv(g_R,g_T);
% plotting commands follow
```

Figure 6.25(a) illustrates $g_T\left(n - \frac{n-1}{2}\right)$, $n = 0, 1, \ldots, N - 1$, for $N = 31$. The corresponding frequency response characteristic is shown in Figure 6.25(b). Note that the frequency response characteristic is no longer zero for $|f| > W$ because the digital filter has finite duration. However, the sidelobes in the spectrum are relatively small. Finally, in Figure 6.26, we show the impulse response of the cascade of the transmitter and receiver FIR filters. This impulse response may be compared with the ideal impulse response obtained by sampling $x(t)$ given by (6.5.17) at a rate $F_s = 4/T = 8W$.

6.5.3   Precoding for Detection of Partial Response Signals

For the duobinary signal pulse, $x(nT) = 1$ for $n = 0, 1$ and $x(nT) = 0$ otherwise. Hence, the samples of the output of the receiver filter $G_R(f)$ are expressed as

$$y_k = a_k + a_{k-1} + v_k \tag{6.5.22}$$
$$= b_k + v_k$$

where $\{a_k\}$ is the transmitted sequence of amplitudes, $\{v_k\}$ is a sequence of additive Gaussian noise samples, and $b_k = a_k + a_{k-1}$. Let us ignore the noise for the moment and consider the binary case where $a_k = \pm 1$ with equal probability. Then b_k takes one of three possible values—namely, $b_k = -2, 0, 2$ with corresponding probabilities $\frac{1}{4}$, $\frac{1}{2}$, and $\frac{1}{4}$. If a_{k-1} is the detected signal from the $(k-1)$st signaling interval, its effect on b_k, the received signal in the kth signaling interval, can be eliminated by subtraction, thus allowing a_k to be detected. The process can be repeated sequentially for every received symbol.

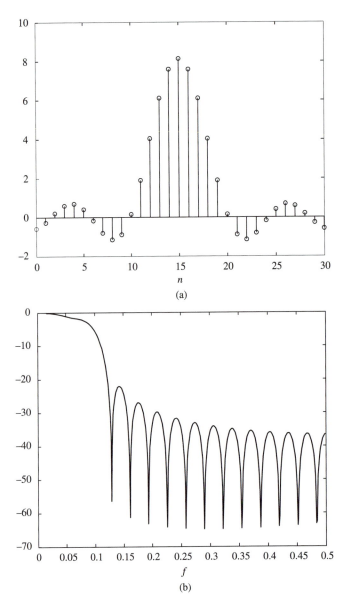

Figure 6.25   (a) Impulse response and (b) frequency response of truncated discrete-time
duobinary FIR filter at the transmitter

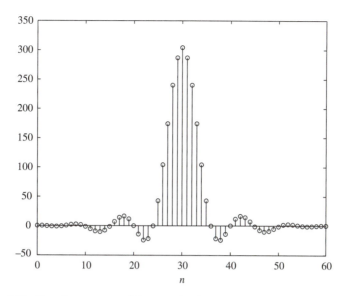

Figure 6.26   Impulse response of the cascade of the transmitter filter with the matched filter at the receiver

The major problem with this procedure is that errors arising from the additive noise tend to propagate. For example, if a_{k-1} is detected in error, its effect on a_k is not eliminated; in effect, it is reinforced by the incorrect subtraction. Consequently, the detection of a_k is also likely to be in error.

Error propagation can be prevented by *precoding* the data at the transmitter instead of eliminating the controlled ISI by subtraction at the receiver. The precoding is performed on the binary data sequence prior to modulation. From the data sequence $\{D_k\}$ of 1's and 0's that is to be transmitted, a new sequence $\{p_k\}$, called the *precoded sequence*, is generated. For the duobinary signal, the precoded sequence is defined as

$$p_k = D_k \ominus p_{k-1}, \qquad k = 1, 2, \ldots \tag{6.5.23}$$

where $\ominus$ denotes modulo-2 subtraction.[1] Then the transmitted signal amplitude is $a_k = -1$ if $p_k = 0$ and $a_k = 1$ if $p_k = 1$; that is,

$$a_k = 2p_k - 1 \tag{6.5.24}$$

The noise-free samples at the output of the receiving filter are given by

$$
\begin{aligned}
b_k &= a_k + a_{k-1} \\
&= (2p_k - 1) + (2p_{k-1} - 1) \\
&= 2(p_k + p_{k-1} - 1) \tag{6.5.25}
\end{aligned}
$$

---

[1]Although this operation is identical to modulo-2 addition, it is convenient to view the precoding operation for duobinary in terms of modulo-2 subtraction.

Table 6.1 Binary signaling with duobinary pulses

Data sequence D_k	—	1	1	1	0	1	0	0	1	0	0	0	1
Precoded sequence p_k	0	1	0	1	1	0	0	0	1	1	1	1	0
Transmitted sequence a_k	−1	1	−1	1	1	−1	−1	−1	1	1	1	1	−1
Received sequence b_k	—	0	0	0	2	0	−2	−2	0	2	2	2	0
Decoded sequence D_k	—	1	1	1	0	1	0	0	1	0	0	0	1

Consequently,

$$p_k + p_{k-1} = \frac{1}{2}b_k + 1 \tag{6.5.26}$$

Since $D_k = p_k \oplus p_{k-1}$, it follows that the data sequence $\{D_k\}$ is obtained from $\{b_k\}$ using the relation

$$D_k = \frac{1}{2}b_k + 1 \quad (\text{mod } 2) \tag{6.5.27}$$

Therefore, if $b_k = \pm 2$, then $D_k = 0$, and if $b_k = 0$, then $D_k = 1$. An example that illustrates the precoding and decoding operations is given in Table 6.1.

In the presence of additive noise, the sampled outputs from the receiving filter are given by (6.5.22). In this case, $y_k = b_k + v_k$ is compared with the two thresholds set at $+1$ and -1. The data sequence $\{D_k\}$ is obtained according to the detection rule

$$D_k = \begin{cases} 1, & |y_k| < 1 \\ 0, & |y_k| \geq 1 \end{cases} \tag{6.5.28}$$

Thus, precoding the data allows us to perform symbol-by-symbol detection at the receiver without the need for subtraction of previously detected symbols.

The extension from binary PAM to multilevel PAM using duobinary pulses is straightforward. The M-level transmitted sequence $\{a_k\}$ results in a (noise-free) received sequence

$$b_k = a_k + a_{k-1}, \qquad k = 1, 2, 3, \dots \tag{6.5.29}$$

which has $2M - 1$ possible equally spaced amplitude levels. The amplitude levels for the sequence $\{a_k\}$ are determined from the relation

$$a_k = 2p_k - (M - 1) \tag{6.5.30}$$

where $\{p_k\}$ is the precoded sequence that is obtained from an M-level data sequence $\{D_k\}$ according to the relation

$$p_k = D_k \ominus p_{k-1} \quad (\text{mod } M) \tag{6.5.31}$$

where the possible values of the data sequence $\{D_k\}$ are $0, 1, 2, \dots, M$.

In the absence of noise, the samples at the output of the receiving filter may be expressed as

$$b_k = a_k + a_{k-1}$$
$$= [2p_k - (M-1)] + [2p_{k-1} - (M-1)]$$
$$= 2[p_k + p_{k-1} - (M-1)] \tag{6.5.32}$$

Hence,

$$p_k + p_{k-1} = \frac{1}{2}b_k + (M-1) \tag{6.5.33}$$

Since $D_k = p_k + p_{k-1} \pmod{M}$, it follows that the transmitted data $\{D_k\}$ are recovered from the received sequence $\{b_k\}$ by means of the relation

$$D_k = \frac{1}{2}b_k + (M-1) \pmod{M} \tag{6.5.34}$$

In the case of the modified duobinary pulse, the received signal samples at the output of the receiving filter $G_R(f)$ are expressed as

$$y_k = a_k - a_{k-2} + v_k$$
$$= b_k + v_k \tag{6.5.35}$$

The precoder for the modified duobinary pulse produces the sequence $\{p_k\}$ from the data sequence $\{D_k\}$ according to the relation

$$p_k = D_k \oplus p_{k-2} \pmod{M} \tag{6.5.36}$$

From these relations, it is easy to show that the detection rule for the recovering of the data sequence $\{D_k\}$ from $\{b_k\}$ in the absence of noise is

$$D_k = \frac{1}{2}b_k \pmod{M} \tag{6.5.37}$$

───(ILLUSTRATIVE PROBLEM)───

Illustrative Problem 6.9 [Duobinary precoding] Write a MATLAB program that takes a data sequence $\{D_k\}$, precodes it for a duobinary pulse transmission system to produce $\{p_k\}$, and maps the precoded sequence into the transmitted amplitude levels $\{a_k\}$. Then, from the transmitted sequence $\{a_k\}$, form the received noise-free sequence $\{b_k\}$ and, using the relation in (6.5.34), recover the data sequence $\{D_k\}$.

───── **SOLUTION** ───

The MATLAB script is given next. By using this program, we can verify the results in Table 6.1 for the case where $M = 2$.

───── **M-FILE** ───

```
% MATLAB script for Illustrative Problem 6.9.
echo on
d=[1 1 1 0 1 0 0 1 0 0 0 1];
p(1)=0;
for  i=1:length(d)
  p(i+1)=rem(p(i)+d(i),2);
  echo off ;
end
echo on ;
a=2.*p−1;
b(1)=0;
dd(1)=0;
for  i=1:length(d)
  b(i+1)=a(i+1)+a(i);
  d_out(i+1)=rem(b(i+1)/2+1,2);
  echo off ;
end
echo on ;
d_out=d_out(2:length(d)+1);
```

6.6   Linear Equalizers

The most common type of channel equalizer used in practice to reduce ISI is a linear FIR filter with adjustable coefficients $\{c_i\}$, as shown in Figure 6.27.

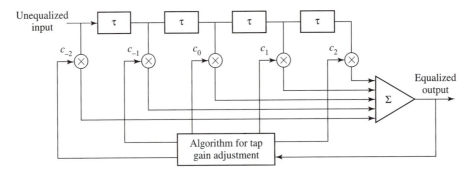

Figure 6.27   Linear transversal filter

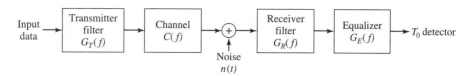

Figure 6.28   Block diagram of a system with an equalizer

On channels whose frequency response characteristics are unknown but time-invariant, we may measure the channel characteristics and adjust the parameters of the equalizer; once adjusted, the parameters remain fixed during the transmission of data. Such equalizers are called *preset equalizers.* On the other hand, *adaptive equalizers* update their parameters on a periodic basis during the transmission of data, so they are capable of tracking a slowly time-varying channel response.

First, let us consider the design characteristics for a linear equalizer from a frequency domain viewpoint. Figure 6.28 shows a block diagram of a system that employs a linear filter as a channel equalizer.

The demodulator consists of a receiver filter with frequency response $G_R(f)$ in cascade with a channel equalizing filter that has a frequency response $G_E(f)$. As indicated in the previous section, the receiver filter response $G_R(f)$ is matched to the transmitter response—that is, $G_R(f) = G_T^*(f)$—and the product $G_R(f)G_T(f)$ is usually designed so that there is either zero ISI at the sampling instants, as, for example, when $G_R(t)G_T(f) = X_{rc}(f)$, or controlled ISI for partial response signals.

For the system shown in Figure 6.28, in which the channel frequency response is not ideal, the desired condition for zero ISI is

$$G_T(f)C(f)G_R(f)G_E(f) = X_{rc}(f) \tag{6.6.1}$$

where $X_{rc}(f)$ is the desired raised-cosine spectral characteristic. Since $G_T(f)G_R(f) = X_{rc}(f)$ by design, the frequency response of the equalizer that compensates for the channel distortion is

$$G_E(f) = \frac{1}{C(f)} = \frac{1}{|C(f)|}e^{-j\theta_c(f)} \tag{6.6.2}$$

Thus, the amplitude response of the equalizer is $|G_E(f)| = 1/|C(f)|$, and its phase response is $\theta_E(f) = -\theta_C(f)$. In this case, the equalizer is said to be the *inverse channel filter* to the channel response.

We note that the inverse channel filter completely eliminates the ISI caused by the channel. Since it forces the ISI to be zero at the sampling instants $t = kT$ for $k = 0, 1, \ldots,$ the equalizer is called a *zero-forcing equalizer.* Hence, the input to the detector is simply

$$z_k = a_k + \eta_k, \qquad k = 0, 1, \ldots \tag{6.6.3}$$

where η_k represents the additive noise and a_k is the desired symbol.

In practice, the ISI caused by channel distortion is usually limited to a finite number of symbols on either side of the desired symbol. Hence, the number of terms that constitute the ISI in the summation given by (6.4.7) is finite. As a consequence, in practice the channel equalizer is implemented as a finite-duration impulse response (FIR) filter, or transversal filter, with adjustable tap coefficients $\{c_n\}$, as illustrated in Figure 6.27. The time delay τ between adjacent taps may be selected as large as T, the symbol interval, in which case the FIR equalizer is called a *symbol-spaced equalizer*. In this case, the input to the equalizer is the sampled sequence given by (6.4.6). However, we note that when the symbol rate $1/T < 2W$, frequencies in the received signal above the folding frequency $1/T$ are aliased into frequencies below $1/T$. In this case, the equalizer compensates for the aliased channel-distorted signal.

On the other hand, when the time delay τ between adjacent taps is selected such that $1/\tau \geq 2W > 1/T$, no aliasing occurs; hence, the inverse channel equalizer compensates for the true channel distortion. Since $\tau < T$, the channel equalizer is said to have *frac-tionally spaced taps*, and it is called a *fractionally spaced equalizer*. In practice, τ is often selected at $\tau = T/2$. Notice that, in this case, the sampling rate at the input to the filter $G_E(f)$ is $2/T$.

The impulse response of the FIR equalizer is

$$g_E(t) = \sum_{n=-K}^{K} c_n \delta(t - n\tau) \tag{6.6.4}$$

and the corresponding frequency response is

$$G_E(f) = \sum_{n=-K}^{K} c_n e^{-j2\pi f n\tau} \tag{6.6.5}$$

where $\{c_n\}$ are the $2K + 1$ equalizer coefficients, and K is chosen sufficiently large so that the equalizer spans the length of the ISI—that is, $2K + 1 \geq L$, where L is the number of signal samples spanned by the ISI. Since $X(f) = G_T(f)C(f)G_R(f)$, and $x(t)$ is the signal pulse corresponding to $X(f)$, the equalized output signal pulse is

$$q(t) = \sum_{n=-K}^{K} c_n x(t - n\tau) \tag{6.6.6}$$

The zero-forcing condition can now be applied to the samples of $q(t)$ taken at times $t = mT$. These samples are

$$q(mT) = \sum_{n=-K}^{K} c_n x(mT - n\tau), \qquad m = 0, \pm 1, \ldots, \pm K \tag{6.6.7}$$

Since there are $2K + 1$ equalizer coefficients, we can control only $2K + 1$ sampled values of $q(t)$. Specifically, we may force the conditions

$$q(mT) = \sum_{n=-K}^{K} c_n x(mT - n\tau)$$

$$= \begin{cases} 1, & m = 0 \\ 0, & m = \pm 1, \pm 2, \ldots, \pm K \end{cases} \tag{6.6.8}$$

which may be expressed in matrix form as $Xc = q$, where X is a $(2K + 1) \times (2K + 1)$ matrix with elements $x(mT - n\tau)$, c is the $(2K+1)$ coefficient vector, and q is the $(2K+1)$ column vector with one nonzero element. Thus, we obtain a set of $2K + 1$ linear equations for the coefficients of the zero-forcing equalizer.

We should emphasize that the FIR zero-forcing equalizer does not completely eliminate the ISI because it has a finite length. However, as K is increased, the residual ISI can be reduced, and in the limit of $K \to \infty$, the ISI is completely eliminated.

─────(ILLUSTRATIVE PROBLEM)─────────────────────────

Illustrative Problem 6.10 [Equalizer design] Consider a channel-distorted pulse, $x(t)$, at the input to the equalizer, given by the expression

$$x(t) = \frac{1}{1 + (2t/T)^2}$$

where $1/T$ is the symbol rate. The pulse is sampled at the rate $2/T$ and is equalized by a zero-forcing equalizer. Determine the coefficients of a five-tap zero-forcing equalizer.

─────(SOLUTION)─────────────────────────────────────

According to (6.6.8), the zero-forcing equalizer must satisfy the equation

$$q(mT) = \sum_{n=-2}^{2} c_n x\left(mT - \frac{nT}{2}\right) = \begin{cases} 1, & m = 0 \\ 0, & m = \pm 1, \pm 2 \end{cases}$$

The matrix X with elements $x(mT - nT/2)$ is given as

$$X = \begin{bmatrix} \frac{1}{5} & \frac{1}{10} & \frac{1}{17} & \frac{1}{26} & \frac{1}{37} \\ 1 & \frac{1}{2} & \frac{1}{5} & \frac{1}{10} & \frac{1}{17} \\ \frac{1}{5} & \frac{1}{2} & 1 & \frac{1}{2} & \frac{1}{5} \\ \frac{1}{17} & \frac{1}{10} & \frac{1}{5} & \frac{1}{2} & 1 \\ \frac{1}{37} & \frac{1}{26} & \frac{1}{17} & \frac{1}{10} & \frac{1}{5} \end{bmatrix} \tag{6.6.9}$$

The coefficient vector c and the vector q are given as

$$c = \begin{bmatrix} c_{-2} \\ c_{-1} \\ c_0 \\ c_1 \\ c_2 \end{bmatrix} \qquad q = \begin{bmatrix} 0 \\ 0 \\ 1 \\ 0 \\ 0 \end{bmatrix} \tag{6.6.10}$$

Then the linear equations $Xc = q$ can be solved by inverting the matrix X. Thus, we obtain

$$c_{\text{opt}} = X^{-1}q = \begin{bmatrix} -2.2 \\ 4.9 \\ -3 \\ 4.9 \\ -2.2 \end{bmatrix} \tag{6.6.11}$$

Figure 6.29 illustrates the original pulse $x(t)$ and the equalized pulse. Note the small amount of residual ISI in the equalized pulse.

The MATLAB script for this computation is given next.

M-FILE

```
% MATLAB script for Illustrative Problem 6.10.
echo on
T=1;
Fs=2/T;
Ts=1/Fs;
c_opt=[−2.2 4.9 −3 4.9 −2.2];
t=−5*T:T/2:5*T;
x=1./(1+((2/T)*t).^2);                    % sampled pulse
equalized_x=filter(c_opt,1,[x 0 0]);      % since there will be a delay of two samples at the output
% to take care of the delay
equalized_x=equalized_x(3:length(equalized_x));
% Now, let us downsample the equalizer output
for i=1:2:length(equalized_x),
   downsampled_equalizer_output((i+1)/2)=equalized_x(i);
   echo off;
end;
echo on ;
% plotting commands follow
```

One drawback to the zero-forcing equalizer is that it ignores the presence of additive noise. As a consequence, its use may result in significant noise enhancement. This is easily seen by noting that in a frequency range where $C(f)$ is small, the channel equalizer $G_E(f) = 1/C(f)$ compensates by placing a large gain in that frequency range. Consequently, the noise in that frequency range is greatly enhanced. An alternative is to relax the zero ISI condition and select the channel equalizer characteristic such that the combined power in the residual ISI and the additive noise at the output of the equalizer is minimized. A channel equalizer that is optimized based on the minimum mean-square error (MMSE) criterion accomplishes the desired goal.

To elaborate, let us consider the noise-corrupted output of the FIR equalizer, which is

$$z(t) = \sum_{n=-K}^{K} c_n y(t - n\tau) \tag{6.6.12}$$

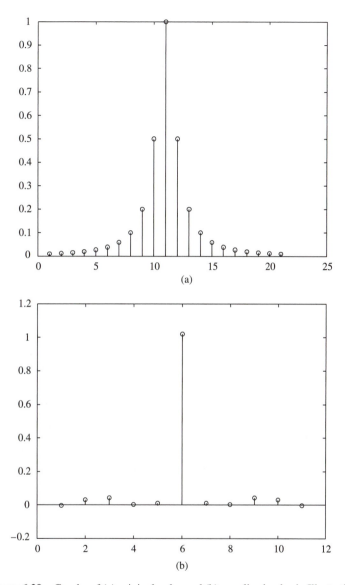

Figure 6.29   Graphs of (a) original pulse and (b) equalized pulse in Illustrative Problem 6.10

where $y(t)$ is the input to the equalizer, given by (6.4.3). The equalizer output is sampled at times $t = mT$. Thus, we obtain

$$z(mT) = \sum_{n=-K}^{K} c_n y(mT - n\tau) \qquad (6.6.13)$$

The desired response at the output of the equalizer at $t = mT$ is the transmitted symbol a_m. The error is defined as the difference between a_m and $z(mT)$. Then the mean-square error (MSE) between the actual output sample $z(mT)$ and the desired values a_m is as follows:[2]

$$
\begin{aligned}
\text{MSE} &= E\,|z(mT) - a_m|^2 \\
&= E\left[\left|\sum_{n=-K}^{K} c_n y(mT - n\tau) - a_m\right|^2\right] \\
&= \sum_{n=-K}^{K}\sum_{k=-K}^{K} c_n c_k R_y(n - k) - 2\sum_{k=-K}^{K} c_k R_{ay}(k) + E(|a_m|^2) \quad (6.6.14)
\end{aligned}
$$

where the correlations are defined as

$$
\begin{aligned}
R_y(n - k) &= E[y^*(mT - n\tau)y(mT - k\tau)] \\
R_{ay}(k) &= E[y(mT - k\tau)a_m^*] \quad\quad (6.6.15)
\end{aligned}
$$

and the expectation is taken with respect to the random information sequence $\{a_m\}$ and the additive noise.

The minimum MSE solution is obtained by differentiating (6.6.14) with respect to the equalizer coefficients $\{c_n\}$. Thus, we obtain the necessary conditions for the minimum MSE as

$$
\sum_{n=-K}^{K} c_n R_y(n - k) = R_{ay}(k), \qquad k = 0, \pm 1, \pm 2, \dots, \pm K \quad\quad (6.6.16)
$$

These are the $2K + 1$ linear equations for the equalizer coefficients. In contrast to the zero-forcing solution described previously, these equations depend on the statistical properties (the autocorrelation) of the noise as well as the ISI through the autocorrelation $R_y(n)$.

In practice, the autocorrelation matrix $R_y(n)$ and the cross-correlation vector $R_{ay}(n)$ are unknown a priori. However, these correlation sequences can be estimated by transmitting a test signal over the channel and using the time-average estimates

$$
\hat{R}_y(n) = \frac{1}{K}\sum_{k=1}^{K} y^*(kT - n\tau)y(kT)
$$

$$
\hat{R}_{ay}(n) = \frac{1}{K}\sum_{k=1}^{K} y(kT - n\tau)a_k^* \quad\quad (6.6.17)
$$

in place of the ensemble averages to solve for the equalizer coefficients given by (6.6.16).

---

[2]In this development, we allow the signals $z(t)$ and $y(t)$ to be complex-valued and the data sequence to be complex-valued.

─────(ILLUSTRATIVE PROBLEM)─────────────────────

Illustrative Problem 6.11 [Equalizer design] Consider the same channel-distorted pulse $x(t)$ as in Illustrative Problem 6.10, but now design the five-tap equalizer based on the minimum MSE criterion. The information symbols have zero mean and unit variance and are uncorrelated—that is,

$$E(a_n) = 0$$
$$E(a_n a_m) = 0, \qquad n \neq m$$
$$E(|a_n|^2) = 1$$

The additive noise $v(t)$ has zero mean and autocorrelation

$$\varphi_{vv}(\tau) = \frac{N_0}{2}\delta(\tau)$$

─────(SOLUTION)────────────────────────────────

The equalizer tap coefficients are obtained by solving (6.6.16) with $K = 2$ and $\tau = T/2$. The matrix with elements $R_y(n - k)$ is simply

$$R_y = X^t X + \frac{N_0}{2} I$$

where X is given by (6.6.9) and I is the identity matrix. The vector with elements $R_{ay}(k)$ is given as

$$R_{ay} = \begin{bmatrix} \frac{1}{5} \\ \frac{1}{2} \\ 1 \\ \frac{1}{2} \\ \frac{1}{5} \end{bmatrix}$$

The equalizer coefficients obtained by solving (6.6.16) are as follows:

$$c_{\text{opt}} = \begin{bmatrix} 0.0956 \\ -0.7347 \\ 1.6761 \\ -0.7347 \\ 0.0956 \end{bmatrix}$$

A plot of the equalized pulse is shown in Figure 6.30.

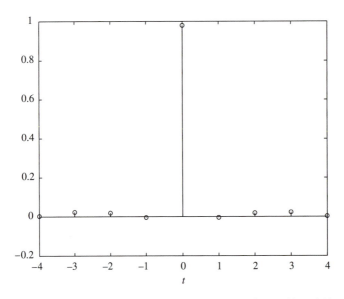

Figure 6.30   Plot of the equalized pulse in Illustrative Problem 6.11

The MATLAB script for this computation is given next.

M-FILE

```
% MATLAB script for Illustrative Problem 6.11.
echo on
T=1;
for n=−2:2,
  for k=−2:2,
    temp=0;
    for i=−2:2, temp=temp+(1/(1+(n−i)^2))*(1/(1+(k−i)^2)); end;
    X(k+3,n+3)=temp;
    echo off ;
  end;
end;
echo on;
N0=0.01;                              % assuming that N0=0.01
Ry=X+(N0/2)*eye(5);
Riy=[1/5 1/2 1 1/2 1/5].' ;
c_opt=inv(Ry)*Riy;                    % optimal tap coefficients
% find the equalized pulse...
t=−3:1/2:3;
x=1./(1+(2*t/T).^2);                  % sampled pulse
equalized_pulse=conv(x,c_opt);
% decimate the pulse to get the samples at the symbol rate
decimated_equalized_pulse=equalized_pulse(1:2:length(equalized_pulse));
% plotting command follows
```

6.6.1 Adaptive Linear Equalizers

We have shown that the tap coefficients of a linear equalizer can be determined by solving a set of linear equations. In the zero-forcing optimization criterion, the linear equations are given by (6.6.8). On the other hand, if the optimization criterion is based on minimizing the MSE, the optimum equalizer coefficients are determined by solving the set of linear equations given by (6.6.16).

In both cases, we may express the set of linear equations in the general matrix form

$$Bc = d \qquad (6.6.18)$$

where B is a $(2K + 1) \times (2K + 1)$ matrix, c is a column vector representing the $2K + 1$ equalizer coefficients, and d is a $(2K + 1)$-dimensional column vector. The solution of (6.6.18) yields

$$c_{\text{opt}} = B^{-1} d \qquad (6.6.19)$$

In practical implementations of equalizers, the solution of (6.6.18) for the optimum coefficient vector is usually obtained by an iterative procedure that avoids the explicit computation of the inverse of the matrix B. The simplest iterative procedure is the method of steepest descent, in which one begins by choosing arbitrarily the coefficient vector c, say, c_0. This initial choice of coefficient vector c_0 corresponds to a point on the criterion function that is being optimized. For example, in the case of the MSE criterion, the initial guess c_0 corresponds to a point on the quadratic MSE surface in the $(2K + 1)$-dimensional space of coefficients. The gradient vector, defined as g_0, which is the derivative of the MSE with respect to the $2K + 1$ filter coefficients, is then computed at this point on the criterion surface, and each tap coefficient is changed in the direction opposite to its corresponding gradient component. The change in the jth tap coefficient is proportional to the size of the jth gradient component.

For example, the gradient vector, denoted as g_k, for the MSE criterion, found by taking the derivatives of the MSE with respect to each of the $2K + 1$ coefficients, is

$$g_k = Bc_k - d, \qquad k = 0, 1, 2, \ldots \qquad (6.6.20)$$

Then the coefficient vector c_k is updated according to the relation

$$c_{k+1} = c_k - \Delta g_k \qquad (6.6.21)$$

where Δ is the *step-size parameter* for the iterative procedure. To ensure convergence of the iterative procedure, Δ is chosen to be a small positive number. In such a case, the gradient vector g_k converges toward zero—that is, $g_k \to 0$ as $k \to \infty$—and the coefficient vector $c_k \to c_{\text{opt}}$, as illustrated in Figure 6.31 based on two-dimensional optimization. In general, convergence of the equalizer tap coefficients to c_{opt} cannot be attained in a finite number of iterations with the steepest-descent method. However, the optimum solution c_{opt} can be approached as closely as desired in a few hundred iterations. In digital communication systems that employ channel equalizers, each iteration corresponds to a time interval for sending one symbol; hence, a few hundred iterations to achieve convergence to c_{opt} correspond to a fraction of a second.

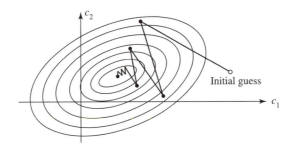

Figure 6.31   Example of the convergence characteristics of a gradient algorithm

Adaptive channel equalization is required for channels whose characteristics change with time. In such a case, the ISI varies with time. The channel equalizer must track such time variations in the channel response and adapt its coefficients to reduce the ISI. In the context of the above discussion, the optimum coefficient vector c_{opt} varies with time due to time variations in the matrix $\boldsymbol{B}$ and, for the case of the MSE criterion, time variations in the vector $\boldsymbol{d}$. Under these conditions, the iterative method described above can be modified to use estimates of the gradient components. Thus, the algorithm for adjusting the equalizer tap coefficients may be expressed as

$$\hat{c}_{k+1} = \hat{c}_k - \Delta\hat{g}_k \tag{6.6.22}$$

where $\hat{g}_k$ denotes an estimate of the gradient vector $\boldsymbol{g}_k$, and $\hat{c}_k$ denotes the estimate of the tap coefficient vector.

In the case of the MSE criterion, the gradient vector $\boldsymbol{g}_k$ given by (6.6.20) may also be expressed as

$$\boldsymbol{g}_k = -E(e_k \boldsymbol{y}_k^*)$$

An estimate $\hat{g}_k$ of the gradient vector at the kth iteration is computed as

$$\hat{g}_k = -e_k \boldsymbol{y}_k^* \tag{6.6.23}$$

where e_k denotes the difference between the desired output from the equalizer at the kth time instant and the actual output $z(kT)$, and $\boldsymbol{y}_k$ denotes the column vector of $2K + 1$ received signal values contained in the equalizer at time instant k. The *error signal* e_k is expressed as

$$e_k = a_k - z_k \tag{6.6.24}$$

where $z_k = z(kT)$ is the equalizer output given by (6.6.13) and a_k is the desired symbol. Hence, by substituting (6.6.23) into (6.6.22), we obtain the adaptive algorithm for optimizing the taps coefficients (based on the MSE criterion) as

$$\hat{c}_{k+1} = \hat{c}_k + \Delta e_k \boldsymbol{y}_k^* \tag{6.6.25}$$

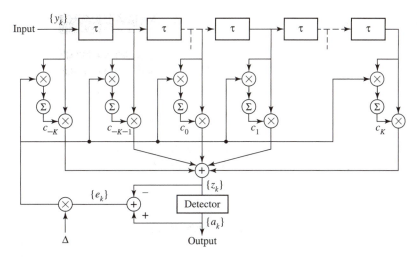

Figure 6.32   Linear adaptive equalizer based on the MSE criterion

Since an estimate of the gradient vector is used in (6.6.25), the algorithm is called a *stochastic gradient algorithm*. It is also known as the *LMS algorithm*.

A block diagram of an adaptive equalizer that adapts its tap coefficients according to (6.6.25) is illustrated in Figure 6.32. Note that the difference between the desired output a_k and the actual output z_k from the equalizer is used to form the error signal e_k. This error is scaled by the step-size parameter Δ, and the scaled error signal Δe_k multiples the received signal values $\{y(kT - n\tau)\}$ at the $2K + 1$ taps. The products $\Delta e_k y^*(kT - n\tau)$ at the $2K + 1$ taps are then added to the previous values of the tap coefficients to obtain the updated tap coefficients, according to (6.6.25). This computation is repeated as each new signal sample is received. Thus, the equalizer coefficients are updated at the symbol rate.

Initially, the adaptive equalizer is trained by the transmission of a known pseudorandom sequence $\{a_m\}$ over the channel. At the demodulator, the equalizer employs the known sequence to adjust its coefficients. Upon initial adjustment, the adaptive equalizer switches from a *training mode* to a *decision-directed mode*, in which case the decisions at the output of the detector are sufficiently reliable so that the error signal is formed by computing the difference between the detector output and the equalizer output—that is,

$$e_k = \hat{a}_k - z_k \tag{6.6.26}$$

where $\hat{a}_k$ is the output of the detector. In general, decision errors at the output of the detector occur infrequently; consequently, such errors have little effect on the performance of the tracking algorithm given by (6.6.25).

A rule of thumb for selecting the step-size parameter in order to ensure convergence and good tracking capabilities in slowly varying channels is

$$\Delta = \frac{1}{5(2K + 1)P_R} \tag{6.6.27}$$

where P_R denotes the received signal-plus-noise power, which can be estimated from the received signal.

ILLUSTRATIVE PROBLEM

Illustrative Problem 6.12 [Adaptive equalizer] Implement an adaptive equalizer based on the LMS algorithm given in (6.6.25). The channel number of taps selected for the equalizer is $2K + 1 = 11$. The received signal-plus-noise power P_R is normalized to unity. The channel characteristic is given by the vector x as

$$x = [0.05 \ -0.063 \ 0.088 \ -0.126 \ -0.25 \ 0.9047 \ 0.25 \ 0 \ 0.126 \ 0.038 \ 0.088]$$

SOLUTION

The convergence characteristics of the stochastic gradient algorithm in (6.6.25) are illustrated in Figure 6.33. These graphs were obtained from a computer simulation of the 11-tap adaptive equalizer. The graphs represent the mean-square error averaged over several realizations. As shown, when Δ is decreased, the convergence is slowed somewhat, but a lower MSE is achieved, indicating that the estimated coefficients are closer to c_{opt}.

The MATLAB script for this example is given next.

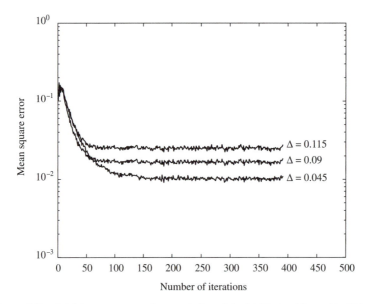

Figure 6.33 Initial convergence characteristics of the LMS algorithm with different step sizes

━━━ **M-FILE** ━━━

```
% MATLAB script for Illustrative Problem 6.12.
echo on
N=500;                                    % length of the information sequence
K=5;
actual_isi=[0.05 −0.063 0.088 −0.126 −0.25 0.9047 0.25 0 0.126 0.038 0.088];
sigma=0.01;
delta=0.115;
Num_of_realizations=1000;
mse_av=zeros(1,N−2*K);
for j=1:Num_of_realizations,              % compute the average over a number of realizations
  % the information sequence
  for i=1:N,
    if (rand<0.5),
      info(i)=−1;
    else
      info(i)=1;
    end;
    echo off ;
  end;
  if (j==1) ; echo on ; end
  % the channel output
  y=filter(actual_isi,1,info);
  for i=1:2:N, [noise(i) noise(i+1)]=gngauss(sigma); end;
  y=y+noise;
  % now the equalization part follows
  estimated_c=[0 0 0 0 0 1 0 0 0 0 0];   % initial estimate of ISI
  for k=1:N−2*K,
    y_k=y(k:k+2*K);
    z_k=estimated_c*y_k.' ;
    e_k=info(k)−z_k;
    estimated_c=estimated_c+delta*e_k*y_k;
    mse(k)=e_k^2;
    echo off ;
  end;
  if (j==1) ; echo on ; end
  mse_av=mse_av+mse;
  echo off ;
end;
echo on ;
mse_av=mse_av/Num_of_realizations;        % mean squared error versus iterations
% plotting commands follow
```

───

Although we have described in some detail the operation of an adaptive equalizer that is optimized on the basis of the MSE criterion, the operation of an adaptive equalizer based on the zero-forcing method is very similar. The major difference lies in the method for estimating the gradient vectors g_k at each iteration. A block diagram of an adaptive zero-forcing equalizer is shown in Figure 6.34.

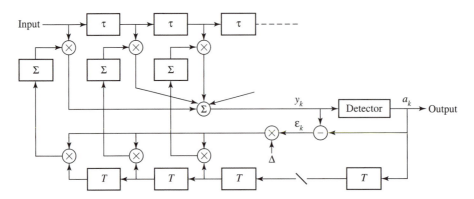

Figure 6.34   An adaptive zero-forcing equalizer

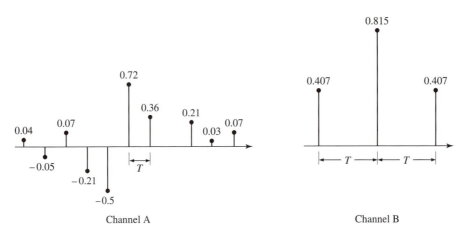

Channel A

Channel B

Figure 6.35   Two channels with ISI

6.7   Nonlinear Equalizers

The linear filter equalizers described earlier are very effective on channels, such as wire line telephone channels, where the ISI is not severe. The severity of the ISI is directly related to the spectral characteristics of the channel and not necessarily to the time span of the ISI. For example, consider the ISI resulting from two channels, illustrated in Figure 6.35. The time span for the ISI in channel A is five symbol intervals on each side of the desired signal component, which has a value of 0.72. On the other hand, the time span for the ISI in channel B is one symbol interval on each side of the desired signal component, which has a value of 0.815. The energy of the total response is normalized to unity for both channels.

In spite of the shorter ISI span, channel B results in more severe ISI. This is evidenced in the frequency response characteristics of these channels, which are shown in Figure 6.36. We observe that channel B has a spectral null [the frequency response $C(f) = 0$ for some

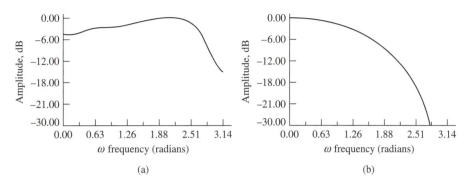

Figure 6.36   Amplitude spectra for (a) channel A and (b) channel B from Figure 6.35

frequencies in the band $|f| \leq W$ at $f = 1/2T$, whereas this does not occur in the case of channel A. Consequently, a linear equalizer will introduce a large gain in its frequency response to compensate for the channel null. Thus, the noise in channel B will be enhanced much more than in channel A. This implies that the performance of the linear equalizer for channel B will be sufficiently poorer than that for channel A. In general, the basic limitation of a linear equalizer is that it performs poorly on channels having spectral nulls. Such channels are often encountered in radio communications, such as ionospheric transmission at frequencies below 30 MHz, and mobile radio channels, such as those used for cellular radio communications.

A *decision-feedback equalizer* (DFE) is a nonlinear equalizer that employs previous decisions to eliminate the ISI caused by previously detected symbols on the current symbol to be detected. A simple block diagram for a DFE is shown in Figure 6.37. The DFE consists of two filters. The first filter is called a *feedforward filter*; it is generally a fractionally spaced FIR filter with adjustable tap coefficients. This filter is identical in form to the linear equalizer described earlier. Its input is the received filtered signal $y(t)$ sampled at some rate that is a multiple of the symbol rate—for example, at rate $2/T$. The second filter is a *feedback filter*. It is implemented as an FIR filter with symbol-spaced taps having adjustable coefficients. Its input is the set of previously detected symbols. The output of the feedback filter is subtracted from the output of the feedforward filter to form the input to the detector. Thus, we have

$$z_m = \sum_{n=1}^{N_1} c_n y(mT - n\tau) - \sum_{n=1}^{N_2} b_n \tilde{a}_{m-n}$$

where $\{c_n\}$ and $\{b_n\}$ are the adjustable coefficients of the feedforward and feedback filters, respectively; $\tilde{a}_{m-n}$, $n = 1, 2, \ldots, N_2$, are the previously detected symbols; N_1 is the length of the feedforward filter; and N_2 is the length of the feedback filter. Based on the input z_m, the detector determines which of the possible transmitted symbols is closest in distance to the input signal a_m. Thus, it makes its decision and outputs $\tilde{a}_m$. What makes the DFE nonlinear is the nonlinear characteristic of the detector that provides the input to the feedback filter.

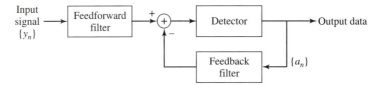

Figure 6.37   Block diagram of a DFE

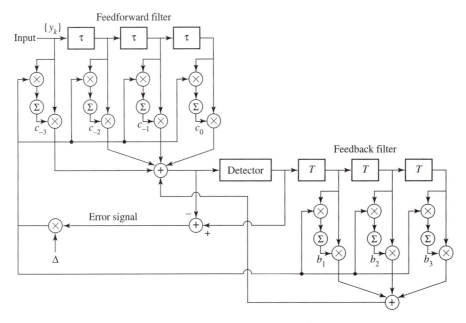

Figure 6.38   Block diagram of an adaptive DFE

The tap coefficients of the feedforward and feedback filters are selected to optimize some desired performance measure. For mathematical simplicity, the MSE criterion is usually applied, and a stochastic gradient algorithm is commonly used to implement an adaptive DFE. Figure 6.38 illustrates the block diagram of an adaptive DFE whose tap coefficients are adjusted by means of the LMS stochastic gradient algorithm.

We should mention that decision errors from the detector that are fed to the feedback filter have a small effect on the performance of the DFE. In general, a small loss in performance of 1 to 2 dB is possible at error rates below 10^{-2}, but the decision errors in the feedback filters are not catastrophic.

Although the DFE outperforms a linear equalizer, it is not the optimum equalizer from the viewpoint of minimizing the probability of error in the detection of the information sequence $\{a_k\}$ from the received signal samples $\{y_k\}$ given in (6.4.6). In a digital communication system that transmits information over a channel that causes ISI, the optimum

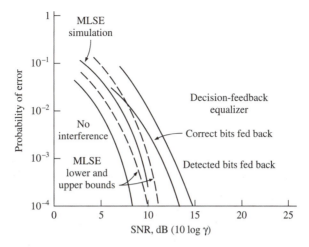

Figure 6.39   Error probability of the Viterbi algorithm for a binary PAM signal transmitted through channel B in Figure 6.35

detector is a maximum-likelihood symbol sequence detector that produces at its output the most probable symbol sequence $\{\tilde{a}_k\}$ for the given received sampled sequence $\{y_k\}$. That is, the detector finds the sequence $\{\tilde{a}_k\}$ that maximizes the *likelihood function*

$$\Lambda(\{a_k\}) = \ln p(\{y_k\} \mid \{a_k\})$$

where $p(\{y_k\} \mid \{a_k\})$ is the joint probability of the received sequence $\{y_k\}$ conditioned on $\{a_k\}$. The sequence of symbols $\{\tilde{a}_k\}$ that maximizes this joint conditional probability is called *the maximum-likelihood sequence detector.*

An algorithm that implements maximum-likelihood sequence detection (MLSD) is the Viterbi algorithm, which was originally devised for decoding convolutional codes as described in Section 8.3.2. For a description of this algorithm in the context of sequence detection in the presence of ISI, the reader is referred to [3,4].

The major drawback of MLSD for channels with ISI is the exponential behavior in computational complexity as a function of the span of the ISI. Consequently, MLSD is practical only for channels where the ISI spans only a few symbols and the ISI is severe, in the sense that it causes a severe degradation in the performance of a linear equalizer or a decision-feedback equalizer. For example, Figure 6.39 illustrates the error probability performance of the Viterbi algorithm for a binary PAM signal transmitted through channel B (see Figure 6.35). For purposes of comparison, we also illustrate the probability of error for a DFE. Both results were obtained by computer simulation. We observe that the performance of the MLSD is about 4.5 dB better than that of the DFE at an error probability of 10^{-4}. Hence, this is one example where the MLSD provides a significant performance gain on a channel with a relatively short ISI span.

In conclusion, channel equalizers are widely used in digital communication systems to mitigate the effects of the ISI caused by channel distortion. Linear equalizers are generally used for high-speed modems that transmit data over telephone channels. For wireless (radio) transmission, such as mobile cellular communications and interoffice communications, the multipath propagation of the transmitted signal results in severe ISI. Such channels require more powerful equalizers to combat the severe ISI. The decision-feedback equalizer and the MLSD are two nonlinear channel equalizers that are suitable for radio channels with severe ISI.

Problems

6.1 The Fourier transform of the rectangular pulse in Illustrative Problem 6.1 and the power spectrum $S_v(f)$ can be computed numerically with MATLAB by using the discrete Fourier transform (DFT) or the FFT algorithm. Normalize $T = 1$ and $\sigma_a^2 = 1$. Then sample the rectangular pulse $g(t)$ at $t = k/10$ for $k = 0, 1, 2, \ldots, 127$. This yields the sequence $\{g_k\}$ of sample values of $g(t)$. Use MATLAB to compute the 128-point DFT of $\{g_k\}$ and plot the values $|G_m|^2$ for $m = 0, 1, \ldots, 127$. Also plot the exact spectrum $|G(f)|^2$ given in (6.2.13), and compare the two results.

6.2 Repeat the computation in Problem 6.1 when the pulse $g(t)$ given as

$$g(t) = \begin{cases} \dfrac{1}{2}\left(1 - \cos\dfrac{2\pi t}{T}\right), & 0 \le t \le T \\ 0, & \text{otherwise} \end{cases}$$

Let $T = 1$ for this computation.

6.3 Write a MATLAB program to compute the power spectrum $S_v(f)$ of the signal $V(t)$ when the pulse $g(t)$ is

$$g(t) = \begin{cases} \dfrac{1}{2}\left(1 - \cos\dfrac{2\pi t}{T}\right), & 0 \le t \le T \\ 0, & \text{otherwise} \end{cases}$$

and the sequence $\{a_n\}$ of signal amplitudes has the correlation function given by (6.2.14).

6.4 Use MATLAB to design an FIR linear phase filter that models a lowpass bandlimited channel that has a $\frac{1}{2}$-dB ripple in the passband $|f| \le 3000$ Hz and a stopband attenuation of -40 dB for $|f| \ge 3500$. Plot the impulse response and the frequency response.

6.5 Write a MATLAB program to design an FIR linear phase filter that models a lowpass bandlimited channel with desired amplitude response

$$A(f) = \begin{cases} 1, & |f| \le 3000 \\ 0, & f > 3000 \end{cases}$$

via the window method, using a Hanning window.

6.6 Write a MATLAB program that generates the impulse response for the two-path (multipath) channel in Illustrative Problem 6.5, and plot the impulse response for $p = 0.95$ and a delay of five samples.

6.7 Write a MATLAB simulation program that implements channel 1 in Illustrative Problem 6.6, and measure the error rate when 10,000 binary data bits $\{\pm 1\}$ are transmitted through this channel. The channel is corrupted by AWGN with variance $\sigma^2 = 0$, $\sigma^2 = 0.1$, $\sigma^2 = 0.2$, $\sigma^2 = 0.5$, and $\sigma^2 = 1.0$.

6.8 Repeat Problem 6.7 for the following channel:

$$x_n = \begin{cases} 1, & n = 0 \\ 0.25, & n = \pm 1 \\ 0, & \text{otherwise} \end{cases}$$

6.9 Write a MATLAB program that generates the sampled version of the transmit filter impulse response $g_T(t)$ given by (6.5.9) for an arbitrary value of the roll-off factor α. Evaluate and plot $g_T(n)$ for $\alpha = \frac{1}{2}$ and $N = 31$. Also, evaluate and plot the magnitude of the frequency response characteristic for this filter. [Use a $4N$-point DFT of $g_T(n)$ by padding $g_T(n)$ with $3N$ zeros.]

6.10 Write a MATLAB program that computes the overall impulse response of the cascade of any transmit filter $g_T(n)$ with its matched filter at the receiver. This computation may be performed by use of the DFT as follows. Pad $g_T(n)$ with $N - 1$ (or more) 0's, and compute the $(2N - 1)$-point (or more) DFT. This yields $G_T(k)$. Then form $|G_T(k)|^2$, and compute the $(2N - 1)$-point inverse DFT of $|G_T(k)|^2$. Evaluate this overall impulse response for the filter in Problem 6.9, and compare this result with the ideal impulse response obtained by sampling $x_{rc}(t)$ at a rate $F_s = 4/T$.

6.11 Repeat Problem 6.9 for $N = 21$ and $N = 41$. Plot and compare the frequency response of the discrete-time filters with those in Problem 6.9. Describe the major differences.

6.12 Write a MATLAB program that takes a data sequence $\{D_k\}$, precodes it for a modified duobinary pulse transmission system to produce $\{p_k\}$, and maps the precoded sequence into the transmitted amplitude levels $\{a_k\}$. Then, from the transmitted sequence, form the received noise-free sequence $\{b_k = a_k - a_{k-2}\}$ and, using the relation given by (6.5.37), recover the data signal $\{D_k\}$. Run the program for any pseudorandom data sequence $\{D_k\}$ for $M = 2$ and $M = 4$ transmitted amplitude levels, and check the results.

6.13 Write a MATLAB program that performs a Monte Carlo simulation of a binary PAM communication system that employs duobinary signal pulse, where the precoding and amplitude sequence $\{a_k\}$ are performed as in Illustrative Problem 6.9. Add Gaussian noise to the received sequence $\{b_k\}$ as indicated in (6.5.22) to form the input to the detector, and use the detection rule in (6.5.28) to recover the data. Perform the simulation for 10,000 bits, and measure the bit-error probability for $\sigma^2 = 0.1$, $\sigma^2 = 0.5$, and $\sigma^2 = 1$. Plot the theoretical error probability for binary PAM with no ISI, and compare the Monte Carlo simulation results with this ideal performance. You should observe some small degradation in the performance of the duobinary system.

6.14 Repeat Problem 6.9 for a sampling rate $F_s = 8/T$, $\alpha = \frac{1}{2}$, and $N = 61$. Does the higher sampling rate result in a better frequency response characteristic—that is, a closer match to $X_{rc}(f)$?

6.15 For the filter designed in Problem 6.14, compute and plot the output of the cascade of this filter with its matched filter using the procedure described in Problem 6.10. Compare this sampled impulse response with the ideal impulse response obtained by sampling $x_{rc}(t)$ at a rate $F_s = 8/T$. Does this higher sampling rate result in a better approximation of the discrete-time filter impulse response to the ideal filter impulse response?

6.16 Write a MATLAB program that generates the sampled version of the transmit filter impulse response $g_T(t)$ given by (6.5.21) for the modified duobinary pulse specified by (6.5.18). Evaluate and plot $g_T(n)$ for $N = 31$. Also, evaluate and plot the magnitude of the frequency response of this filter.

6.17 Repeat Problem 6.10 for the filter designed in Problem 6.16.

6.18 Repeat Problem 6.16 for $N = 21$ and $N = 41$. Compare the frequency responses of these filters with the frequency response of the filter designed in Problem 6.16. What are the major differences in these frequency response characteristics?

6.19 Consider the channel-distorted pulse $x(t)$ given in Illustrative Problem 6.10. The pulse is sampled at the rate $2/T$ and equalized by a zero-forcing equalizer with $2K + 1 = 11$ taps. Write a MATLAB program to solve for the coefficients of the zero-forcing equalizer. Evaluate and plot the output of this zero-forcing equalizer for 50 sampled values.

6.20 Repeat Problem 6.19 for a MSE equalizer with $\sigma^2 = 0.01$, $\sigma^2 = 0.1$, and $\sigma^2 = 1.0$. Compare these equalizer coefficients with those obtained in Problem 6.19, and comment on the results as σ^2 is varied.

6.21 Write a general MATLAB program for computing the tap coefficients of an FIR equalizer of arbitrary length $2K + 1$ based on the MSE criterion, given as input the sampled values of the pulse $x(t)$ taken at the symbol rate and the spectral density of the additive noise σ^2. Use the program to evaluate the coefficients of an 11-tap equalizer when the sampled values of $x(t)$ are

$$ x(nT) = \begin{cases} 1, & n = 0 \\ 0.5, & n = \pm 1 \\ 0.3, & n = \pm 3 \\ 0.1, & n = \pm 4 \end{cases} $$

$\sigma^2 = 0.01$, and $\sigma^2 = 0.1$. Also, evaluate the minimum MSE for the optimum equalizer coefficients.

6.22 For the channel characteristics given in Problem 6.21, evaluate the coefficients of the MSE equalizer and the minimum MSE when the number of equalizer taps is 21. Compare these equalizer coefficients with the values of the coefficients obtained in Problem 6.21, and comment on whether the reduction in the MSE obtained with the longer equalizer is sufficiently large to justify its use.

6.23 The amount of residual ISI at the output of an equalizer can be evaluated by convolving the channel sampled response with the equalizer coefficients and observing the resulting output sequence. Write a MATLAB program that computes the output of the equalizer of a specified length when the input is the sampled channel characteristic. For simplicity, consider the case where the equalizer is a symbol-spaced equalizer and the channel sampled response also consists of symbol-spaced samples. Use the program to evaluate the output of a zero-forcing, symbol-spaced equalizer for the channel response given in Problem 6.21.

6.24 Write a MATLAB Monte Carlo simulation program that simulates the digital communication system that is modeled in Figure P6.24. The channel is modeled as an FIR filter with symbol-spaced values. The MSE equalizer is also an FIR filter with symbol-spaced tap coefficients. Training symbols are transmitted initially to train the equalizer. In the data mode, the equalizer employs the output of the detector in forming the error signal. Perform a Monte Carlo simulation of the system using 1000 training (binary) symbols and 10,000 binary data symbols for the channel model given in Problem 6.21. Use $\sigma^2 = 0.01$, $\sigma^2 = 0.1$, and $\sigma^2 = 1$. Compare the measured error rate with that of an ideal channel with no ISI.

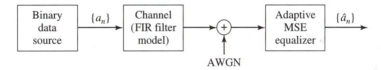

Figure P6.24

Chapter 7

Digital Transmission via Carrier Modulation

7.1 Preview

In the two preceding chapters, we considered the transmission of digital information through baseband channels. In such a case, the information-bearing signal is transmitted directly through the channel without the use of a sinusoidal carrier. However, most communication channels are bandpass channels; hence, the only way to transmit signals through such channels is by shifting the frequency of the information-bearing signal to the frequency band of the channel.

In this chapter, we consider four types of carrier-modulated signals that are suitable for bandpass channels: amplitude-modulated signals, quadrature-amplitude-modulated signals, phase-shift keying, and frequency-shift keying.

7.2 Carrier-Amplitude Modulation

In baseband digital PAM, the signal waveforms have the form

$$s_m(t) = A_m g_T(t) \tag{7.2.1}$$

where A_m is the amplitude of the mth waveforms and $g_T(t)$ is a pulse whose shape determines the spectral characteristics of the transmitted signal. The spectrum of the baseband signals is assumed to be contained in the frequency band $|f| \leq W$, where W is the bandwidth of $|G_T(f)|^2$, as illustrated in Figure 7.1. Recall that the signal amplitude takes the discrete values

$$A_m = (2m - 1 - M)d, \qquad m = 1, 2, \ldots, M \tag{7.2.2}$$

where $2d$ is the Euclidean distance between two adjacent signal points.

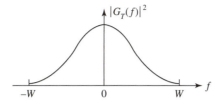

Figure 7.1   Energy density spectrum of the transmitted signal $g_T(t)$

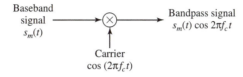

Figure 7.2   Amplitude modulation of a sinusoidal carrier by the baseband PAM signal

To transmit the digital signal waveforms through a bandpass channel, the baseband signal waveforms $s_m(t)$, $m = 1, 2, \ldots, M$, are multiplied by a sinusoidal carrier of the form $\cos 2\pi f_c t$, as shown in Figure 7.2, where f_c is the carrier frequency ($f_c > W$) and corresponds to the center frequency in the passband of the channel. Hence, the transmitted signal waveforms are expressed as

$$u_m(t) = A_m g_T(t) \cos 2\pi f_c t, \qquad m = 1, 2, \ldots, M \tag{7.2.3}$$

In the special case when the transmitted pulse shape $g_T(t)$ is rectangular—that is,

$$g_T(t) = \begin{cases} \sqrt{\dfrac{2}{T}}, & 0 \leq t \leq T \\ 0, & \text{otherwise} \end{cases}$$

the amplitude-modulated carrier signal is usually called *amplitude-shift keying* (ASK). In this case, the PAM signal is not bandlimited.

Amplitude modulation of the carrier $\cos 2\pi f_c t$ by the baseband signal waveforms $s_m(t)$ shifts the spectrum of the baseband signal by an amount f_c and thus places the signal into the passband of the channel. Recall that the Fourier transform of the carrier is $[\delta(f - f_c) + \delta(f + f_c)]/2$. Since multiplication of two signals in the time domain corresponds to the convolution of their spectra in the frequency domain, the spectrum of the amplitude-modulated signal is

$$U_m(f) = \frac{A_m}{2}[G_T(f - f_c) + G_T(f + f_c)] \tag{7.2.4}$$

Thus, the spectrum of the baseband signal $s_m(t) = A_m g_T(t)$ is shifted in frequency by the carrier frequency f_c. The bandpass signal is a double-sideband suppressed-carrier (DSB-SC) AM signal, as illustrated in Figure 7.3.

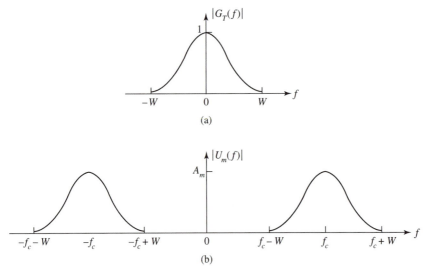

Figure 7.3   Spectra of (a) baseband and (b) amplitude-modulated signals

We note that impressing the baseband signal $s_m(t)$ onto the amplitude of the carrier signal $\cos 2\pi f_c(t)$ does not change the basic geometric representation of the digital PAM signal waveforms. The bandpass PAM signal waveforms may be represented in general as

$$u_m(t) = s_m \psi(t) \tag{7.2.5}$$

where the signal waveform $\psi(t)$ is defined as

$$\psi(t) = g_T(t) \cos 2\pi f_c t \tag{7.2.6}$$

and

$$s_m = A_m, \qquad m = 1, 2, \ldots, M \tag{7.2.7}$$

denotes the signal points that take the M values on the real line, as shown in Figure 7.4. The signal waveform $\psi(t)$ is normalized to unit energy; that is,

$$\int_{-\infty}^{\infty} \psi^2(t)\, dt = 1 \tag{7.2.8}$$

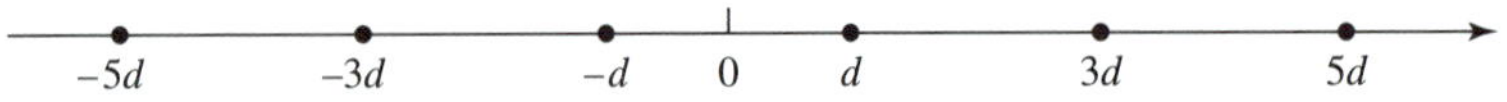

Figure 7.4   Signal point constellation for PAM signal

Consequently,

$$\int_{-\infty}^{\infty} g_T^2(t) \cos^2 2\pi f_c t \, dt = \frac{1}{2} \int_{-\infty}^{\infty} g_T^2(t) \, dt + \frac{1}{2} \int_{-\infty}^{\infty} g_T^2(t) \cos 4\pi f_c t \, dt$$
$$= 1 \qquad (7.2.9)$$

But

$$\int_{-\infty}^{\infty} g_T^2(t) \cos 4\pi f_c t \, dt = 0 \qquad (7.2.10)$$

because the bandwidth W of $g_T(t)$ is much smaller than the carrier frequency—that is, $f_c \gg W$. In such a case, $g_T(t)$ is essentially constant within any one cycle of $\cos 4\pi f_c t$; hence, the integral in (7.2.10) is equal to zero for each cycle of the integrand. In view of (7.2.10), it follows that

$$\frac{1}{2} \int_{-\infty}^{\infty} g_T^2(t) \, dt = 1 \qquad (7.2.11)$$

Therefore, $g_T(t)$ must be appropriately scaled so that (7.2.8) and (7.2.11) are satisfied.

7.2.1   Demodulation of PAM Signals

The demodulation of a bandpass digital PAM signal may be accomplished in one of several ways by means of correlation or matched filtering. For illustrative purposes, we consider a correlation-type demodulator.

The received signal may be expressed as

$$r(t) = A_m g_T(t) \cos 2\pi f_c t + n(t) \qquad (7.2.12)$$

where $n(t)$ is a bandpass noise process, which is represented as

$$n(t) = n_c(t) \cos 2\pi f_c t - n_s(t) \sin 2\pi f_c t \qquad (7.2.13)$$

and where $n_c(t)$ and $n_s(t)$ are the quadrature components of the noise. By cross-correlating the received signal $r(t)$ with $\psi(t)$ given by (7.2.6), as shown in Figure 7.5, we obtain the output

$$\int_{-\infty}^{\infty} r(t)\psi(t) \, dt = A_m + n = s_m + n \qquad (7.2.14)$$

where n represents the additive noise component at the output of the correlator.

The noise component has a zero mean. Its variance can be expressed as

$$\sigma_n^2 = \int_{-\infty}^{\infty} |\Psi(f)|^2 S_n(f) \, df \qquad (7.2.15)$$

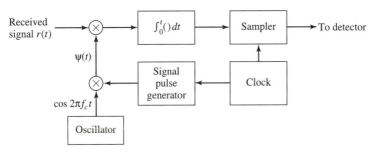

Figure 7.5   Demodulation of bandpass digital PAM signal

where $\Psi(f)$ is the Fourier transform of $\psi(t)$, and $S_n(f)$ is the power-spectral density of the additive noise. The Fourier transform of $\psi(t)$ is

$$\Psi(f) = \frac{1}{2}[G_T(f - f_c) + G_T(f + f_c)] \tag{7.2.16}$$

and the power-spectral density of the bandpass additive noise process is

$$S_n(f) = \begin{cases} \dfrac{N_0}{2}, & |f - f_c| \le W \\ 0, & \text{otherwise} \end{cases} \tag{7.2.17}$$

By substituting (7.2.16) and (7.2.17) into (7.2.15) and evaluating the integral, we obtain $\sigma_n^2 = N_0/2$.

It is apparent from (7.2.14), which is the input to the amplitude detector, that the probability of error of the optimum detector for the carrier-modulated PAM signal is identical to that of baseband PAM. That is,

$$P_M = \frac{2(M - 1)}{M} Q\left(\sqrt{\frac{6(\log_2 M)\mathcal{E}_{avb}}{(M^2 - 1)N_0}}\right) \tag{7.2.18}$$

where $\mathcal{E}_{avb}$ is the average energy per bit.

────────────(ILLUSTRATIVE PROBLEM)────────────

Illustrative Problem 7.1 [PAM signal spectrum] In an amplitude-modulated digital PAM system, the transmitter filter with impulse response $g_T(t)$ has a square-root raised-cosine spectral characteristic as described in Illustrative Problem 6.7, with a roll-off factor $\alpha = 0.5$. The carrier frequency is $f_c = 40/T$. Evaluate and graph the spectrum of the baseband signal and the spectrum of the amplitude-modulated signal.

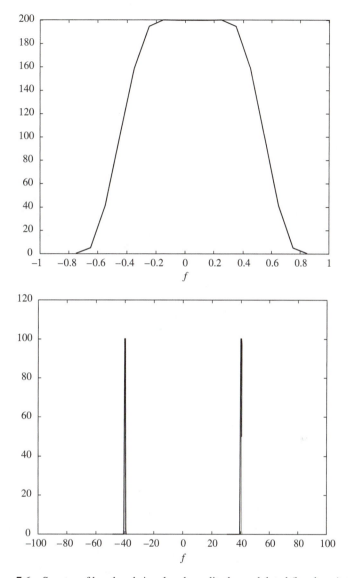

Figure 7.6   Spectra of baseband signal and amplitude-modulated (bandpass) signal

---

SOLUTION

Figure 7.6 illustrates these two spectral characteristics. The MATLAB script for this computation is given next.

```
             ( M-FILE )
```

```
% MATLAB script for Illustrated Problem 7.1.
echo on
T=1;
delta_T=T/200;                        % sampling interval
alpha=0.5;                            % roll-off factor
fc=40/T;                              % carrier frequency
A_m=1;                                % amplitude
t=−5*T+delta_T:delta_T:5*T;           % time axis
N=length(t);
for i=1:N,
  if (abs(t(i))~=T/(2*alpha)),
    g_T(i) = sinc(t(i)/T)*(cos(pi*alpha*t(i)/T)/(1−4*alpha^2*t(i)^2/T^2));
  else
    g_T(i) = 0;                       % the value of g_T is 0 at t=T/(2*alpha)
  end;                                % and at t=-T/(2*alpha)
    echo off ;
end;
echo on;
G_T=abs(fft(g_T));                    % spectrum of g_T
u_m=A_m*g_T.*cos(2*pi*fc*t);          % the modulated signal
U_m=abs(fft(u_m));                    % spectrum of the modulated signal
% actual frequency scale
f=−0.5/delta_T:1/(delta_T*(N−1)):0.5/delta_T;
% plotting commands follow
figure(1);
plot(f,fftshift(G_T));
axis([−1/T  1/T  0  max(G_T)]);
figure(2);
plot(f,fftshift(U_m));
```

7.3 Carrier-Phase Modulation

In carrier-phase modulation, the information that is transmitted over a communication channel is impressed on the phase of the carrier. Since the range of the carrier phase is $0 \leq \theta < 2\pi$, the carrier phases used to transmit digital information via digital-phase modulation are $\theta_m = 2\pi m/M$, for $m = 0, 1, \ldots, M - 1$. Thus, for binary phase modulation ($M = 2$), the two carrier phases are $\theta_0 = 0$ and $\theta_1 = \pi$ rad. For M-ary phase modulation, $M = 2^k$, where k is the number of information bits per transmitted symbol.

The general representation of a set of M carrier-phase-modulated signal waveforms is

$$u_m(t) = Ag_T(t) \cos\left(2\pi f_c t + \frac{2\pi m}{M}\right), \qquad m = 0, 1, \ldots, M - 1 \qquad (7.3.1)$$

where $g_T(t)$ is the transmitting filter pulse shape, which determines the spectral characteristics of the transmitted signal, and A is the signal amplitude. This type of digital phase modulation is called *phase-shift keying* (PSK). We note that PSK signals have equal energy; that is,

$$\mathcal{E}_m = \int_{-\infty}^{\infty} u_m^2(t)\, dt \tag{7.3.2}$$

$$= \int_{-\infty}^{\infty} A^2 g_T^2(t) \cos^2\left(2\pi f_c t + \frac{2\pi m}{M}\right) dt$$

$$= \frac{1}{2}\int_{-\infty}^{\infty} A^2 g_T^2(t)\, dt + \frac{1}{2}\int_{-\infty}^{\infty} A^2 g_T^2(t) \cos\left(4\pi f_c t + \frac{4\pi m}{M}\right) dt$$

$$= \frac{A^2}{2}\int_{-\infty}^{\infty} g_T^2(t)\, dt \tag{7.3.3}$$

$$\equiv \mathcal{E}_s, \quad \text{for all } m \tag{7.3.4}$$

where $\mathcal{E}_s$ denotes the energy per transmitted symbol. The term involving the double-frequency component in (7.3.4) averages out to zero when $f_c \gg W$, where W is the bandwidth of $g_T(t)$.

When $g_T(t)$ is a rectangular pulse, it is defined as

$$g_T(t) = \sqrt{\frac{2}{T}}, \qquad 0 \le t \le T \tag{7.3.5}$$

In this case, the transmitted signal waveforms in the symbol interval $0 \le t \le T$ may be expressed as (with $A = \sqrt{\mathcal{E}_s}$)

$$u_m(t) = \sqrt{\frac{2\mathcal{E}_s}{T}} \cos\left(2\pi f_c t + \frac{2\pi m}{M}\right), \qquad m = 0, 1, \ldots, M-1 \tag{7.3.6}$$

Note that the transmitted signals given by (7.3.6) have a constant envelope, and the carrier phase changes abruptly at the beginning of each signal interval. Figure 7.7 illustrates a four-phase ($M = 4$) PSK signal waveform.

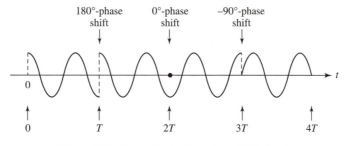

Figure 7.7   Example of a four-phase PSK signal

By viewing the angle of the cosine function in (7.3.6) as the sum of two angles, we may express the waveforms in (7.3.1) as

$$u_m(t) = \sqrt{\mathcal{E}_s}\, g_T(t) \cos\left(\frac{2\pi m}{M}\right) \cos 2\pi f_c t - \sqrt{\mathcal{E}_s}\, g_T(t) \sin\left(\frac{2\pi m}{M}\right) \sin 2\pi f_c t$$

$$= s_{mc}\psi_1(t) + s_{ms}\psi_2(t) \tag{7.3.7}$$

where

$$s_{mc} = \sqrt{\mathcal{E}_s}\, \cos\frac{2\pi m}{M}$$

$$s_{ms} = \sqrt{\mathcal{E}_s}\, \sin\frac{2\pi m}{M} \tag{7.3.8}$$

and $\psi_1(t)$ and $\psi_2(t)$ are orthogonal basis functions defined as

$$\psi_1(t) = g_T(t) \cos 2\pi f_c t$$

$$\psi_2(t) = -g_T(t) \sin 2\pi f_c t \tag{7.3.9}$$

By appropriately normalizing the pulse shape $g_T(t)$, we can normalize the energy of these two basis functions to unity. Thus, a phase-modulated signal may be viewed as two quadrature carriers with amplitudes that depend on the transmitted phase in each signal interval. Hence, digital phase-modulated signals are represented geometrically as two-dimensional vectors with components s_{mc} and s_{ms}—that is,

$$s_m = \left(\sqrt{\mathcal{E}_s}\, \cos\frac{2\pi m}{M} \quad \sqrt{\mathcal{E}_s}\, \sin\frac{2\pi m}{M}\right) \tag{7.3.10}$$

Signal point constellations for $M = 2$, 4, and 8 are illustrated in Figure 7.8. We observe that binary phase modulation is identical to binary PAM (binary antipodal signals).

The mapping, or assignment, of k information bits into the $M = 2^k$ possible phases may be done in a number of ways. The preferred assignment is to use *Gray encoding*, in which adjacent phases differ by one binary digit, as illustrated in Figure 7.8. Consequently, only a single bit error occurs in the k-bit sequence with Gray encoding when noise causes the erroneous selection of an adjacent phase to the transmitted phase.

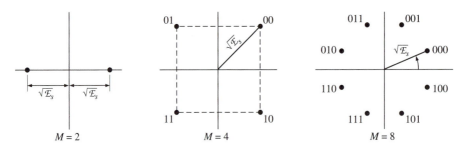

Figure 7.8   PSK signal constellations

ILLUSTRATIVE PROBLEM

Illustrative Problem 7.2 [PSK waveform] Generate the constant-envelope PSK signal waveforms given by (7.3.6) for $M = 8$. For convenience, the signal amplitude is normalized to unity.

SOLUTION

Figure 7.9 illustrates the eight waveforms for the case in which $f_c = 6/T$. The MATLAB script for this computation is given next.

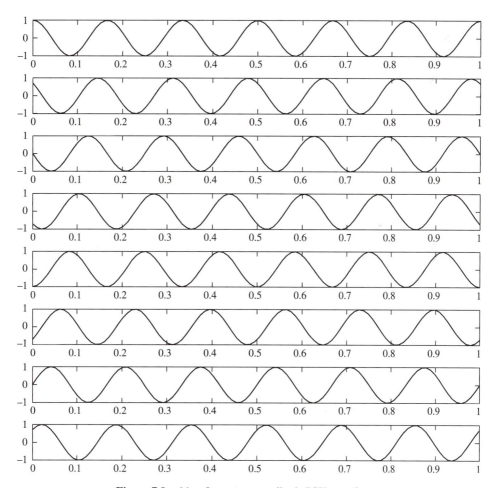

Figure 7.9 $M = 8$ constant-amplitude PSK waveforms

M-FILE

```
% MATLAB script for Illustrative Problem 7.2.
echo on
T=1;
M=8;
Es=T/2;
fc=6/T;                              % carrier frequency
N=100;                               % number of samples
delta_T=T/(N−1);
t=0:delta_T:T;
u0=sqrt(2*Es/T)*cos(2*pi*fc*t);
u1=sqrt(2*Es/T)*cos(2*pi*fc*t+2*pi/M);
u2=sqrt(2*Es/T)*cos(2*pi*fc*t+4*pi/M);
u3=sqrt(2*Es/T)*cos(2*pi*fc*t+6*pi/M);
u4=sqrt(2*Es/T)*cos(2*pi*fc*t+8*pi/M);
u5=sqrt(2*Es/T)*cos(2*pi*fc*t+10*pi/M);
u6=sqrt(2*Es/T)*cos(2*pi*fc*t+12*pi/M);
u7=sqrt(2*Es/T)*cos(2*pi*fc*t+14*pi/M);
% plotting commands follow
subplot(8,1,1);
plot(t,u0);
subplot(8,1,2);
plot(t,u1);
subplot(8,1,3);
plot(t,u2);
subplot(8,1,4);
plot(t,u3);
subplot(8,1,5);
plot(t,u4);
subplot(8,1,6);
plot(t,u5);
subplot(8,1,7);
plot(t,u6);
subplot(8,1,8);
plot(t,u7);
```

7.3.1   Phase Demodulation and Detection

The received bandpass signal in a signaling interval from an AWGN channel may be expressed as

$$r(t) = u_m(t) + n(t)$$
$$= u_m(t) + n_c(t) \cos 2\pi f_c t - n_s(t) \sin 2\pi f_c t \qquad (7.3.11)$$

where $n_c(t)$ and $n_s(t)$ are the two quadrature components of the additive noise.

The received signal may be correlated with $\psi_1(t)$ and $\psi_2(t)$ given by (7.3.9). The outputs of the two correlators yield the noise-corrupted signal components, which may be expressed as

$$r = s_m + n$$

$$= \left(\sqrt{\mathcal{E}_s} \cos \frac{2\pi m}{M} + n_c \quad \sqrt{\mathcal{E}_s} \sin \frac{2\pi m}{M} + n_s \right) \tag{7.3.12}$$

where n_c and n_s are defined as

$$n_c = \frac{1}{2} \int_{-\infty}^{\infty} g_T(t) n_c(t) \, dt$$

$$n_s = \frac{1}{2} \int_{-\infty}^{\infty} g_T(t) n_s(t) \, dt \tag{7.3.13}$$

The quadrature noise components $n_c(t)$ and $n_s(t)$ are zero-mean Gaussian random processes that are uncorrelated. As a consequence, $E(n_c) = E(n_s) = 0$ and $E(n_c n_s) = 0$. The variance of n_c and n_s is

$$E(n_c^2) = E(n_s^2) = \frac{N_0}{2} \tag{7.3.14}$$

The optimum detector projects the received signal vector r onto each of the M possible transmitted signal vectors $\{s_m\}$ and selects the vector corresponding to the largest projection. Thus, we obtain the correlation metrics

$$C(r, s_m) = r \cdot s_m, \qquad m = 0, 1, \ldots, M - 1 \tag{7.3.15}$$

Because all signals have equal energy, an equivalent detector metric for digital phase-modulation is to compute the phase of the received signal vector $r = (r_c, r_s)$ as

$$\theta_r = \tan^{-1} \frac{r_s}{r_c} \tag{7.3.16}$$

and select the signal from the set $\{s_m\}$ whose phase is closest to θ_r.

The probability of error at the detector for phase modulation in an AWGN channel may be found in any textbook that treats digital communications. Since binary phase modulation is identical to binary PAM, the probability of error is

$$P_2 = Q\left(\sqrt{\frac{2\mathcal{E}_b}{N_0}} \right) \tag{7.3.17}$$

where $\mathcal{E}_b$ is the energy per bit. Four-phase modulation may be viewed as two binary phase-modulation systems on quadrature (orthogonal) carriers. Consequently, the probability of a bit error is identical to that for binary phase modulation. For $M > 4$, there is no simple closed-form expression for the probability of a symbol error. A good approximation for P_M is

$$P_M \approx 2Q\left(\sqrt{\frac{2\mathcal{E}_s}{N_0}} \sin \frac{\pi}{M} \right)$$

$$\approx 2Q\left(\sqrt{\frac{2k\mathcal{E}_b}{N_0}} \sin \frac{\pi}{M} \right) \tag{7.3.18}$$

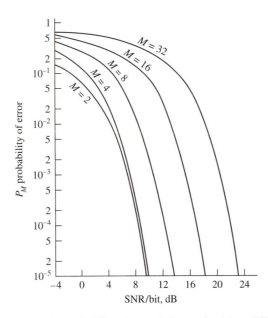

Figure 7.10   Probability of a symbol error for M-ary PSK

where $k = \log_2 M$ bits per symbol. Figure 7.10 illustrates the symbol-error probability as a function of the SNR $\mathscr{E}_b/N_0$.

The equivalent bit-error probability for M-ary phase modulation is also difficult to derive due to the dependence of the mapping of k-bit symbols into the corresponding signal phases. When a Gray code is used in the mapping, two k-bit symbols corresponding to adjacent signal phases differ in only a single bit. Because the most probable errors due to noise result in the erroneous selection of an adjacent phase to the true phase, most k-bit symbol errors contain only a single bit error. Hence, the equivalent bit-error probability for M-ary phase modulation is well approximated as

$$P_b \approx \frac{1}{k} P_M \tag{7.3.19}$$

⬤⬤⬤ ILLUSTRATIVE PROBLEM ⬤⬤⬤

Illustrative Problem 7.3 [PSK simulation] Perform a Monte Carlo simulation of an $M = 4$ PSK communication system that models the detector as the one that computes the correlation metrics given in (7.3.15). The model for the system to be simulated is shown in Figure 7.11.

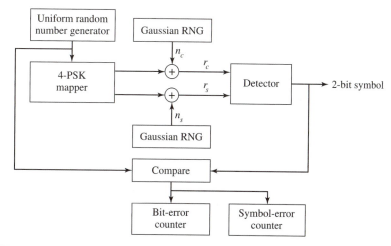

Figure 7.11   Block diagram of an $M = 4$ PSK system for a Monte Carlo simulation

SOLUTION

As shown, we simulate the generation of the random vector r given by (7.3.12), which is the output of the signal correlator and the input to the detector. We begin by generating a sequence of quaternary (2-bit) symbols that are mapped into the corresponding four-phase signal points, as shown in Figure 7.8 for $M = 4$. To accomplish this task, we use a random number generator that generates a uniform random number in the range (0, 1). This range is subdivided into four equal intervals, (0, 0.25), (0.25, 0.5), (0.5, 0.75), and (0.75, 1.0), where the subintervals correspond to the pairs of information bits 00, 01, 11, and 10, respectively. These pairs of bits are used to select the signal phase vector s_m.

The additive noise components n_c and n_s are statistically independent zero-mean Gaussian random variables with variance σ^2. For convenience, we may normalize the variance to $\sigma^2 = 1$ and control the SNR in the received signal by scaling the signal energy parameter $\mathcal{E}_s$, or vice versa.

The detector observes the received signal vector $r = s_m + n$ as given in (7.3.12) and computes the projection (dot product) of r onto the four possible signal vectors s_m. Its decision is based on selecting the signal point corresponding in the largest projection. The output decisions from the detector are compared with the transmitted symbols, and symbol errors and bit errors are counted.

Figure 7.12 illustrates the results of the Monte Carlo simulation for the transmission of $N=10,000$ symbols at different values of the SNR parameter $\mathcal{E}_b/N_0$, where $\mathcal{E}_b = \mathcal{E}_s/2$ is the bit energy. Also shown in Figure 7.12 is the bit-error rate, which is defined as $P_b \approx P_M/2$, and the corresponding theoretical error probability, given by (7.3.18).

The MATLAB scripts for this Monte Carlo simulation are given next.

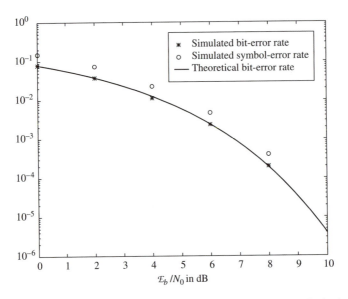

Figure 7.12   Performance of a four-phase PSK system from the Monte Carlo simulation

M-FILE

```
% MATLAB script for Illustrative Problem 7.3.
echo on
SNRindB1=0:2:10;
SNRindB2=0:0.1:10;
for i=1:length(SNRindB1),
    [pb,ps]=cm_sm32(SNRindB1(i));          % simulated bit and symbol error rates
    smld_bit_err_prb(i)=pb;
    smld_symbol_err_prb(i)=ps;
    echo off ;
end;
echo on;
for i=1:length(SNRindB2),
    SNR=exp(SNRindB2(i)*log(10)/10);       % signal to noise ratio
    theo_err_prb(i)=Qfunct(sqrt(2*SNR));   % theoretical bit error rate
    echo off ;
end;
echo on ;
% Plotting commands follow
semilogy(SNRindB1,smld_bit_err_prb,' * ');
hold
semilogy(SNRindB1,smld_symbol_err_prb,' o ');
semilogy(SNRindB2,theo_err_prb);
```

M-FILE

```
function [pb,ps]=cm_sm32(snr_in_dB)
% [pb,ps]=cm_sm32(snr_in_dB)
%              CM_SM32  finds the probability of bit error and symbol error for the
%              given value of snr_in_dB, signal to noise ratio in dB.
N=10000;
E=1;                              % energy per symbol
snr=10^(snr_in_dB/10);           % signal to noise ratio
sgma=sqrt(E/snr)/2;              % noise variance
% the signal mapping
s00=[1  0];
s01=[0  1];
s11=[-1  0];
s10=[0  -1];
% generation of the data source
for  i=1:N,
  temp=rand;                     % a uniform random variable between 0 and 1
  if (temp<0.25),                % with probability 1/4, source output is "00"
    dsource1(i)=0;
    dsource2(i)=0;
  elseif (temp<0.5),             % with probability 1/4, source output is "01"
    dsource1(i)=0;
    dsource2(i)=1;
  elseif (temp<0.75),            % with probability 1/4, source output is "10"
    dsource1(i)=1;
    dsource2(i)=0;
  else                           % with probability 1/4, source output is "11"
    dsource1(i)=1;
    dsource2(i)=1;
  end;
end;
% detection and the probability of error calculation
numofsymbolerror=0;
numofbiterror=0;
for  i=1:N,
  % the received signal at the detector, for the ith symbol, is:
  n(1)=gngauss(sgma);
  n(2)=gngauss(sgma);
  if ((dsource1(i)==0) & (dsource2(i)==0)),
    r=s00+n;
  elseif ((dsource1(i)==0) & (dsource2(i)==1)),
    r=s01+n;
  elseif ((dsource1(i)==1) & (dsource2(i)==0)),
    r=s10+n;
  else
    r=s11+n;
  end;
  % The correlation metrics are computed below
  c00=dot(r,s00);
  c01=dot(r,s01);
  c10=dot(r,s10);
  c11=dot(r,s11);
```

```
% The decision on the ith symbol is made next
c_max=max([c00 c01 c10 c11]);
if (c00==c_max),
   decis1=0; decis2=0;
elseif (c01==c_max),
   decis1=0; decis2=1;
elseif (c10==c_max),
   decis1=1; decis2=0;
else
   decis1=1; decis2=1;
end;
% increment the error counter, if the decision is not correct
symbolerror=0;
if (decis1~=dsource1(i)),
   numofbiterror=numofbiterror+1;
   symbolerror=1;
end;
if (decis2~=dsource2(i)),
   numofbiterror=numofbiterror+1;
   symbolerror=1;
end;
if (symbolerror==1),
   numofsymbolerror = numofsymbolerror+1;
end;
end;
ps=numofsymbolerror/N;            % since there are totally N symbols
pb=numofbiterror/(2*N);           % since 2N bits are transmitted
```

7.3.2 Differential Phase Modulation and Demodulation

The demodulation of the phase-modulated carrier signal, as described above, requires that the carrier phase components $\psi_1(t)$ and $\psi_2(t)$ be phase-locked to the received carrier-modulated signal. In general, this means that the receiver must estimate the carrier-phase offset of the received signal caused by a transmission delay through the channel and compensate for this carrier-phase offset in the cross-correlation of the received signal with the two reference components $\psi_1(t)$ and $\psi_2(t)$. The estimation of the carrier-phase offset is usually performed by use of a phase-locked loop (PLL). Thus, we achieve coherent phase demodulation.

Another type of carrier-phase modulation is differential phase modulation, in which the transmitted data are differentially encoded prior to the modulator. In differential encoding, the information is conveyed by phase shifts relative to the previous signal interval. For example, in binary phase modulation, the information bit 1 may be transmitted by shifting the phase of the carrier by $180°$ relative to the previous carrier phase, whereas the information bit 0 is transmitted by a zero-phase shift relative to the phase in the preceding signaling interval. In four-phase modulation, the relative phase shifts between successive intervals are $0°$, $90°$, $180°$, and $270°$, corresponding to the information bits 00, 01, 11, and 10, respectively. The generalization of differential encoding for $M > 4$ is straightforward.

The phase-modulated signals resulting from this encoding process are described as *differentially encoded*. The encoding is performed by a relatively simple logic circuit preceding the modulator.

Demodulation and detection of the differentially encoded phase-modulated signal may be performed as follows. The received signal phase $\theta_r = \tan^{-1} r_s/r_c$ at the detector is mapped into one of the M possible transmitted signal phases $\{\theta_m\}$ that is closest to θ_r. Following the detector is a relatively simple phase comparator that compares the phases of the detected signal over two consecutive intervals to extract the transmitted information.

We observe that the demodulation of a differentially encoded phase-modulated signal does not require the estimation of the carrier phase. To elaborate, suppose that we demodulate the differentially encoded signal by cross-correlating $r(t)$ with $g_T(t) \cos 2\pi f_c t$ and $-g_T(t) \sin 2\pi f_c t$. At the kth signaling interval, the two components of the demodulator output may be represented in complex-valued form as

$$r_k = \sqrt{\mathcal{E}_s}\, e^{j(\theta_k - \phi)} + n_k \qquad (7.3.20)$$

where θ_k is the phase angle of the transmitted signal at the kth signaling interval, ϕ is the carrier phase, and $n_k = n_{kc} + jn_{ks}$ is the noise. Similarly, the received signal vector at the output of the demodulator in the preceding signaling interval is the complex-valued quantity

$$r_{k-1} = \sqrt{\mathcal{E}_s}\, e^{j(\theta_{k-1} - \phi)} + n_{k-1} \qquad (7.3.21)$$

The decision variable for the phase detector is the phase difference between these two complex numbers. Equivalently, we can project r_k onto r_{k-1} and use the phase of the resulting complex number; that is,

$$r_k r_{k-1}^* = \mathcal{E}_s e^{j(\theta_k - \theta_{k-1})} + \sqrt{\mathcal{E}_s}\, e^{j(\theta_k - \phi)} n_{k-1} + \sqrt{\mathcal{E}_s}\, e^{-j(\theta_{k-1} - \phi)} n_k + n_k n_{k-1}^* \quad (7.3.22)$$

which, in the absence of noise, yields the phase difference $\theta_k - \theta_{k-1}$. Thus, the mean value of $r_k r_{k-1}^*$ is independent of the carrier phase. Differentially encoded PSK signaling that is demodulated and detected as described above is called *differential* PSK (DPSK). The demodulation and detection of DPSK are illustrated in Figure 7.13.

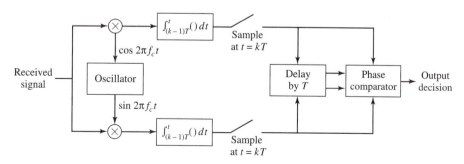

Figure 7.13   Block diagram of DPSK demodulator

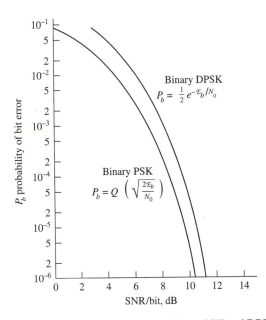

Figure 7.14   Probability of error for binary PSK and DPSK

The probability of error for DPSK in an AWGN channel is relatively simple to derive for binary ($M = 2$) phase modulation. The result is

$$P_2 = \frac{1}{2}e^{-\mathcal{E}_b/N_0} \tag{7.3.23}$$

The graph of (7.3.23) is shown in Figure 7.14. Also shown in this figure is the probability of error for binary PSK. We observe that at error probabilities below 10^{-4}, the difference in SNR between binary PSK and binary DPSK is less than 1 dB.

For $M > 2$, the error probability performance of a DPSK demodulator and detector is extremely difficult to evaluate exactly. The major difficulty is encountered in the determination of the probability density function for the phase of the random variable $r_k r_{k-1}^*$, given by (7.3.22). However, an approximation to the performance of DPSK is easily obtained, as we now demonstrate.

Without loss of generality, suppose the phase difference $\theta_k - \theta_{k-1} = 0$. Furthermore, the exponential factors $e^{-j(\theta_{k-1}-\phi)}$ and $e^{j(\theta_k-\phi)}$ in (7.3.22) can be absorbed into the Gaussian noise components n_{k-1} and n_k without changing their statistical properties. Therefore, $r_k r_{k-1}^*$ in (7.3.22) can be expressed as

$$r_k r_{k-1}^* = \mathcal{E}_s + \sqrt{\mathcal{E}_s}\,(n_k + n_{k-1}^*) + n_k n_{k-1}^* \tag{7.3.24}$$

The complication in determining the probability density function of the phase is the term $n_k n_{k-1}^*$. However, at SNRs of practical interest, the term $n_k n_{k-1}^*$ is small relative to the dominant noise term $\sqrt{\mathcal{E}_s}\,(n_k + n_{k-1}^*)$. If we neglect the term $n_k n_{k-1}^*$ and we also normalize $r_k r_{k-1}^*$ by dividing through by $\sqrt{\mathcal{E}_s}$, the new set of decision metrics becomes

$$x = \sqrt{\mathcal{E}_s} + \mathrm{Re}(n_k + n_{k-1}^*)$$
$$y = \mathrm{Im}(n_k + n_{k-1}^*) \tag{7.3.25}$$

The variables x and y are uncorrelated Gaussian random variables with identical variances $\sigma_n^2 = N_0$. The phase is

$$\theta_r = \tan^{-1} \frac{y}{x} \tag{7.3.26}$$

At this stage, we have a problem that is identical to the one for phase-coherent demodulation. The only difference is that the noise variance is now twice as large as in the case of PSK. Thus, we conclude that the performance of DPSK is 3 dB poorer than that for PSK. This result is relatively good for $M \geq 4$, but it is pessimistic for $M = 2$ in the sense that the loss in binary DPSK relative to binary PSK is less than 3 dB at large SNR.

─────(**ILLUSTRATIVE PROBLEM**)─────────────

Illustrative Problem 7.4 [DPSK encoder] Implement a differential encoder for the case of $M = 8$ DPSK.

─────(**SOLUTION**)──────────────────

The signal points are the same as those for PSK shown in Figure 7.8. However, for DPSK, these signal points represent the phase change relative to the phase of the previous transmitted signal points.

The MATLAB script for implementing the differential encoder is given next.

─────(**M-FILE**)────────────────────

```
% MATLAB script for Illustrative Problem 7.4.
mapping=[0 1 3 2 7 6 4 5];          % For Gray mapping
M=8;
E=1;
sequence=[0 1 0  0 1 1  0 0 1  1 1 1  1 1 0  0 0 0];
[e]=cm_dpske(E,M,mapping,sequence);      % e is the differential encoder output
```

─────(**ILLUSTRATIVE PROBLEM**)─────────────

Illustrative Problem 7.5 Perform a Monte Carlo simulation of an $M = 4$ DPSK communication system. The model for the system to be simulated is shown in Figure 7.15.

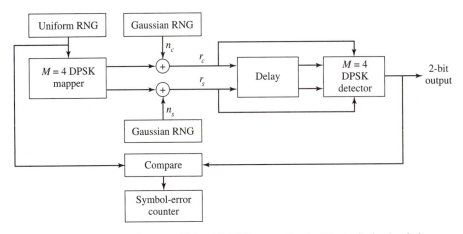

Figure 7.15 Block diagram of $M = 4$ DPSK system for the Monte Carlo simulation

---

SOLUTION

The uniform random number generator (RNG) is used to generate the pairs of bits {00, 01, 11, 10}, as described in Illustrative Problem 7.3. Each 2-bit symbol is mapped into one of the four signal points $s_m = [\cos \pi m/2 \quad \sin \pi m/2]$, $m = 0, 1, 2, 3$, by the differential encoder. Two Gaussian RNG are used to generate the noise components $[n_c \quad n_s]$. Then the received signal-plus-noise vector is

$$r = \left[\cos \frac{\pi m}{2} + n_c \quad \sin \frac{\pi m}{2} + n_s\right]$$
$$= [r_c \quad r_s]$$

The differential detector basically computes the phase difference between r_k and r_{k-1}. Mathematically, this computation can be performed as in (7.3.22)—that is,

$$r_k r_{k-1}^* = (r_{ck} + j r_{sk})(r_{c\,k-1} - j r_{s\,k-1})$$
$$= r_{ck} r_{c\,k-1} + r_{sk} r_{s\,k-1} + j(r_{sk} r_{c\,k-1} - r_{ck} r_{s\,k-1})$$
$$= x_k + j\, y_k$$

and $\theta_k = \tan^{-1}(y_k/x_k)$ is the phase difference. The value of θ_k is compared with the possible phase differences $\{0°, \ 90°, \ 180°, \ 270°\}$, and a decision is made in favor of the phase that is closest to θ_k. The detected phase is then mapped into the pair of information bits. The error counter counts the symbol errors in the detected sequence.

Figure 7.16 illustrates the results of the Monte Carlo simulation for the transmission of $N=10,000$ symbols at different values of the SNR parameter $\mathcal{E}_b/N_0$, where $\mathcal{E}_b = \mathcal{E}_s/2$ is the bit energy. Also shown in Figure 7.16 is the theoretical value of the symbol-error rate based on the approximation that the term $n_k n_{k-1}^*$ is negligible. We observe from Figure 7.16 that the approximation results in an upper bound to the error probability.

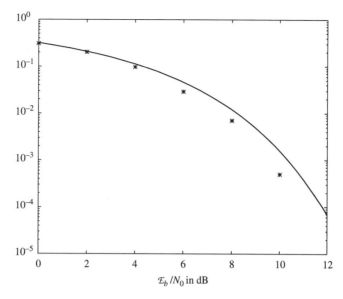

Figure 7.16   Performance of four-phase DPSK system from Monte Carlo simulation. (The solid curve is an upper bound based on approximation that neglects the noise term $n_l n^*_{k-1}$.)

The MATLAB scripts for this Monte Carlo simulation are given next.

M-FILE

```
% MATLAB script for Illustrative Problem 7.5.
echo on
SNRindB1=0:2:12;
SNRindB2=0:0.1:12;
for i=1:length(SNRindB1),
   smld_err_prb(i)=cm_sm34(SNRindB1(i));     % simulated error rate
   echo off ;
end;
echo on ;
for i=1:length(SNRindB2),
   SNR=exp(SNRindB2(i)*log(10)/10);          % signal to noise ratio
   theo_err_prb(i)=2*Qfunct(sqrt(SNR));      % theoretical symbol error rate
   echo off ;
end;
echo on ;
% Plotting commands follow
semilogy(SNRindB1,smld_err_prb,' * ');
hold
semilogy(SNRindB2,theo_err_prb);
```

M-FILE

```
function [p]=cm_sm34(snr_in_dB)
% [p]=cm_sm34(snr_in_dB)
%               CM_SM34  finds the probability of error for the given
%               value of snr_in_dB, signal to noise ratio in dB.
N=10000;
E=1;                                    % energy per symbol
snr=10^(snr_in_dB/10);                  % signal to noise ratio
sgma=sqrt(E/(4*snr));                   % noise variance
% generation of the data source follows
for i=1:2*N,
  temp=rand;                            % a uniform random variable between 0 and 1
  if (temp<0.5),
    dsource(i)=0;                       % with probability 1/2, source output is "0"
  else
    dsource(i)=1;                       % with probability 1/2, source output is "1"
  end;
end;
% Differential encoding of the data source follows
mapping=[0 1 3 2];
M=4;
[diff_enc_output] = cm_dpske(E,M,mapping,dsource);
% received signal is then
for i=1:N,
  [n(1) n(2)]=gngauss(sgma);
  r(i,:)=diff_enc_output(i,:)+n;
end;
% detection and the probability of error calculation
numoferr=0;
prev_theta=0;
for i=1:N,
  theta=angle(r(i,1)+j*r(i,2));
  delta_theta=mod(theta-prev_theta,2*pi);
  if ((delta_theta<pi/4) | (delta_theta>7*pi/4)),
    decis=[0 0];
  elseif (delta_theta<3*pi/4),
    decis=[0 1];
  elseif (delta_theta<5*pi/4)
    decis=[1 1];
  else
    decis=[1 0];
  end;
  prev_theta=theta;
  % increase the error counter, if the decision is not correct
  if ((decis(1)~=dsource(2*i-1)) | (decis(2)~=dsource(2*i))),
    numoferr=numoferr+1;
  end;
end;
p=numoferr/N;
```

M-FILE

```
function [enc_comp] = cm_dpske(E,M,mapping,sequence);
% [enc_comp] = cm_dpske(E,M,mapping,sequence)
%            CM_DPSKE differentially encodes a sequence.
%            E is the average energy, M is the number of constellation points
%            and mapping is the vector defining how the constellation points are
%            allocated. Finally, "sequence" is the uncoded binary data sequence.
k=log2(M);
N=length(sequence);
% If N is not divisible by k, append zeros, so that it is...
remainder=rem(N,k);
if (remainder~=0),
   for i=N+1:N+k-remainder,
      sequence(i)=0;
   end;
   N=N+k-remainder;
end;
theta=0;                                % Initially, assume that theta=0
for i=1:k:N,
   index=0;
   for j=i:i+k-1,
      index=2*index+sequence(j);
   end;
   index=index+1;
   theta=mod(2*pi*mapping(index)/M+theta,2*pi);
   enc_comp((i+k-1)/k,1)=sqrt(E)*cos(theta);
   enc_comp((i+k-1)/k,2)=sqrt(E)*sin(theta);
end;
```

7.4   Quadrature Amplitude Modulation

A quadrature-amplitude-modulated (QAM) signal employs two quadrature carriers, $\cos 2\pi f_c t$ and $\sin 2\pi f_c t$, each of which is modulated by an independent sequence of information bits. The transmitted signal waveforms have the form

$$u_m(t) = A_{mc} g_T(t) \cos 2\pi f_c t + A_{ms} g_T(t) \sin 2\pi f_c t, \qquad m = 1, 2, \ldots, M \quad (7.4.1)$$

where $\{A_{mc}\}$ and $\{A_{ms}\}$ are the sets of amplitude levels that are obtained by mapping k-bit sequences into signal amplitudes. For example, Figure 7.17 illustrates a 16-QAM signal constellation that is obtained by amplitude modulating each quadrature carrier by $M = 4$ PAM. In general, rectangular signal constellations result when two quadrature carriers are each modulated by PAM.

More generally, QAM may be viewed as a form of combined digital-amplitude and digital-phase modulation. Thus, the transmitted QAM signal waveforms may be expressed as

$$u_{mn}(t) = A_m g_T(t) \cos(2\pi f_c t + \theta_n), \qquad m = 1, 2, \ldots, M_1, \quad n = 1, 2, \ldots, M_2 \quad (7.4.2)$$

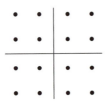

Figure 7.17 $M = 16$-QAM signal constellation

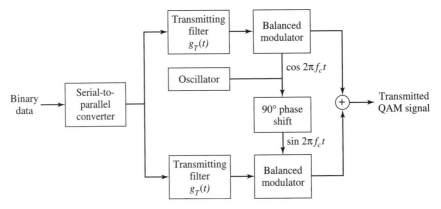

Figure 7.18   Functional block diagram of modulator for QAM

If $M_1 = 2^{k_1}$ and $M_2 = 2^{k_2}$, the combined amplitude- and phase-modulation method results in the simultaneous transmission of $k_1 + k_2 = \log_2 M_1 M_2$ binary digits occurring at a symbol rate $R_b/(k_1 + k_2)$. Figure 7.18 illustrates the functional block diagram of a QAM modulator.

It is clear that the geometric signal representation of the signals given by (7.4.1) and (7.4.2) is in terms of two-dimensional signal vectors of the form

$$s_m = \left(\sqrt{\mathcal{E}_s}\, A_{mc} \quad \sqrt{\mathcal{E}_s}\, A_{ms} \right), \qquad m = 1, 2, \ldots, M \tag{7.4.3}$$

Examples of signal space constellations for QAM are shown in Figure 7.19. Note that $M = 4$ QAM is identical to $M = 4$ PSK.

7.4.1   Demodulation and Detection of QAM

Let us assume that a carrier-phase offset is introduced in the transmission of the signal through the channel. In addition, the received signal is corrupted by additive Gaussian noise. Hence, $r(t)$ may be expressed as

$$r(t) = A_{mc} g_T(t) \cos(2\pi f_c t + \phi) + A_{ms} g_T(t) \sin(2\pi f_c t + \phi) + n(t) \tag{7.4.4}$$

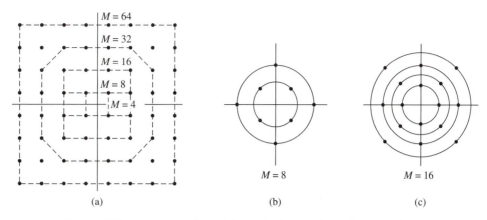

Figure 7.19   (a) Rectangular and (b, c) circular QAM signal constellations

where ϕ is the carrier-phase offset and

$$n(t) = n_c(t) \cos 2\pi f_c t - n_s(t) \sin 2\pi f_c t$$

The received signal is correlated with the two phase-shifted basis functions

$$\psi_1(t) = g_T(t) \cos(2\pi f_c t + \phi)$$
$$\psi_2(t) = g_T(t) \sin(2\pi f_c t + \phi) \tag{7.4.5}$$

as illustrated in Figure 7.20, and the outputs of the correlators are sampled and passed to the detector. The phase-locked loop (PLL) shown in Figure 7.20 estimates the carrier-phase offset ϕ of the received signal and compensates for this phase offset by phase shifting $\psi_1(t)$ and $\psi_2(t)$, as indicated in (7.4.5). The clock shown in Figure 7.20 is assumed to be synchronized to the received signal so that the correlator outputs are sampled at the proper instant in time. Under these conditions, the outputs from the two correlators are

$$r_c = A_{mc} + n_c \cos \phi - n_s \sin \phi$$
$$r_s = A_{ms} + n_c \sin \phi + n_s \cos \phi \tag{7.4.6}$$

where

$$n_c = \frac{1}{2} \int_0^T n_c(t) g_T(t) \, dt$$

$$n_s = \frac{1}{2} \int_0^T n_s(t) g_T(t) \, dt \tag{7.4.7}$$

The noise components are zero-mean, uncorrelated Gaussian random variables with variance $N_0/2$.

The optimum detector computes the distance metrics

$$D(\mathbf{r}, \mathbf{s}_m) = |\mathbf{r} - \mathbf{s}_m|^2, \qquad m = 1, 2, \dots, M \tag{7.4.8}$$

where $\mathbf{r}^t = (r_c, r_s)$, and $\mathbf{s}_m$ is given by (7.4.3).

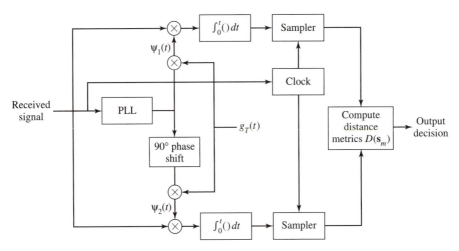

Figure 7.20   Demodulation and detection of QAM signals

7.4.2   Probability of Error for QAM in an AWGN Channel

In this section, we consider the performance of QAM systems that employ rectangular signal constellations. Rectangular QAM signal constellations have the distinct advantage of being easily generated as two PAM signals impressed on phase quadrature carriers. In addition, they are easily demodulated.

For rectangular signal constellations in which $M = 2^k$, where k is even, the QAM signal constellation is equivalent to two PAM signals on quadrature carriers, each having $\sqrt{M} = 2^{k/2}$ signal points. Because the signals in the phase quadrature components are perfectly separated by coherent detection, the probability of error for QAM is easily determined from the probability of error for PAM. Specifically, the probability of a correct decision for the M-ary QAM system is

$$P_c = \left(1 - P_{\sqrt{M}} \right)^2 \tag{7.4.9}$$

where $P_{\sqrt{M}}$ is the probability of error of a $\sqrt{M}$-ary PAM with one-half the average power in each quadrature signal of the equivalent QAM system. By appropriately modifying the probability of error for M-ary PAM, we obtain

$$P_{\sqrt{M}} = 2 \left(1 - \frac{1}{\sqrt{M}} \right) Q \left(\sqrt{\frac{3}{M-1} \frac{\mathcal{E}_{av}}{N_0}} \right) \tag{7.4.10}$$

where $\mathcal{E}_{av}/N_0$ is the average SNR per symbol. Therefore, the probability of a symbol error for the M-ary QAM is

$$P_M = 1 - \left(1 - P_{\sqrt{M}} \right)^2 \tag{7.4.11}$$

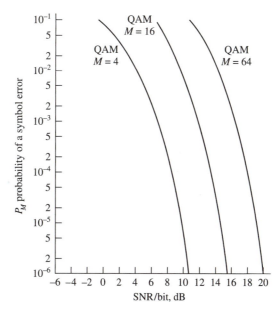

Figure 7.21   Probability of a symbol error for QAM

We note that this result is exact for $M = 2^k$ when k is even. On the other hand, when k is odd, there is no equivalent $\sqrt{M}$-ary PAM system. This is no problem, however, because it is rather easy to determine the error rate for a rectangular signal set. If we employ the optimum detector that bases its decisions on the optimum distance metrics given by (7.4.8), it is relatively straightforward to show that the symbol-error probability is tightly upper-bounded as

$$P_M \leq 1 - \left[1 - 2Q\left(\sqrt{\frac{3\mathcal{E}_{av}}{(M-1)N_0}} \right) \right]^2$$

$$\leq 4Q\left(\sqrt{\frac{3k\mathcal{E}_{avb}}{(M-1)N_0}} \right) \tag{7.4.12}$$

for any $k \geq 1$, where $\mathcal{E}_{avb}/N_0$ is the average SNR per bit. The probability of a symbol error is plotted in Figure 7.21 as a function of the average SNR per bit.

════ ILLUSTRATIVE PROBLEM ════

Illustrative Problem 7.6 [QAM simulation] Perform a Monte Carlo simulation of an $M = 16$-QAM communication system using a rectangular signal constellation. The model of the system to be simulated is shown in Figure 7.22.

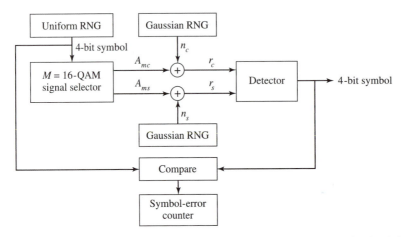

Figure 7.22   Block diagram of an $M = 16$-QAM system for the Monte Carlo simulation

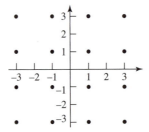

Figure 7.23 $M = 16$-QAM signal constellation for the Monte Carlo simulation

SOLUTION

The uniform random number generator (RNG) is used to generate the sequence of information symbols corresponding to the 16 possible 4-bit combinations of b_1, b_2, b_3, b_4. The information symbols are mapped into the corresponding signal points, as illustrated in Figure 7.23, which have the coordinates $[A_{mc}, A_{ms}]$. Two Gaussian RNG are used to generate the noise components $[n_c, n_s]$. The channel-phase shift ϕ is set to 0 for convenience. Consequently, the received signal-plus-noise vector is

$$r = [A_{mc} + n_c \quad A_{ms} + n_s]$$

The detector computes the distance metrics given by (7.4.8) and decides in favor of the signal point that is closest to the received vector r. The error counter counts the symbol errors in the detected sequence. Figure 7.24 illustrates the results of the Monte Carlo simulation for the transmission of $N = 10{,}000$ symbols at different values of the SNR parameter $\mathscr{E}_b/N_0$, where $\mathscr{E}_b = \mathscr{E}_s/4$ is the bit energy. Also shown in Figure 7.24 is the theoretical value of the symbol-error probability given by (7.4.10) and (7.4.11).

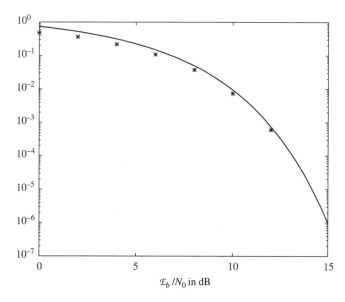

Figure 7.24   Performance of $M = 16$-QAM system from the Monte Carlo simulation

The MATLAB scripts for this problem are given next.

M-FILE

```
% MATLAB script for Illustrative Problem 7.6.
echo on
SNRindB1=0:2:15;
SNRindB2=0:0.1:15;
M=16;
k=log2(M);
for i=1:length(SNRindB1),
    smld_err_prb(i)=cm_sm41(SNRindB1(i));    % simulated error rate
    echo off;
end;
echo on ;
for i=1:length(SNRindB2),
    SNR=exp(SNRindB2(i)*log(10)/10);        % signal to noise ratio
    % theoretical symbol error rate
    theo_err_prb(i)=4*Qfunct(sqrt(3*k*SNR/(M−1)));
    echo off ;
end;
echo on ;
% Plotting commands follow
semilogy(SNRindB1,smld_err_prb,' * ');
hold
semilogy(SNRindB2,theo_err_prb);
```

M-FILE

```
function [p]=cm_sm41(snr_in_dB)
% [p]=cm_sm41(snr_in_dB)
%             CM_SM41   finds the probability of error for the given
%             value of snr_in_dB, SNR in dB.
N=10000;
d=1;                                % min. distance between symbols
Eav=10*d^2;                         % energy per symbol
snr=10^(snr_in_dB/10);              % SNR per bit (given)
sgma=sqrt(Eav/(8*snr));             % noise variance
M=16;
% generation of the data source follows
for i=1:N,
  temp=rand;                        % a uniform R.V. between 0 and 1
  dsource(i)=1+floor(M*temp);       % a number between 1 and 16, uniform
end;
% mapping to the signal constellation follows
mapping=[-3*d  3*d;
          -d   3*d;
           d   3*d;
         3*d   3*d;
        -3*d    d;
          -d    d;
           d    d;
         3*d    d;
        -3*d   -d;
          -d   -d;
           d   -d;
         3*d   -d;
        -3*d  -3*d;
          -d  -3*d;
           d  -3*d;
         3*d  -3*d];
for i=1:N,
  qam_sig(i,:)=mapping(dsource(i),:);
end;
% received signal
for i=1:N,
  [n(1) n(2)]=gngauss(sgma);
  r(i,:)=qam_sig(i,:)+n;
end;
% detection and error probability calculation
numoferr=0;
for i=1:N,
  % metric computation follows
  for j=1:M,
    metrics(j)=(r(i,1)-mapping(j,1))^2+(r(i,2)-mapping(j,2))^2;
  end;
  [min_metric decis] = min(metrics);
  if (decis~=dsource(i)),
    numoferr=numoferr+1;
  end;
```

end;
p=numoferr/(N);

7.5   Carrier-Frequency Modulation

We have described methods for transmitting digital information by modulating either the amplitude of the carrier, the phase of the carrier, or the combined amplitude and phase. Digital information can also be transmitted by modulating the frequency of the carrier.

As we will observe from our treatment below, digital transmission by frequency modulation is a modulation method that is appropriate for channels that lack the phase stability that is necessary to perform carrier-phase estimation. In contrast, the linear modulation methods that we have introduced—namely, PAM, coherent PSK, and QAM—require the estimation of the carrier phase to perform phase-coherent detection.

7.5.1   Frequency-Shift Keying

The simplest form of frequency modulation is binary frequency-shift keying (FSK). In binary FSK we employ two different frequencies, say, f_1 and $f_2 = f_1 + \Delta f$, to transmit a binary information sequence. The choice of frequency separation $\Delta f = f_2 - f_1$ is considered later. Thus, the two signal waveforms may be expressed as

$$u_1(t) = \sqrt{\frac{2\mathcal{E}_b}{T_b}} \cos 2\pi f_1 t, \qquad 0 \le t \le T_b$$

$$u_2(t) = \sqrt{\frac{2\mathcal{E}_b}{T_b}} \cos 2\pi f_2 t, \qquad 0 \le t \le T_b \tag{7.5.1}$$

where $\mathcal{E}_b$ is the signal energy per bit and T_b is the duration of the bit interval.

More generally, M-ary FSK may be used to transmit a block of $k = \log_2 M$ bits per signal waveform. In this case, the M signal waveforms may be expressed as

$$u_m(t) = \sqrt{\frac{2\mathcal{E}_s}{T}} \cos(2\pi f_c t + 2\pi m \Delta f t), \qquad m = 0, 1, \ldots, M-1, \quad 0 \le t \le T \tag{7.5.2}$$

where $\mathcal{E}_s = k\mathcal{E}_b$ is the energy per symbol, $T = kT_b$ is the symbol interval, and Δf is the frequency separation between successive frequencies—that is, $\Delta f = f_m - f_{m-1}$ for all $m = 1, 2, \ldots, M-1$, where $f_m = f_c + m\Delta f$.

Note that the M FSK waveforms have equal energy, $\mathcal{E}_s$. The frequency separation Δf determines the degree to which we can discriminate among the M possible transmitted signals. As a measure of the similarity (or dissimilarity) between a pair of signal waveforms, we use the correlation coefficient γ_{mn}:

$$\gamma_{mn} = \frac{1}{\mathcal{E}_s} \int_0^T u_m(t) u_n(t) \, dt \tag{7.5.3}$$

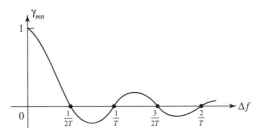

Figure 7.25   Cross-correlation coefficient as a function of frequency separation for FSK signals

By substituting for $u_m(t)$ and $u_m(t)$ in (7.5.3), we obtain

$$
\begin{aligned}
\gamma_{mn} &= \frac{1}{\mathcal{E}_s} \int_0^T \frac{2\mathcal{E}_s}{T} \cos(2\pi f_c t + 2\pi m \Delta f t) \cos(2\pi f_c t + 2\pi n \Delta f t)\, dt \\
&= \frac{1}{T} \int_0^T \cos 2\pi (m-n) \Delta f t\, dt + \frac{1}{T} \int_0^T \cos[4\pi f_c t + 2\pi (m+n) \Delta f t]\, dt \\
&= \frac{\sin 2\pi (m-n) \Delta f T}{2\pi (m-n) \Delta f T}
\end{aligned}
\tag{7.5.4}
$$

where the second integral vanishes when $f_c \gg 1/T$. A plot of γ_{mn} as a function of the frequency separation Δf is given in Figure 7.25. We observe that the signal waveforms are orthogonal when Δf is a multiple of $1/2T$. Hence, the minimum frequency separation between successive frequencies for orthogonality is $1/2T$. We also note that the minimum value of the correlation coefficient is $\gamma_{mn} = -0.217$, which occurs at the frequency separation $\Delta f = 0.715/T$.

M-ary orthogonal FSK waveforms have a geometric representation as M M-dimensional orthogonal vectors, given as

$$
s_0 = (\sqrt{\mathcal{E}_s}, 0, \ldots, 0)
\tag{7.5.5}
$$

$$
s_1 = (0, \sqrt{\mathcal{E}_s}, 0, \ldots, 0)
\tag{7.5.6}
$$

$$
\vdots
\tag{7.5.7}
$$

$$
s_{M-1} = (0, 0, \ldots, 0, \sqrt{\mathcal{E}_s})
\tag{7.5.8}
$$

where the basis functions are $\psi_m(t) = \sqrt{2/T} \cos 2\pi (f_c + m \Delta f) t$. The distance between pairs of signal vectors is $d = \sqrt{2\mathcal{E}_s}$ for all m, n, which is also the minimum distance among the M signals. Note that these signals are equivalent to the M-ary baseband orthogonal signals described in Section 5.4.

The demodulation and detection of the M-ary FSK signals are considered next.

7.5.2   Demodulation and Detection of FSK Signals

Let us assume that the FSK signals are transmitted through an additive white Gaussian noise channel. Furthermore, we assume that each signal is delayed in the transmission through the channel. Consequently, the filtered received signal at the input to the demodulator may be expressed as

$$r(t) = \sqrt{\frac{2\mathcal{E}_s}{T}} \, \cos(2\pi f_c t + 2\pi m \Delta f t + \phi_m) + n(t) \tag{7.5.9}$$

where ϕ_m denotes the phase shift of the mth signal (due to the transmission delay) and $n(t)$ represents the additive bandpass noise, which may be expressed as

$$n(t) = n_c(t) \cos 2\pi f_c t - n_s(t) \sin 2\pi f_c t \tag{7.5.10}$$

The demodulation and detection of the M FSK signals may be accomplished by one of two methods. One approach is to estimate the M carrier-phase shifts $\{\phi_m\}$ and perform *phase-coherent demodulation and detection*. As an alternative method, the carrier phases may be ignored in the demodulation and detection.

In phase-coherent demodulation, the received signal $r(t)$ is correlated with each of the M possible received signals $\cos\left(2\pi f_c t + 2\pi m \Delta f t + \hat{\phi}_m\right)$, for $m = 0, 1, \dots, M - 1$, where $\{\hat{\phi}_m\}$ are the carrier-phase estimates. A block diagram illustrating this type of demodulation is shown in Figure 7.26. It is interesting to note that when $\hat{\phi}_m \neq \phi_m$ for $m = 0, 1, \dots, M - 1$ (imperfect phase estimates), the frequency separation required for signal orthogonality at the demodulator is $\Delta f = 1/T$, which is twice the minimum separation for orthogonality when $\phi = \hat{\phi}$.

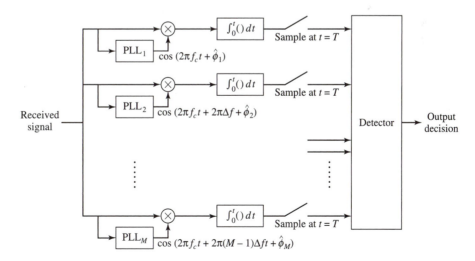

Figure 7.26   Phase-coherent demodulation of M-ary FSK signals

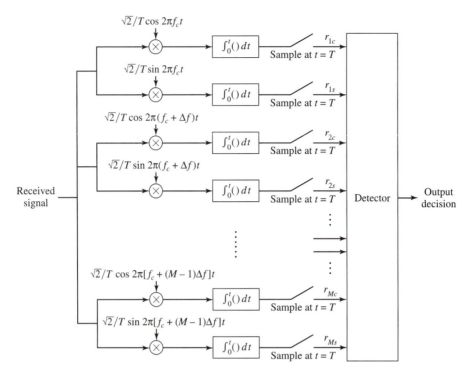

Figure 7.27   Demodulation of M-ary signals for noncoherent detection

The requirement of estimating M carrier phases makes coherent demodulation of FSK signals extremely complex and impractical, especially when the number of signals is large. Therefore, we shall not consider coherent detection of FSK signals.

Instead, we consider a method for demodulation and detection that does not require knowledge of the carrier phases. The demodulation may be accomplished as shown in Figure 7.27. In this case, there are two correlators per signal waveform, or a total of $2M$ correlators, in general. The received signal is correlated with the basis functions (quadrature carriers) $\sqrt{(2/T)}\cos(2\pi f_c t + 2\pi m\Delta f t)$ and $\sqrt{(2/T)}\sin(2\pi f_c t + 2\pi m\Delta f t)$ for $m = 0, 1, \ldots, M-1$. The $2M$ outputs of the correlators are sampled at the end of the signal interval and are passed to the detector. Thus, if the mth signal is transmitted, the $2M$ samples at the detector may be expressed as

$$r_{kc} = \sqrt{\mathcal{E}_s}\left[\frac{\sin 2\pi(k-m)\Delta fT}{2\pi(k-m)\Delta fT}\cos\phi_m - \frac{\cos 2\pi(k-m)\Delta fT - 1}{2\pi(k-m)\Delta fT}\sin\phi_m\right] + n_{kc}$$

$$r_{ks} = \sqrt{\mathcal{E}_s}\left[\frac{\cos 2\pi(k-m)\Delta fT - 1}{2\pi(k-m)\Delta fT}\cos\phi_m + \frac{\sin 2\pi(k-m)\Delta fT}{2\pi(k-m)\Delta fT}\sin\phi_m\right] + n_{ks}$$

$$(7.5.11)$$

where n_{kc} and n_{ks} denote the Gaussian noise components in the sampled outputs.

We observe that when $k = m$, the sampled values to the detector are

$$r_{mc} = \sqrt{\mathcal{E}_s} \cos \phi_m + n_{mc}$$
$$r_{ms} = \sqrt{\mathcal{E}_s} \sin \phi_m + n_{ms} \tag{7.5.12}$$

Furthermore, we observe that when $k \neq m$, the signal components in the samples r_{kc} and r_{ks} will vanish, independent of the values of the phase shift ϕ_k, provided that the frequency separation between successive frequencies is $\Delta f = 1/T$. In such a case, the other $2(M-1)$ correlator outputs consist of noise only—that is,

$$r_{kc} = n_{kc}, \qquad r_{ks} = n_{ks}, \qquad k \neq m \tag{7.5.13}$$

In the following development, we assume that $\Delta f = 1/T$, so that the signals are orthogonal.

It can be shown that the $2M$ noise samples $\{n_{kc}\}$ and $\{n_{ks}\}$ are zero-mean, mutually uncorrelated Gaussian random variables with equal variance $\sigma^2 = N_0/2$. Consequently, the joint probability density function for r_{mc} and r_{ms} conditioned on ϕ_m is

$$f_{r_m}(r_{mc}, r_{ms} \mid \phi_m) = \frac{1}{2\pi\sigma^2} e^{-[(r_{mc}-\sqrt{\mathcal{E}_s}\cos\phi_m)^2 + (r_{ms}-\sqrt{\mathcal{E}_s}\sin\phi_m)^2]/2\sigma^2} \tag{7.5.14}$$

and for $m \neq k$, we have

$$f_{r_k}(r_{kc}, r_{ks}) = \frac{1}{2\pi\sigma^2} e^{-(r_{kc}^2 + r_{ks}^2)/2\sigma^2} \tag{7.5.15}$$

Given the $2M$ observed random variables $\{r_{kc}, r_{ks}\}_{k=0}^{M-1}$, the optimum detector selects the signal that corresponds to the maximum of the posterior probabilities—that is,

$$P\,[s_m \text{ was transmitted} \mid \boldsymbol{r}] \equiv P(s_m \mid \boldsymbol{r}), \qquad m = 0, 1, \ldots, M-1 \tag{7.5.16}$$

where $\boldsymbol{r}$ is the $2M$-dimensional vector with elements $\{r_{kc}, r_{ks}\}_{k=0}^{M-1}$. When the signals are equally probable, the optimum detector specified by (7.5.16) computes the signal envelopes, defined as

$$r_m = \sqrt{r_{mc}^2 + r_{ms}^2}, \qquad m = 0, 1, \ldots, M-1 \tag{7.5.17}$$

and selects the signal corresponding to the largest envelope of the set $\{r_m\}$. In this case, the optimum detector is called an *envelope detector*.

An equivalent detector is one that computes the squared envelopes

$$r_m^2 = r_{mc}^2 + r_{ms}^2, \qquad m = 0, 1, \ldots, M-1 \tag{7.5.18}$$

and selects the signal corresponding to the largest, $\{r_m^2\}$. In this case, the optimum detector is called a *square-law detector*.

─────(ILLUSTRATIVE PROBLEM)──────────────────────────────

Illustrative Problem 7.7 [FSK signaling] Consider a binary communication system that employs the two FSK signal waveforms given as

$$u_1(t) = \cos 2\pi f_1 t, \qquad 0 \leq t \leq T_b$$
$$u_2(t) = \cos 2\pi f_2 t, \qquad 0 \leq t \leq T_b$$

where $f_1 = 1000/T_b$ and $f_2 = f_1 + 1/T_b$. The channel imparts a phase shift of $\phi = 45°$ on each of the transmitted signals, so that the received signal in the absence of noise is

$$r(t) = \cos\left(2\pi f_i t + \frac{\pi}{4}\right), \qquad i = 1, 2, \qquad 0 \leq t \leq T_b$$

Numerically implement the correlation-type demodulator for the FSK signals.

─────(SOLUTION)──────────────────────────────

We sample the received signal $r(t)$ at a rate $F_s = 5000/T_b$ in the bit interval T_b. Thus, the received signal $r(t)$ is represented by the 5000 samples $\{r(n/F_s)\}$. The correlation demodulator multiplies $\{r(n/F_s)\}$ by the sampled version of $u_1(t) = \cos 2\pi f_1 t$, $v_1(t) = \sin 2\pi f_1 t$, $u_2(t) = \cos 2\pi f_2 t$, and $v_2(t) = \sin 2\pi f_2 t$, as illustrated in Figure 7.27. Thus, the correlator outputs are

$$r_{1c}(k) = \sum_{n=0}^{k} r\left(\frac{n}{F_s}\right) u_1\left(\frac{n}{F_s}\right), \qquad k = 1, 2, \ldots, 5000$$

$$r_{1s}(k) = \sum_{n=0}^{k} r\left(\frac{n}{F_s}\right) v_1\left(\frac{n}{F_s}\right), \qquad k = 1, 2, \ldots, 5000$$

$$r_{2c}(k) = \sum_{n=0}^{k} r\left(\frac{n}{F_s}\right) u_2\left(\frac{n}{F_s}\right), \qquad k = 1, 2, \ldots, 5000$$

$$r_{2s}(k) = \sum_{n=0}^{k} r\left(\frac{n}{F_s}\right) v_2\left(\frac{n}{F_s}\right), \qquad k = 1, 2, \ldots, 5000$$

The detector is a square-law detector that computes the two decision variables

$$r_1 = r_{1c}^2(5000) + r_{1s}^2(5000)$$
$$r_2 = r_{2c}^2(5000) + r_{2s}^2(5000)$$

and selects the information bit corresponding to the larger decision variable.

A MATLAB program that implements the correlations numerically is given next. The graphs of the correlator outputs are shown in Figure 7.28, based on the signal $u_1(t)$ being transmitted.

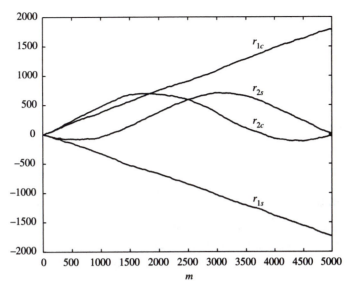

Figure 7.28   Output of correlators for binary FSK demodulation

---

M-FILE

```
% MATLAB script for Illustrative Problem 7.7.
echo on
Tb=1;
f1=1000/Tb;
f2=f1+1/Tb;
phi=pi/4;
N=5000;                              % number of samples
t=0:Tb/(N−1):Tb;
u1=cos(2*pi*f1*t);
u2=cos(2*pi*f2*t);
% assuming that u1 is transmitted, the received signal r is
sgma=1;                              % noise variance
for i=1:N,
  r(i)=cos(2*pi*f1*t(i)+phi)+gngauss(sgma);
  echo off;
end;
echo on ;
% the correlator outputs are computed next
v1=sin(2*pi*f1*t);
v2=sin(2*pi*f2*t);
r1c(1)=r(1)*u1(1);
r1s(1)=r(1)*v1(1);
r2c(1)=r(1)*u2(1);
r2s(1)=r(1)*v2(1);
for k=2:N,
  r1c(k)=r1c(k−1)+r(k)*u1(k);
```

```
    r1s(k)=r1s(k−1)+r(k)*v1(k);
    r2c(k)=r2c(k−1)+r(k)*u2(k);
    r2s(k)=r2s(k−1)+r(k)*v2(k);
    echo off;
end;
echo on;
% decision variables
r1=r1c(5000)^2+r1s(5000)^2;
r2=r2c(5000)^2+r2s(5000)^2;
% plotting commands follow
```

7.5.3 Probability of Error for Noncoherent Detection of FSK

The derivation of the performance of the optimum envelope detector for the M-ary FSK signals can be found in most texts on digital communication. The probability of a symbol error may be expressed as

$$P_M = \sum_{n=1}^{M-1} (-1)^{n+1} \binom{M-1}{n} \frac{1}{n+1} e^{-nk\mathscr{E}_b/N_0(n+1)} \tag{7.5.19}$$

When $M = 2$, this expression reduces to the probability of error for binary FSK, which is

$$P_2 = \frac{1}{2} e^{-\mathscr{E}_b/2N_0} \tag{7.5.20}$$

For $M > 2$, the bit-error probability may be obtained from the symbol-error probability by means of the relation

$$P_b = \frac{2^{k-1}}{2^k - 1} P_M \tag{7.5.21}$$

The bit-error probability is plotted in Figure 7.29 as a function of the SNR per bit for $M = 2, 4, 8, 16, 32$. We observe that for any given bit-error probability, the SNR per bit decreases as M increases. In the limit as $M \to \infty$, the error probability can be made arbitrarily small provided that the SNR per bit exceeds -1.6 dB. This is the channel capacity limit, or *Shannon limit*, for any digital communication system transmitting information through an AWGN channel.

The cost of increasing M is the bandwidth required to transmit the signals. Since the frequency separation between adjacent frequencies is $\Delta f = 1/T$ for signal orthogonality, the bandwidth required for the M signals is $W = M/T$. The bit rate is $R = k/T$, where $k = \log_2 M$. Therefore, the bit-rate-to-bandwidth ratio is

$$\frac{R}{W} = \frac{\log_2 M}{M} \tag{7.5.22}$$

We observe that $R/W \to 0$ as $M \to \infty$.

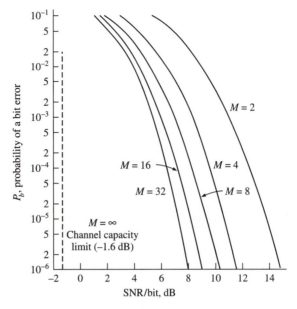

Figure 7.29   Probability of a bit error for noncoherent detection of orthogonal FSK signals

---

ILLUSTRATIVE PROBLEM

Illustrative Problem 7.8 [Binary FSK simulation] Perform a Monte Carlo simulation of a binary FSK communication system in which the signal waveforms are given by (7.5.1), where $f_2 = f_1 + 1/T_b$ and the detector is a square-law detector. The block diagram of the binary FSK system to be simulated is shown in Figure 7.30.

SOLUTION

Since the signals are orthogonal, when $u_1(t)$ is transmitted, the first demodulator output is

$$r_{1c} = \sqrt{\mathcal{E}_b}\,\cos\phi + n_{1c}$$
$$r_{1s} = \sqrt{\mathcal{E}_b}\,\sin\phi + n_{1s}$$

and the second demodulator output is

$$r_{2c} = n_{2c}$$
$$r_{2s} = n_{2s}$$

where n_{1c}, n_{1s}, n_{2c}, and n_{2s} are mutually statistically independent, zero-mean Gaussian random variables with variance σ^2 and ϕ represents the channel-phase shift.

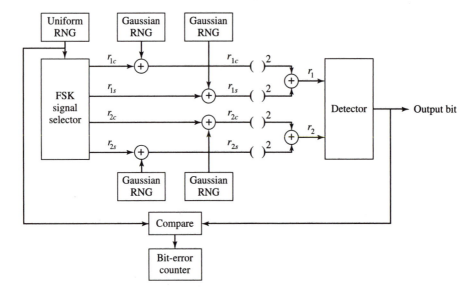

Figure 7.30   Block diagram of a binary FSK system for the Monte Carlo simulation

In the above expression, the channel-phase shift ϕ may be set to zero for convenience. The square-law detector computes

$$r_1 = r_{1c}^2 + r_{1s}^2$$
$$r_2 = r_{2c}^2 + r_{2s}^2$$

and selects the information bit corresponding to the larger of these two decision variables. An error counter measures the error rate by comparing the transmitted sequence to the output of the detector.

The MATLAB programs that implement the Monte Carlo simulation are given next. Figure 7.31 illustrates the measured error rate and compares it with the theoretical error probability given by (7.5.20).

M-FILE

```
% MATLAB script for Illustrative Problem 7.8.
echo on
SNRindB1=0:2:15;
SNRindB2=0:0.1:15;
for i=1:length(SNRindB1),
  smld_err_prb(i)=cm_sm52(SNRindB1(i));    % simulated error rate
  echo off ;
end;
echo on ;
for i=1:length(SNRindB2),
```

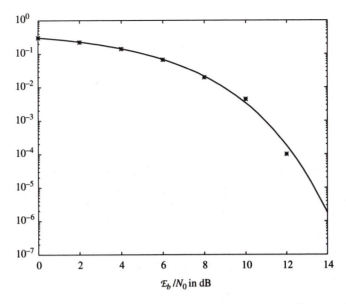

Figure 7.31   Performance of a binary FSK system from the Monte Carlo simulation

```
    SNR=exp(SNRindB2(i)*log(10)/10);           % signal to noise ratio
    theo_err_prb(i)=(1/2)*exp(−SNR/2);         % theoretical symbol error rate
    echo off;
end;
echo on;
% Plotting commands follow
semilogy(SNRindB1,smld_err_prb,' * ');
hold
semilogy(SNRindB2,theo_err_prb);
```

M-FILE

```
function [p]=cm_sm52(snr_in_dB)
% [p]=cm_sm52(snr_in_dB)
%               CM_SM52  Returns the probability of error for the given
%               value of snr_in_dB, signal to noise ratio in dB.
N=10000;
Eb=1;
d=1;
snr=10^(snr_in_dB/10);              % signal to noise ratio per bit
sgma=sqrt(Eb/(2*snr));             % noise variance
phi=0;
% generation of the data source follows
for i=1:N,
  temp=rand;                        % a uniform random variable between 0 and 1
```

```
    if (temp<0.5),
       dsource(i)=0;
    else
       dsource(i)=1;
    end;
 end;
 % detection and the probability of error calculation
 numoferr=0;
 for i=1:N,
    % demodulator output
    if (dsource(i)==0),
       r0c=sqrt(Eb)*cos(phi)+gngauss(sgma);
       r0s=sqrt(Eb)*sin(phi)+gngauss(sgma);
       r1c=gngauss(sgma);
       r1s=gngauss(sgma);
    else
       r0c=gngauss(sgma);
       r0s=gngauss(sgma);
       r1c=sqrt(Eb)*cos(phi)+gngauss(sgma);
       r1s=sqrt(Eb)*sin(phi)+gngauss(sgma);
    end;
    % square law detector outputs
    r0=r0c^2+r0s^2;
    r1=r1c^2+r1s^2;
    % decision is made next
    if (r0>r1),
       decis=0;
    else
       decis=1;
    end;
    % Let's increment the error counter, if the decision is not correct
    if (decis~=dsource(i)),
       numoferr=numoferr+1;
    end;
 end;
 p=numoferr/(N);
```

7.6   Synchronization in Communication Systems

In Section 3.3, we described the demodulation process for AM signals. In particular, we
showed that we can classify the demodulation schemes into *coherent* and *noncoherent*
schemes. In coherent demodulation, the AM signal is multiplied by a sinusoid with the
same frequency and phase of the carrier and then demodulated. In noncoherent demodulation, which can be applied only to the conventional AM scheme, envelope demodulation
is employed, and there is no need for precise tracking of the phase and the frequency of
the carrier at the receiver. Furthermore, we showed in Illustrative Problem 3.6 that correct
phase synchronization in coherent demodulation is essential, and phase errors can result in
considerable performance degradation.

In this chapter, we discussed the demodulation schemes for digital carrier-modulated systems. In the demodulation of PAM, PSK, and QAM, we assumed that we have complete knowledge of the carrier frequency and phase.

In this section, we discuss methods to generate sinusoids with the same frequency and phase of the carrier at the demodulator. These methods are studied under the title of *carrier synchronization* and apply to both analog and digital carrier modulation systems discussed in this chapter and Chapter 3. Another type of synchronization, called *timing synchronization, clock synchronization*, or *timing recovery*, is encountered only in digital communication systems. We briefly discuss this type of synchronization problem in this section as well.

7.6.1 Carrier Synchronization

A carrier-synchronization system consists of a local oscillator whose phase is controlled to be in synch with the carrier signal. This is achieved by employing a *phase-locked loop* (PLL). A phase-locked loop is a nonlinear feedback-control system for controlling the phase of the local oscillator. In the following discussion, for the sake of simplicity, we consider only binary PSK modulation systems.

The PLL is driven by a sinusoidal signal at the carrier frequency (or a multiple of it). In order to obtain the sinusoidal signal to drive the PLL, the DSB-modulated signal

$$u(t) = A_c m(t) \cos(2\pi f_c t - \phi(t)) \tag{7.6.1}$$

where $m(t) = \pm 1$, is squared to obtain

$$
\begin{aligned}
u^2(t) &= A_c^2 m^2(t) \cos^2(2\pi f_c t - \phi(t)) \\
&= \frac{A_c^2}{2} m^2(t) + \frac{A_c^2}{2} m^2(t) \cos(4\pi f_c t - 2\phi(t)) \\
&= \frac{A_c^2}{2} + \frac{A_c^2}{2} \cos(4\pi f_c t - 2\phi(t)) \tag{7.6.2}
\end{aligned}
$$

Obviously, this signal has a component at $2f_c$. The reason that we do not deal directly with $u(t)$ is that usually the process $m(t)$ is zero-mean, so the power content of $u(t)$ at f_c is zero. Now, if the signal $u^2(t)$ is passed through a bandpass filter tuned to $2f_c$, the output will be a sinusoidal signal with the central frequency $2f_c$, a phase of $-2\phi(t)$, and an amplitude of $A_c^2 H(2f_c)/2$. Without loss of generality, we can assume that the amplitude is unity; that is, the input to the PLL is

$$r(t) = \cos(4\pi f_c t - 2\phi(t)) \tag{7.6.3}$$

The PLL consists of a multiplier, a loop filter, and a voltage-controlled oscillator (VCO), as shown in Figure 7.32. If we assume that the output of the VCO is $\sin(4\pi f_c t - 2\hat{\phi}(t))$, then at the input of the loop filter we have

$$
\begin{aligned}
e(t) &= \cos(4\pi f_c t - 2\phi(t)) \sin(4\pi f_c t - 2\hat{\phi}(t)) \\
&= \frac{1}{2} \sin(2\phi(t) - 2\hat{\phi}(t)) + \frac{1}{2} \sin(8\pi f_c t - 2\phi(t) - 2\hat{\phi}(t)) \tag{7.6.4}
\end{aligned}
$$

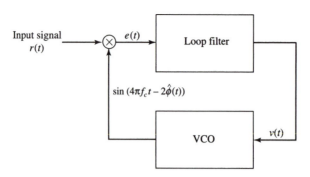

Figure 7.32   The phase-locked loop

Note that $e(t)$ has a high- and a low-frequency component. The role of the loop filter is to remove the high-frequency component and to make sure that $\hat{\phi}(t)$ follows closely the variations in $\phi(t)$. A simple loop filter is a first-order lowpass filter with a transfer function of

$$G(s) = \frac{1 + \tau_1 s}{1 + \tau_2 s} \tag{7.6.5}$$

where $\tau_2 \gg \tau_1$. If we denote the input of the VCO as $v(t)$, then the output of the VCO will be a sinusoid whose instantaneous frequency deviation from $2f_c$ is proportional to $v(t)$. But the instantaneous frequency of the VCO output is

$$2f_c + \frac{1}{\pi} \frac{d}{dt} \hat{\phi}(t)$$

therefore,

$$\frac{d}{dt} \hat{\phi}(t) = \frac{K}{2} v(t) \tag{7.6.6}$$

or, equivalently,

$$2\hat{\phi}(t) = K \int_{-\infty}^{t} v(\tau) \, d\tau \tag{7.6.7}$$

where K is some proportionality constant. After removal of the second and fourth harmonics, the PLL reduces to the one shown in Figure 7.33.

Assuming that $\hat{\phi}(t)$ closely follows changes in $\phi(t)$, the difference $2\phi(t) - 2\hat{\phi}(t)$ is very small, and we can use the approximation

$$\frac{1}{2} \sin\left(2\phi(t) - 2\hat{\phi}(t)\right) \approx \phi(t) - \hat{\phi}(t) \tag{7.6.8}$$

With this approximation, the only nonlinear component in Figure 7.33 is replaced by a linear component, resulting in the *linearized PLL model* shown in Figure 7.34. Note that

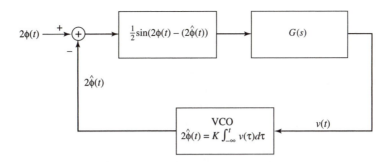

Figure 7.33   The phase-locked loop after removal of high-frequency components

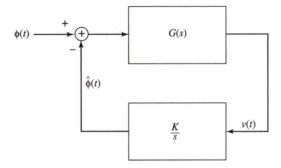

Figure 7.34   The linearized model for a phase-locked loop

this model is represented in the transform domain, and the integrator is replaced by its transform-domain equivalent, $1/s$.

The model shown in Figure 7.34 is a linear control system with a forward gain of $G(s)$ and a feedback gain of K/s; therefore, the transfer function of the system is given by

$$H(s) = \frac{\hat{\Phi}(s)}{\Phi(s)} = \frac{KG(s)/s}{1 + KG(s)/s} \qquad (7.6.9)$$

and with the first-order model for $G(s)$ assumed before,

$$G(s) = \frac{1 + \tau_1 s}{1 + \tau_2 s} \qquad (7.6.10)$$

$H(s)$ is given as

$$H(s) = \frac{1 + \tau_1 s}{1 + (\tau_1 + 1/K)s + \tau_2 s^2/K} \qquad (7.6.11)$$

With $H(s)$ as the transfer function, if the input to the PLL is $\Phi(s)$, the error will be

$$
\begin{aligned}
\Delta\Phi(s) &= \Phi(s) - \hat{\Phi}(s) \\
&= \Phi(s) - \Phi(s)H(s) \\
&= [1 - H(s)]\,\Phi(s) \\
&= \frac{(1 + \tau_2 s)s}{K + (1 + K\tau_1)s + \tau_2 s^2}\Phi(s)
\end{aligned}
\tag{7.6.12}
$$

Now let us assume that up to a certain time $\phi(t) \approx \hat{\phi}(t)$, so $\Delta\phi(t) \approx 0$. At this time, some abrupt change causes a jump in $\phi(t)$ that can be modeled as a step—that is, $\Phi(s) = K_1/s$. With this change, we have

$$
\begin{aligned}
\Delta\Phi(s) &= \frac{(1 + \tau_2 s)s}{K + (1 + K\tau_1)s + \tau_2 s^2}\frac{K_1}{s} \\
&= \frac{K_1(1 + \tau_2 s)}{K + (1 + K\tau_1)s + \tau_2 s^2}
\end{aligned}
\tag{7.6.13}
$$

Now, by using the final value theorem of the Laplace transform, which states that

$$
\lim_{t \to \infty} f(t) = \lim_{s \to 0} sF(s)
\tag{7.6.14}
$$

as long as all poles of $sF(s)$ have negative real parts, we conclude that

$$
\begin{aligned}
\lim_{t \to \infty} \Delta\phi(t) &= \lim_{s \to 0} s\Phi(s) \\
&= \lim_{s \to 0} \frac{K_1 s(1 + \tau_2 s)}{K + (1 + K\tau_1)s + \tau_2 s^2} \\
&= 0
\end{aligned}
\tag{7.6.15}
$$

In other words, a simple first-order loop filter results in a PLL that can track jumps in the input phase.

The transfer function (7.6.11) can be written in the standard form

$$
H(s) = \frac{(2\zeta\omega_n - \omega_n^2/K)s + \omega_n^2}{s^2 + 2\zeta\omega_n s + \omega_n^2}
\tag{7.6.16}
$$

Here,

$$
\omega_n = \sqrt{\frac{K}{\tau_2}}
$$

$$
\zeta = \frac{\omega_n(\tau_1 + 1/K)}{2}
$$

where ω_n is the natural frequency and ζ is the damping factor.

ILLUSTRATIVE PROBLEM

Illustrative Problem 7.9 [First-order PLL] Assuming that

$$G(s) = \frac{1 + 0.01s}{1 + s}$$

and $K = 1$, determine and plot the response of the PLL to an abrupt change of height 1 to the input phase.

SOLUTION

Here, $\tau_1 = 0.01$ and $\tau_2 = 1$; therefore,

$$\omega_n = 1$$
$$\zeta = 0.505$$

which results in

$$H(s) = \frac{0.01s + 1}{s^2 + 1.01s + 1}$$

Thus, the response to $\phi(t) = u(t)$—that is, $\Phi(s) = 1/s$—is given by

$$\hat{\Phi}(s) = \frac{0.01s + 1}{s^3 + 1.01s^2 + s + 1}$$

In order to determine and plot the time response $\hat{\phi}(t)$ to the input $u(t)$, we note that we have to determine the output of a system with transfer function $H(s)$ to the input $u(t)$. This can be done most easily by using state-space techniques. We employ the MATLAB function tf2ss.m, which returns the state-space model of a system described by its transfer function. After determining the state-space representation of the system, we obtain the step response numerically.

The function tf2ss.m takes the numerator and the denominator of the transfer function $H(s)$ and returns A, B, C, and D, its state-space representation, in the form

$$\begin{cases} \dfrac{d}{dt}x(t) = Ax(t) + Bu(t) \\ \quad y(t) = Cx(t) + Du(t) \end{cases}$$

This representation can be approximated by

$$\begin{cases} x(t + \Delta t) = x(t) + Ax(t)\Delta t + Bu(t)\Delta t \\ \quad y(t) = Cx(t) + Du(t) \end{cases}$$

or, equivalently,

$$\begin{cases} x(i + 1) = x(i) + Ax(i)\Delta t + Bu(i)\Delta t \\ \quad y(i) = Cx(i) + Du(i) \end{cases}$$

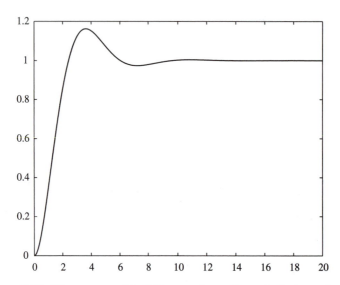

Figure 7.35   The response of the PLL to an abrupt change in the input phase in Illustrative Problem 7.9

For this problem, it is sufficient to choose $u(t)$ to be a step function and the numerator and the denominator vectors of $H(s)$ to be [0.01 1] and [1 1.01 1], respectively. With this choice of numerator and denominator vectors, the state-space parameters of the system will be

$$A = \begin{bmatrix} -1.01 & -1 \\ 1 & 0 \end{bmatrix}$$

$$B = \begin{bmatrix} 1 \\ 0 \end{bmatrix}$$

$$C = \begin{bmatrix} 0.01 & 1 \end{bmatrix}$$

$$D = 0$$

The plot of the output of the PLL is given in Figure 7.35.

As we can see from Figure 7.35, the output of the PLL eventually follows the input; however, the speed by which it follows the input depends on the loop filter parameters and K, the VCO proportionality constant.

The MATLAB script for this problem is given next.

M-FILE

```
% MATLAB script for Illustrative Problem 7.9.
echo on
num=[0.01 1];
```

```
den=[1 1.01 1];
[a,b,c,d]=tf2ss(num,den);
dt=0.01;
u=ones(1,2000);
x=zeros(2,2001);
for i=1:2000
    x(:,i+1)=x(:,i)+dt.*a*x(:,i)+dt.*b*u(i);
    y(i)=c*x(:,i);
    echo off;
end
echo on;
t=[0:dt:20];
plot(t(1:2000),y)
```

7.6.2   Clock Synchronization

In Chapter 5 and in this chapter, we have seen that a popular implementation of the optimal receiver makes use of matched filters and samplers at the matched filter output. In all these cases, we assumed that the receiver has complete knowledge of the sampling instant and can sample perfectly at this time. Systems that achieve this type of synchronization between the transmitter and the receiver are called *timing recovery, clock-synchronization,* or *symbol-synchronization* systems.

A simple implementation of clock synchronization employs an *early-late gate.* The operation of an early-late gate is based on the fact that in a PAM communication system, the output of the matched filter is the autocorrelation function of the basic pulse signal used in the PAM system (possibly with some shift). The autocorrelation function is maximized at the optimum sampling time and is symmetric. This means that, in the absence of noise, at sampling times $T^+ = T + \delta$ and $T^- = T - \delta$, the output of the sampler will be equal—that is,

$$y(T^+) = y(T^-) \tag{7.6.17}$$

In this case, the optimum sampling time is obviously the midpoint between the early and late sampling times:

$$T = \frac{T^+ + T^-}{2} \tag{7.6.18}$$

Now let us assume that we are not sampling at the optimal sampling time T, but instead we are sampling at T_1. If we take two extra samples at $T^+ = T_1 + \delta$ and $T^- = T_1 - \delta$, these samples are not symmetric with respect to the optimum sampling time T and, therefore, will not be equal. A typical autocorrelation function for positive and negative incoming pulses and the three samples is shown in Figure 7.36.

Here

$$T^- = T - \delta_1$$
$$T^+ = T + \delta_2$$

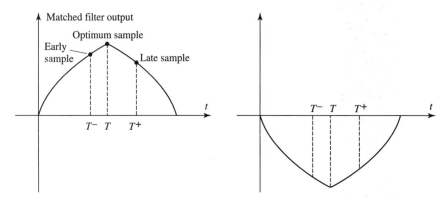

Figure 7.36   The matched filter output and early and late samples

where

$$\delta_1 < \delta_2 \qquad (7.6.19)$$

and, as the figure shows, this results in

$$\left| y(T^-) \right| > \left| y(T^+) \right| \qquad (7.6.20)$$

Also, in this case,

$$T < T_1 = \frac{T^- + T^+}{2} \qquad (7.6.21)$$

Therefore, when $\left| y(T^-) \right| > \left| y(T^+) \right|$, the correct sampling time is before the assumed sampling time, and the sampling should be done earlier. Conversely, when $\left| y(T^-) \right| < \left| y(T^+) \right|$, the sampling time should be delayed. Obviously, when $\left| y(T^-) \right| = \left| y(T^+) \right|$, the sampling time is correct, and no correction is necessary.

The early-late gate synchronization system therefore takes three samples at T_1, $T^- = T_1 - \delta$, and $T^+ = T_1 + \delta$ and then compares $\left| y(T^-) \right|$ and $\left| y(T^+) \right|$ and, depending on their relative values, generates a signal to correct the sampling time.

ILLUSTRATIVE PROBLEM

Illustrative Problem 7.10 [Clock synchronization] A binary PAM communication systems uses a raised-cosine waveform with a roll-off factor of 0.4. The system transmission rate is 4800 bits/s. Write a MATLAB file that simulates the operation of an early-late gate for this system.

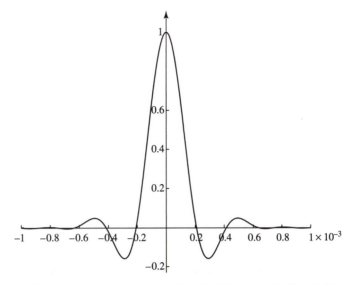

Figure 7.37   The raised-cosine signal in Illustrative Problem 7.10

SOLUTION

Since the rate is 4800 bits/s, we have

$$T = \frac{1}{4800} \tag{7.6.22}$$

and with $\alpha = 0.4$, the expression for a raised-cosine waveform becomes

$$x(t) = \mathrm{sinc}(4800t)\frac{\cos(4800 \times 0.4\pi t)}{1 - 4 \times 0.16 \times 4800^2 t^2}$$

$$= \mathrm{sinc}(4800t)\frac{\cos 1920\pi t}{1 - 1.4746 \times 10^7 t^2} \tag{7.6.23}$$

This signal obviously extends from $-\infty$ to $+\infty$. The plot of this signal is given in Figure 7.37.

From Figure 7.37, it is clear that, for all practical purposes, it is sufficient to consider only the interval $|t| \le 0.6 \times 10^{-3}$, which is roughly $[-3T, 3T]$. Truncating the raised-cosine pulse to this interval and computing the autocorrelation function result in the waveform shown in Figure 7.38.

In the MATLAB script given next, the raised-cosine signal and the autocorrelation function are first computed and plotted. In this particular example, the length of the autocorrelation function is 1201, and the maximum (i.e., the optimum sampling time) occurs at the 600th component. Two cases are examined: one when the incorrect sampling time is 700 and one when it is 500. In both cases, the early-late gate corrects the sampling time to the optimum 600.

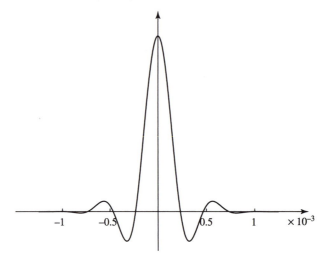

Figure 7.38   The autocorrelation function of the raised-cosine signal

M-FILE

```
% MATLAB script for Illustrative Problem 7.10.
echo on
alpha=0.4;
T=1/4800;
t=[−3*T:1.001*T/100:3*T];
x=sinc(t./T).*(cos(pi*alpha*t./T)./(1−4*alpha^2*t.^2/T^2));
pause % Press any key to see a plot of x(t).
plot(t,x)
y=xcorr(x);
ty=[t−3*T,t(2:length(t))+3*T];
pause % Press any key to see a plot of the autocorrelation of x(t).
plot(ty,y);
d=60;                          % Early and late advance and delay
ee=0.01;                       % Precision
e=1;                           % Step size
n=700;                         % The incorrect sampling time
while  abs(abs(y(n+d))−abs(y(n−d)))>=ee
  if  abs(y(n+d))−abs(y(n−d))>0
    n=n+e;
  elseif  abs(y(n+d))−abs(y(n−d))<0
    n=n−e;
  end
  echo off ;
end
echo on ;
pause % Press any key to see the corrected sampling time
n
n=500;                                  % Another incorrect sampling time
```

```
while abs(abs(y(n+d))−abs(y(n−d)))>=ee
    if abs(y(n+d))−abs(y(n−d))>0
        n=n+e;
    elseif abs(y(n+d))−abs(y(n−d))<0
        n=n−e;
    end
    echo off ;
end
echo on ;
pause % Press any key to see the corrected sampling time
n
```

Problems

7.1 In a carrier-amplitude-modulated PAM system, the transmitter filter has a square-root raised-cosine spectral characteristic with roll-off factor $\alpha = 1$. The carrier frequency is $f_c = 40/T$. Evaluate and graph the spectrum of the baseband signal and the amplitude-modulated PAM signal.

7.2 Repeat Problem 7.1 when the carrier frequency is $f_c = 80/T$.

7.3 Repeat Problem 7.1 when the transmitter has a square-root duobinary spectral characteristic.

7.4 The purpose of this problem is to demonstrate that (7.2.9) and (7.2.10) hold by evaluating (7.2.9) numerically using MATLAB. The pulse $g_T(t)$ may be assumed to be rectangular—that is,

$$g_T(t) = \begin{cases} 1, & 0 \le t \le 2 \\ 0, & \text{otherwise} \end{cases}$$

Let the carrier frequency $f_c = 2000$ Hz. Use a sampling rate of $F_s = 20,000$ samples per second on the signal waveform $\psi(t)$ given by (7.2.6), and compute the energy of $\psi(t)$ by approximating the integral in (7.2.8) by the summation

$$\frac{1}{N} \sum_{n=0}^{N-1} \psi^2(nT_s) = \frac{1}{N} \sum_{n=0}^{N-1} \psi^2\left(\frac{n}{F_s}\right)$$

where $N = 40,000$ samples. Write a MATLAB program to generate the samples $\psi(n/F_s)$, and perform the computation of the signal energy as described above.

7.5 The cross-correlation of the received signal $r(t)$ with $\psi(t)$ as given by (7.2.14) may be performed numerically using MATLAB. Write a MATLAB program that computes the correlator output

$$y(n) = \sum_{k=0}^{n} r\left(\frac{k}{F_s}\right) \psi\left(\frac{k}{F_s}\right), \qquad n = 0, 1, \ldots, N-1$$

where F_s is the sampling frequency. Evaluate and graph $y(n)$ when $r(t) = \psi(t)$, where $\psi(t)$ is the waveform described in Problem 7.4 and $F_s = 20{,}000$ Hz.

7.6 Evaluate and graph the correlation $\{y(n)\}$ in Problem 7.5 when the signal $g_T(t)$ is

$$g_T(t) = \begin{cases} \frac{1}{2}(1 - \cos \pi t), & 0 \le t \le 2 \\ 0, & \text{otherwise} \end{cases}$$

for the same parameters given in Problem 7.4.

7.7 In Illustrative Problem 7.2, the eight PSK waveforms had a constant amplitude. Instead of the rectangular pulse $g_T(t)$, suppose that the signal pulse shape is

$$g_T(t) = \begin{cases} \frac{1}{2}(1 - \cos 2\pi t/T), & 0 \le t \le T \\ 0, & \text{otherwise} \end{cases}$$

Write a MATLAB program to compute and graph the $M = 8$-PSK signal waveforms for the case in which $f_c = 6/T$.

7.8 Write a MATLAB program that numerically computes the cross-correlation of the received signal $r(t)$ for a PSK signal with the two basis functions given by (7.3.9). That is, compute

$$y_c(n) = \sum_{k=0}^{n} r\left(\frac{k}{F_s}\right) \psi_1\left(\frac{k}{F_s}\right), \qquad n = 0, 1, \ldots, N-1$$

$$y_s(n) = \sum_{k=0}^{n} r\left(\frac{k}{F_s}\right) \psi_2\left(\frac{k}{F_s}\right), \qquad n = 0, 1, \ldots, N-1$$

where N is the number of samples of $r(t)$, $\psi_1(t)$, and $\psi_2(t)$. Evaluate and plot these correlation sequences when

$$r(t) = s_{mc}\psi_1(t) + s_{ms}\psi_2(t)$$

$$g_T(t) = \begin{cases} 2, & 0 \le t \le 2 \\ 0, & \text{otherwise} \end{cases}$$

$f_c = 1000$ Hz, $F_s = 10{,}000$ samples per second, and the transmitted signal point is as given.

  a. $s_m = (s_{mc}, s_{ms}) = (1, 0)$

  b. $s_m = (-1, 0)$

  c. $s_m = (0, 1)$

7.9 Write a MATLAB program that performs a Monte Carlo simulation of an $M = 4$-PSK communication system, as described in Illustrative Problem 7.3, but modify the detector so that it computes the received signal phase θ_r as given by (7.3.16) and selects the signal point whose phase is closest to θ_r.

7.10 Write a MATLAB program that implements a differential encoder and a differential decoder for a $M = 4$ DPSK system. Check the operation of the encoder and decoder by passing a sequence of 2-bit symbols through the cascade of the encoder and decoder, and verify that the output sequence is identical to the input sequence.

7.11 Write a MATLAB program that performs a Monte Carlo simulation of a binary DPSK communication system. In this case, the transmitted signal phases are $\theta = 0$ and $\theta = 180°$. A $\theta = 0$ phase change corresponds to the transmission of a zero. A $\theta = 180°$ phase change corresponds to the transmission of a one. Perform the simulation for $N=10,000$ bits at different values of the SNR parameter $\mathcal{E}_b/N_0$. It is convenient to normalize $\mathcal{E}_b$ to unity. Then, with $\sigma^2 = N_0/2$, the SNR is $\mathcal{E}_b/N_0 = 1/2\sigma^2$, where σ^2 is the variance of the additive noise component. Hence, the SNR can be controlled by scaling the variance of the additive noise component. Plot and compare the measured error rate of the binary DPSK with the theoretical error probability given by (7.3.23).

7.12 Write a MATLAB program that generates and graphs the $M = 8$-QAM signal waveforms given by (7.4.2) for the signal constellation shown in Figure P7.12. Assume that the pulse waveform $g_T(t)$ is rectangular—that is,

$$g_T(t) = \begin{cases} 1, & 0 \leq t \leq T \\ 0, & \text{otherwise} \end{cases}$$

and the carrier frequency is $f_c = 8/T$.

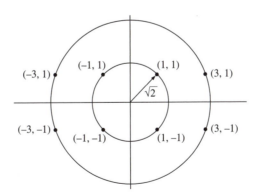

Figure P7.12

7.13 Repeat Problem 7.12 when the pulse waveform $g_T(t)$ is given as

$$g_T(t) = \begin{cases} \frac{1}{2}(1 - \cos 2\pi t/T), & 0 \leq t \leq T \\ 0, & \text{otherwise} \end{cases}$$

7.14 Write a MATLAB program that performs a Monte Carlo simulation of an $M = 8$-QAM communication system for the signal constellation shown in Figure P7.12. Perform the simulation for $N = 10,000$ (3-bit) symbols at different values of the SNR parameter $\mathcal{E}_{avb}/N_0$. It is convenient to normalize $\mathcal{E}_{avb}$ to unity. Then, with $\sigma^2 = N_0/2$, the SNR is $\mathcal{E}_{avb}/N_0 = 1/2\sigma^2$, where σ^2 is the variance of each of the two additive noise components. Plot and compare the measured symbol-error rate of the QAM system with the upper bound on the theoretical error probability given by (7.4.12).

7.15 Repeat the simulation in Problem 7.14 for the $M = 8$-signal constellation shown in Figure P7.15. Compare the error probabilities for the two $M = 8$-QAM signal constellation, and indicate which constellation gives the better performance.

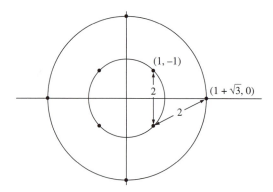

Figure P7.15

7.16 Consider the binary FSK signals of the form

$$u_1(t) = \sqrt{\frac{2\mathcal{E}_b}{T_b}} \cos 2\pi f_1 t, \qquad 0 \le t \le T_b$$

$$u_2(t) = \sqrt{\frac{2\mathcal{E}_b}{T_b}} \cos 2\pi f_2 t, \qquad 0 \le t \le T_b$$

$$f_2 = f_1 + \frac{1}{2T_b}$$

Let $f_1 = 1000/T_b$. By sampling the two waveforms at the rate $F_s = 5000/T_b$, we obtain 5000 samples in the bit interval $0 \le t \le T_b$. Write a MATLAB program that generates the 5000 samples for each of $u_1(t)$ and $u_2(t)$, and compute the cross-correlation

$$\sum_{n=0}^{N-1} u_1\left(\frac{n}{F_s}\right) u_2\left(\frac{n}{F_s}\right)$$

and, thus, verify numerically the orthogonality condition for $u_1(t)$ and $u_2(t)$.

7.17 Use the MATLAB program given in Illustrative Problem 7.7 to compute and graph the correlator outputs when the received signal is

$$r(t) = \cos\left(2\pi f_1 t + \frac{\pi}{2}\right), \qquad 0 \le t \le T_b$$

7.18 Use the MATLAB program given in Illustrative Problem 7.7 to compute and graph the correlator outputs when the transmitted signal is $u_2(t)$ and the received signal is

$$r(t) = \cos\left(2\pi f_2 t + \frac{\pi}{4}\right), \qquad 0 \le t \le T$$

7.19 Write a MATLAB program that performs a Monte Carlo simulation of a quaternary ($M = 4$) FSK communication system and employs the frequencies

$$f_k = f_1 + \frac{k}{T}, \qquad k = 0, 1, 2, 3$$

The detector is a square-law detector. Perform the simulation for N=10,000 (2-bit) symbols at different values of the SNR parameter $\mathcal{E}_b/N_0$, and record the number of symbol errors and bit errors. Plot and compare the measured symbol- and bit-error rates with the theoretical symbol- and bit-error probabilities given by (7.5.19) and (7.5.21).

7.20 In Illustrative Problem 7.9, it was assumed that the input phase has an abrupt jump, and the simulation showed that a first-order loop filter can track such a change. Now assume that the input changes according to a ramp—that is, starts to increase linearly. Simulate the performance of a first-order PLL in this case, and determine whether the loop is capable of tracking such a change.

7.21 Repeat Illustrative Problem 7.10 with a rectangular pulse shape in the presence of AWGN for SNR values of 20, 10, 5, and 0 dB.

Chapter 8

Channel Capacity and Coding

8.1 Preview

The objective of any communication system is to transmit information generated by an information source from one location to another. The medium over which information is transmitted is called the *communication channel*. We have already seen in Chapter 4 that the information content of a source is measured by the entropy of the source, and the most common unit for this quantity is bits. We have also seen that an appropriate mathematical model for an information source is a random process.

In this chapter, we consider appropriate mathematical models for communication channels. We also discuss a quantity called the *channel capacity* that is defined for any communication channel and gives a fundamental limit on the amount of information that can be transmitted through the channel. In particular, we consider two types of channels, the binary symmetric channel (BSC) and the additive white Gaussian noise channel (AWGN).

The second part of this chapter is devoted to coding techniques for reliable communication over communication channels. We discuss the two most commonly used coding methods—namely, block and convolutional coding. Encoding and decoding techniques for these codes and their performance are discussed in detail in the later sections of this chapter.

8.2 Channel Model and Channel Capacity

A communication channel transmits the information-bearing signal to the destination. In this transmission, the information-carrying signal is subject to a variety of changes. Some of these changes are deterministic in nature—for example, attenuation, linear and nonlinear distortion; and some are probabilistic—for example, addition of noise, multipath fading, and so on. Since deterministic changes can be considered as special cases of random changes, in the most general sense the mathematical model for a communication channel is a stochastic dependence between the input and the output signals.

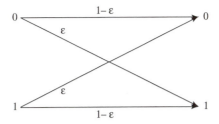

Figure 8.1 A binary symmetric channel (BSC)

8.2.1   Channel Model

In the simplest case a channel can be modeled as a conditional probability relating each output of the channel to its corresponding input. Such a model is called a *discrete-memoryless channel* (DMC) and is completely described by the channel input alphabet $\mathcal{X}$, the channel output alphabet $\mathcal{Y}$, and the *channel transition probability matrix* $p(y|x)$, given for all $x \in \mathcal{X}$, $y \in \mathcal{Y}$. One special case of the discrete-memoryless channel is the *binary symmetric channel* (BSC) that can be considered as a mathematical model for binary transmission over a Gaussian channel with hard decisions at the output. A binary symmetric channel corresponds to the case $\mathcal{X} = \mathcal{Y} = \{0, 1\}$ and $p(y = 0 \mid x = 1) = p(y = 1 \mid x = 0) = \epsilon$. A schematic model for this channel is shown in Figure 8.1. The parameter ϵ is called the *crossover probability* of the channel.

8.2.2   Channel Capacity

By definition, channel capacity is the maximum rate at which reliable transmission of information over the channel is possible. Reliable transmission is possible when there exists a sequence of codes with increasing block length, for which the error probability tends to zero as the block length increases. The channel capacity is denoted by C; by definition, at rates $R<C$, reliable transmission over the channel is possible; at rates $R>C$, reliable transmission is not possible.

Shannon's fundamental result of information theory states that for discrete-memoryless channels, the capacity is given by the following expression:

$$C = \max_{p(x)} I(X; Y) \tag{8.2.1}$$

where $I(X; Y)$ denotes the mutual information between X (channel input) and Y (channel output), and the maximization is carried out over all input probability distributions of the channel.

The mutual information between two random variables X and Y is defined as

$$I(X; Y) = \sum_{x \in \mathcal{X}} \sum_{y \in \mathcal{Y}} p(x) p(y|x) \log \frac{p(x, y)}{p(x) p(y)} \tag{8.2.2}$$

where the mutual information is in bits and the logarithm is in base 2.

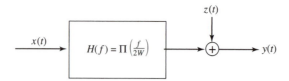

Figure 8.2   Bandlimited additive white Gaussian noise channel

For the case of the binary symmetric channel, the capacity is given by the simple relation

$$C = 1 - H_b(\epsilon) \tag{8.2.3}$$

where ϵ is the crossover probability of the channel and $H_b(\cdot)$ denotes the binary entropy function:

$$H_b(x) = -x \log(x) - (1 - x) \log(1 - x) \tag{8.2.4}$$

Another important channel model is the bandlimited additive white Gaussian noise channel with an input power constraint. This channel is modeled as shown in Figure 8.2.

The channel is bandlimited to $[-W, W]$, the noise is Gaussian and white with a (two-sided) power-spectral density of $N_0/2$, and the channel input is a process that satisfies an input power constraint of P. Shannon showed that the capacity of this channel in bits per second is given by

$$C = W \log \left(1 + \frac{P}{N_0 W} \right) \quad \text{bits/second} \tag{8.2.5}$$

For a discrete-time AWGN channel with input power constraint P and noise variance σ^2, the capacity in bits per transmission is given by

$$C = \frac{1}{2} \log \left(1 + \frac{P}{\sigma^2} \right) \tag{8.2.6}$$

─────────────(ILLUSTRATIVE PROBLEM)─────────────

Illustrative Problem 8.1 [Binary symmetric channel capacity] Binary data are transmitted over an additive white Gaussian noise channel using BPSK signaling and hard-decision decoding at the output using optimal matched filter detection.

1. Plot the error probability of the channel as a function of

$$\gamma = \frac{\mathcal{E}}{N_0} \tag{8.2.7}$$

where $\mathcal{E}$ is the energy in each BPSK signal and $N_0/2$ is the noise power-spectral density. Assume that γ changes from -20 dB to 20 dB.

2. Plot the capacity of the resulting channel as a function of γ.

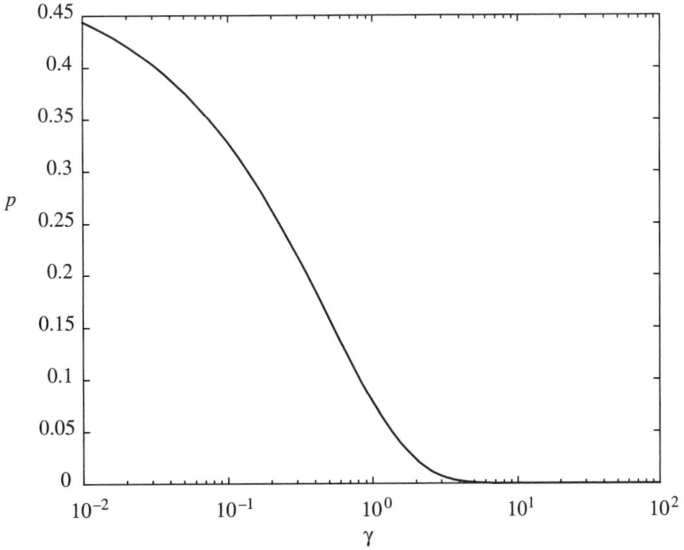

Figure 8.3   BPSK error probability versus $\gamma = \mathcal{E}/N_0$

SOLUTION

1. The error probability of the BPSK with optimal detection is given by

$$p = Q\left(\sqrt{2\gamma}\right) \tag{8.2.8}$$

The corresponding plot is shown in Figure 8.3.

2. Here we use the relation

$$C = 1 - H_b(p)$$
$$= 1 - H_b\left(Q\left(\sqrt{2\gamma}\right)\right) \tag{8.2.9}$$

to obtain a plot of C versus γ. This plot is shown in Figure 8.4.

The MATLAB script for this problem is given next.

M-FILE

```
% MATLAB script for Illustrative Problem 8.1.
echo on
gamma_db=[−20:0.1:20];
gamma=10.^(gamma_db./10);
p_error=q(sqrt(2.*gamma));
```

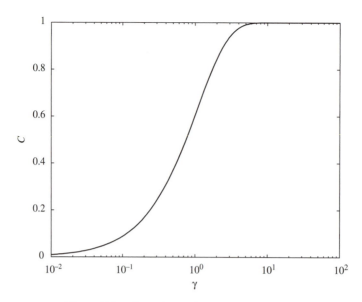

Figure 8.4   Channel capacity versus $\gamma = \mathcal{E}/N_0$

```
capacity=1.−entropy2(p_error);
pause % Press a key to see a plot of error probability vs. SNR/bit
clf
semilogx(gamma,p_error)
xlabel('SNR/bit')
title('Error probability versus SNR/bit')
ylabel('Error Prob.')
pause % Press a key to see a plot of channel capacity vs. SNR/bit
clf
semilogx(gamma,capacity)
xlabel('SNR/bit')
title('Channel capacity versus SNR/bit')
ylabel('Channel capacity')
```

◀─◖ILLUSTRATIVE PROBLEM◗─

Illustrative Problem 8.2 [Gaussian channel capacity]

1. Plot the capacity of an additive white Gaussian noise channel with a bandwidth of $W = 3000$ Hz as a function of P/N_0 for values of P/N_0 between -20 dB and 30 dB.

2. Plot the capacity of an additive white Gaussian noise channel with $P/N_0 = 25$ dB as a function of W. In particular, what is the channel capacity when W increases indefinitely?

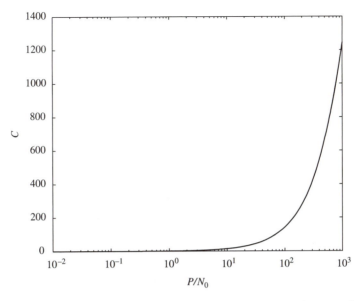

Figure 8.5   Capacity of an AWGN channel with $W = 3000$ Hz as a function of P/N_0

SOLUTION

1. The desired plot is given in Figure 8.5.

2. The capacity as a function of bandwidth is plotted in Figure 8.6. As is seen in the plots, when either P/N_0 or W tend to zero, the capacity of the channel also tends to zero. However, when P/N_0 or W tends to infinity, the capacity behaves differently. When P/N_0 tends to infinity, the capacity also tends to infinity, as shown in Figure 8.5. However, when W tends to infinity, the capacity does go to a certain limit, which is determined by P/N_0. To determine this limiting value, we have

$$\lim_{W \to \infty} W \log_2 \left(1 + \frac{P}{N_0 W} \right) = \frac{P}{N_0 \ln 2} \tag{8.2.10}$$

$$= 1.4427 \frac{P}{N_0} \tag{8.2.11}$$

The MATLAB script for this problem is given next.

M-FILE

```
% MATLAB script for Illustrative Problem 8.2.
echo on
pn0_db=[−20:0.1:30];
pn0=10.^(pn0_db./10);
```

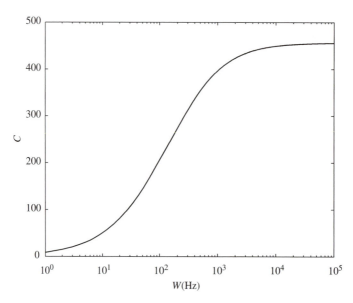

Figure 8.6   Capacity as a function of bandwidth in an AWGN channel ($P/N_0 = 25$dB)

```
capacity=3000.*log2(1+pn0/3000);
pause % Press a key to see a plot of channel capacity vs. P/N0
clf
semilogx(pn0,capacity)
title('Capacity vs. P/N0 in an AWGN channel')
xlabel('P/N0')
ylabel('Capacity (bits/second)')
clear
w=[1:10,12:2:100,105:5:500,510:10:5000,5025:25:20000,20050:50:100000];
pn0_db=25;
pn0=10^(pn0_db/10);
capacity=w.*log2(1+pn0./w);
pause % Press a key to see a plot of channel capacity vs. bandwidth
clf
semilogx(w,capacity)
title('Capacity vs. bandwidth in an AWGN channel')
xlabel('Bandwidth (Hz)')
ylabel('Capacity (bits/second)')
```

⬤ ILLUSTRATIVE PROBLEM ▬

Illustrative Problem 8.3 [Capacity of binary input AWGN channel] A binary input
AWGN channel is modeled by the two binary input levels A and $-A$ and additive zero-
mean Gaussian noise with variance σ^2. In this case, $\mathcal{X} = \{-A, A\}$, $\mathcal{Y} = \mathbb{R}$, $p(y \mid X =$

$A) \sim \mathcal{N}(A, \sigma^2)$, and $p(y \mid X = -A) \sim \mathcal{N}(-A, \sigma^2)$. Plot the capacity of this channel as a function of A/σ.

SOLUTION

Due to the symmetry in the problem, the capacity is achieved for uniform input distribution—that is, for $p(X = A) = p(X = -A) = \frac{1}{2}$. For this input distribution, the output distribution is given by

$$p(y) = \frac{1}{2\sqrt{2\pi\sigma^2}}e^{-(y+A)^2/2\sigma^2} + \frac{1}{2\sqrt{2\pi\sigma^2}}e^{-(y-A)^2/2\sigma^2} \tag{8.2.12}$$

and the mutual information between the input and the output is given by

$$I(X; Y) = \frac{1}{2}\int_{-\infty}^{\infty} p(y \mid X = A) \log_2 \frac{p(y \mid X = A)}{p(y)}\, dy$$
$$+ \frac{1}{2}\int_{-\infty}^{\infty} p(y \mid X = -A) \log_2 \frac{p(y \mid X = -A)}{p(y)}\, dy \tag{8.2.13}$$

Simple integration and change of variables result in

$$I(X; Y) = \frac{1}{2}f\left(\frac{A}{\sigma}\right) + \frac{1}{2}f\left(-\frac{A}{\sigma}\right) \tag{8.2.14}$$

where

$$f(a) = \int_{-\infty}^{\infty} \frac{1}{\sqrt{2\pi}}e^{-(u-a)^2/2} \log_2 \frac{2}{1 + e^{-2au}}\, du \tag{8.2.15}$$

Using these relations, we can calculate $I(X; Y)$ for various values of A/σ and plot the result. A plot of the resulting curve is shown in Figure 8.7.

The MATLAB script for this problem follows.

M-FILE

```
% MATLAB script for Illustrative Problem 8.3.
echo on
a_db=[-20:0.2:20];
a=10.^(a_db/10);
for i=1:201
    f(i)=quad('il3_8fun',a(i)-5,a(i)+5,1e-3,[],a(i));
    g(i)=quad('il3_8fun',-a(i)-5,-a(i)+5,1e-3,[],-a(i));
    c(i)=0.5*f(i)+0.5*g(i);
    echo off ;
end
echo on ;
pause % Press a key to see capacity vs. SNR plot
```

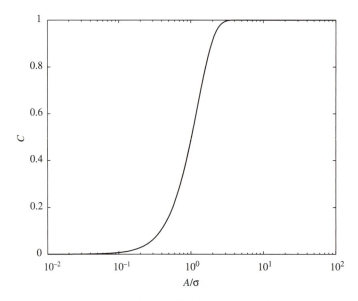

Figure 8.7   Capacity of a binary input AWGN channel as a function of SNR $= A/\sigma$

```
semilogx(a,c)
title('Capacity versus SNR in binary input AWGN channel')
xlabel('SNR')
ylabel('Capacity (bits/transmission)')
```

─────────────────────────────────

── ❨ILLUSTRATIVE PROBLEM❩ ──────────────────────────

Illustrative Problem 8.4 [Comparison of hard- and soft-decision schemes] A binary input channel uses the two input levels A and $-A$. The output of the channel is the sum of the input and an additive white Gaussian noise with mean 0 and variance σ^2. This channel is used under two different conditions. In one case, the output is used directly without quantization (soft decision), and in the other case, an optimal decision is made on each input level (hard decision). Plot the capacity in each case as a function of A/σ.

── ❨SOLUTION❩ ──────────────────────────

The soft-decision part is similar to Illustrative Problem 8.3. For the hard-decision case, the crossover probability of the resulting binary symmetric channel is $Q(A/\sigma)$, and therefore the capacity is given by

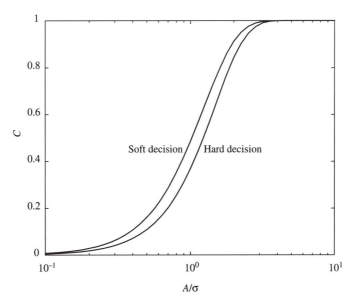

Figure 8.8   Plots of C_H and C_S versus SNR = A/σ

$$C_H = 1 - H_b \left(Q \left(\frac{A}{\sigma} \right) \right)$$

Both C_H and C_S are shown in Figure 8.8. Soft-decision decoding outperforms hard-decision decoding at all values of A/σ, as expected.

The MATLAB script for this problem is given next.

M-FILE

```
% MATLAB script for Illustrative Problem 8.4.
echo on
a_db=[−13:0.5:13];
a=10.^(a_db/10);
c_hard=1−entropy2(Q(a));
for i=1:53
    f(i)=quad('il3_8fun',a(i)−5,a(i)+5,1e−3,[ ],a(i));
    g(i)=quad('il3_8fun',−a(i)−5,−a(i)+5,1e−3,[ ],−a(i));
    c_soft(i)=0.5*f(i)+0.5*g(i);
    echo off ;
end
echo on ;
pause % Press a key to see the capacity curves
semilogx(a,c_soft,a,c_hard)
```

ILLUSTRATIVE PROBLEM

Illustrative Problem 8.5 [Capacity versus bandwidth and SNR] The capacity of a band-limited AWGN channel with input power constraint P and bandwidth W is given by

$$C = W \log_2 \left(1 + \frac{P}{N_0 W} \right)$$

Plot the capacity as a function of both W and P/N_0.

SOLUTION

The desired plot is shown in Figure 8.9. Note that for constant P/N_0, the plot reduces to the curve shown in Figure 8.6. For constant bandwidth, the capacity as a function of P/N_0 is similar to the curve shown in Figure 8.5.

The MATLAB script file for this problem is given next.

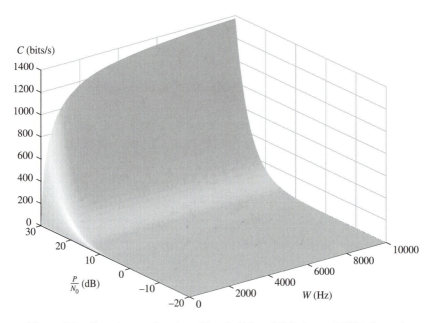

Figure 8.9   Capacity as a function of bandwidth and SNR in an AWGN channel

━━◖ M-FILE ◗━━━━━━━━━━━━━━━━━━━━━━━━━━━━━━━━━━━━

```
% MATLAB script for Illustrative Problem 8.5.
echo off
w=[1:5:20,25:20:100,130:50:300,400:100:1000,1250:250:5000,5500:500:10000];
pn0_db=[-20:1:30];
pn0=10.^(pn0_db/10);
for i=1:45
  for j=1:51
    c(i,j)=w(i)*log2(1+pn0(j)/w(i));
  end
end
echo on
pause % Press a key to see C vs. W and P/N0
k=[0.9,0.8,0.5,0.6];
s=[-70,35];
surfl(w,pn0_db,c',s,k)
title('Capacity vs. bandwidth and SNR')
```

━━◖ ILLUSTRATIVE PROBLEM ◗━━━━━━━━━━━━━━━━━━━━━━

Illustrative Problem 8.6 [Capacity of discrete-time AWGN channel] Plot the capacity of the discrete-time AWGN channel as a function of the input power and the noise variance.

━━◖ SOLUTION ◗━━━━━━━━━━━━━━━━━━━━━━━━━━━━━━━━

The desired plot is given in Figure 8.10.
      The MATLAB script file for this problem is given next.

━━◖ M-FILE ◗━━━━━━━━━━━━━━━━━━━━━━━━━━━━━━━━━━━━

```
% MATLAB script for Illustrative Problem 8.6.
echo on
p_db=[-20:1:20];
np_db=p_db;
p=10.^(p_db/10);
np=p;
for i=1:41
  for j=1:41
    c(i,j)=0.5*log2(1+p(i)/np(j));
    echo off ;
  end
end
echo on ;
pause % Press a key to see the plot
surfl(np_db,p_db,c)
```

Capacity (bits/transmission)

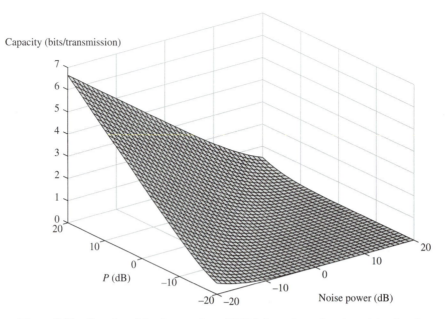

Figure 8.10   Capacity of the discrete-time AWGN channel as a function of the signal power (P) and the noise power (σ^2)

8.3   Channel Coding

Communication through noisy channels is subject to errors. In order to decrease the effect of errors and achieve reliable communication, it is necessary to transmit sequences that are as different as possible so that the channel noise will not change one sequence into another. This means that some redundancy has to be introduced to increase the reliability of communication. The introduction of redundancy results in transmission of extra bits and a reduction of the transmission rate.

Channel coding schemes can be generally divided into two classes: block codes and convolutional codes. In block coding, binary source output sequences of length k are mapped into binary channel input sequences of length n; therefore, the rate of the resulting code is k/n bits per transmission. Such a code is called an (n, k) block code and consists of 2^k codewords of length n, usually denoted by $c_1, c_2, \ldots, c_{2^k}$. Mapping of the information source outputs into channel inputs is done independently, and the output of the encoder depends only on the current input sequence of length k and not on the previous input sequences. In convolutional encoding, source outputs of length k_0 are mapped into n_0 channel inputs, but the channel inputs depend not only on the most recent k_0 source outputs, but also on the last $(L - 1)k_0$ inputs of the encoder.

One of the simplest block codes is the *simple repetition code*, in which there are two messages to be transmitted over a binary symmetric channel, but instead of transmitting a 0 and a 1 for the two messages, two sequences, one consisting of all 0's and one consisting of all 1's, are transmitted. The length of the two sequences is chosen to be some odd number n. The encoding process is shown below:

$$0 \longrightarrow \overbrace{00\ldots00}^{n \text{ odd}} \tag{8.3.1}$$

$$1 \longrightarrow \overbrace{11\ldots11}^{n \text{ odd}} \tag{8.3.2}$$

The decoding is a simple majority vote decoding; that is, if the majority of the received symbols are 1's, the decoder decides in favor of a 1; and if the majority are 0's, the decoder decides in favor of a 0.

An error occurs if at least $(n+1)/2$ of the transmitted symbols are received in error. Since the channel is a binary symmetric channel with crossover probability ϵ, the error probability can be expressed as

$$p_e = \sum_{k=(n+1)/2}^{n} \binom{n}{k} \epsilon^k (1-\epsilon)^{n-k} \tag{8.3.3}$$

For example, with $n = 5$ and $\epsilon = 0.001$, we have

$$p_e = \sum_{k=3}^{5} 0.001^k (0.999)^{5-k} \approx 10^{-9} \tag{8.3.4}$$

This means that by employing the channel five times instead of just once, we can reduce the error probability from 0.001 to 10^{-9}. Of course, a price has been paid for this more reliable performance; that price is a reduction in the rate of transmission and an increase in the complexity of the system. The rate of transmission has been decreased from one binary message per one use of the channel to one binary message per five usages of the channel. The complexity of the system has been increased because now we have to use an encoder (which has a very simple structure) and a decoder, which implements majority vote decoding. More reliable transmission in this problem is achieved if we increase n. For instance, for $n = 9$ we have

$$p_e = \sum_{k=5}^{9} 0.001^k (0.999)^{9-k} \approx 10^{-15} \tag{8.3.5}$$

From above, it seems that if we want to reduce the error probability to zero, we have to increase n indefinitely and, therefore, reduce the transmission rate to zero. This, however, is not the case, and Shannon showed that one can achieve asymptotically reliable communication (i.e., $p_e \to 0$) by keeping the rate of transmission below the channel capacity, which in the above case is

$$C = 1 - H_b(0.001) = 1 - 0.0114 = 0.9886 \quad \text{bits/transmission} \tag{8.3.6}$$

This, however, is achieved by employing encoding and decoding schemes that are much more complex than the simple repetition code.

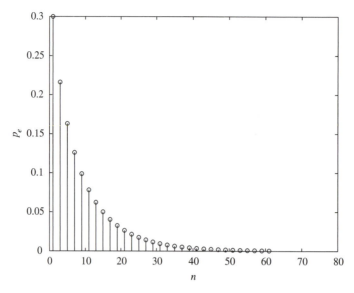

Figure 8.11   Error probability of a simple repetition code for $\epsilon = 0.3$ and
$n = 1, 3, ..., 61$

Illustrative Problem 8.7 [Error probability in simple repetition codes] Assuming that
$\epsilon = 0.3$ in a binary symmetric channel, plot p_e as a function of the block length n.

We derive p_e for values of n from 1 to 61. The error probability is given by

$$p_e = \sum_{k=(n+1)/2}^{n} \binom{n}{k} 0.3^k \times 0.7^{n-k}$$

and the resulting plot is shown in Figure 8.11.
    The MATLAB script file for this problem is given next.

```
% MATLAB script for Illustrative Problem 8.7.
echo on
ep=0.3;
for i=1:2:61
```

```
   p(i)=0;
    for  j=(i+1)/2:i
       p(i)=p(i)+prod(1:i)/(prod(1:j)*prod(1:(i−j)))*ep^j*(1−ep)^(i−j);
       echo off ;
    end
 end
echo on ;
pause % Press a key to see the plot
stem((1:2:61),p(1:2:61))
xlabel('n')
ylabel('pe')
title('Error probability as a function of n in simple repetition code')
```

8.3.1   Linear Block Codes

Linear block codes are the most important and widely used class of block codes. A block code is linear if any linear combination of two codewords is a codeword. In the binary case, this means that the sum of any two codewords is a codeword. In linear block codes, the codewords form a k-dimensional subspace of an n-dimensional space. Linear block codes are described in terms of their *generator matrix* G, which is a $k \times n$ binary matrix such that each codeword c can be written in the form

$$c = uG \tag{8.3.7}$$

where u is the binary data sequence of length k (the encoder input). Obviously, the all-0 sequence of length n is always a codeword of an (n, k) linear block code.

An important parameter in a linear block code, which determines its error-correcting capabilities, is the minimum (Hamming) distance of the code, which is defined as the minimum Hamming distance between any two distinct codewords. The minimum distance of a code is denoted by $d_{\min}$, and we have

$$d_{\min} = \min_{i \neq j} d_H(c_i, c_j) \tag{8.3.8}$$

For linear codes, the minimum distance is equal to the minimum weight of the code, defined by

$$w_{\min} = \min_{c_i \neq 0} w(c_i) \tag{8.3.9}$$

that is, the minimum number of 1's in any nonzero codeword.

---------**ILLUSTRATIVE PROBLEM**--

Illustrative Problem 8.8 [Linear block codes] The generator matrix for a $(10, 4)$ linear block code is given by

$$G = \begin{bmatrix} 1 & 0 & 0 & 1 & 1 & 1 & 0 & 1 & 1 & 1 \\ 1 & 1 & 1 & 0 & 0 & 0 & 1 & 1 & 1 & 0 \\ 0 & 1 & 1 & 0 & 1 & 1 & 0 & 1 & 0 & 1 \\ 1 & 1 & 0 & 1 & 1 & 1 & 1 & 0 & 0 & 1 \end{bmatrix}$$

Determine all the codewords and the minimum weight of the code.

---------**SOLUTION**--

In order to obtain all codewords, we have to use all information sequences of length 4 and find the corresponding encoded sequences. Since there is a total of 16 binary sequences of length 4, there will be 16 codewords. Let U denote a $2^k \times k$ matrix whose rows are all possible binary sequences of length k, starting from the all-0 sequence and ending with the all-1 sequence. The rows are chosen in such a way that the decimal representation of each row is smaller than the decimal representation of all rows below it. For the case of $k = 4$, the matrix U is given by

$$U = \begin{bmatrix} 0 & 0 & 0 & 0 \\ 0 & 0 & 0 & 1 \\ 0 & 0 & 1 & 0 \\ 0 & 0 & 1 & 1 \\ 0 & 1 & 0 & 0 \\ 0 & 1 & 0 & 1 \\ 0 & 1 & 1 & 0 \\ 0 & 1 & 1 & 1 \\ 1 & 0 & 0 & 0 \\ 1 & 0 & 0 & 1 \\ 1 & 0 & 1 & 0 \\ 1 & 0 & 1 & 1 \\ 1 & 1 & 0 & 0 \\ 1 & 1 & 0 & 1 \\ 1 & 1 & 1 & 0 \\ 1 & 1 & 1 & 1 \end{bmatrix}$$

We have

$$C = UG \tag{8.3.10}$$

where C is the matrix of codewords, which in this case is a 16×10 matrix whose rows are the codewords. The matrix of codewords is given by

$$C = \begin{bmatrix} 0 & 0 & 0 & 0 \\ 0 & 0 & 0 & 1 \\ 0 & 0 & 1 & 0 \\ 0 & 0 & 1 & 1 \\ 0 & 1 & 0 & 0 \\ 0 & 1 & 0 & 1 \\ 0 & 1 & 1 & 0 \\ 0 & 1 & 1 & 1 \\ 1 & 0 & 0 & 0 \\ 1 & 0 & 0 & 1 \\ 1 & 0 & 1 & 0 \\ 1 & 0 & 1 & 1 \\ 1 & 1 & 0 & 0 \\ 1 & 1 & 0 & 1 \\ 1 & 1 & 1 & 0 \\ 1 & 1 & 1 & 1 \end{bmatrix} \begin{bmatrix} 1 & 0 & 0 & 1 & 1 & 1 & 0 & 1 & 1 & 1 \\ 1 & 1 & 1 & 0 & 0 & 0 & 1 & 1 & 1 & 0 \\ 0 & 1 & 1 & 0 & 1 & 1 & 0 & 1 & 0 & 1 \\ 1 & 1 & 0 & 1 & 1 & 1 & 1 & 0 & 0 & 1 \end{bmatrix}$$

$$= \begin{bmatrix} 0 & 0 & 0 & 0 & 0 & 0 & 0 & 0 & 0 & 0 \\ 1 & 1 & 0 & 1 & 1 & 1 & 1 & 0 & 0 & 1 \\ 0 & 1 & 1 & 0 & 1 & 1 & 0 & 1 & 0 & 1 \\ 1 & 0 & 1 & 1 & 0 & 0 & 1 & 1 & 0 & 0 \\ 1 & 1 & 1 & 0 & 0 & 0 & 1 & 1 & 1 & 0 \\ 0 & 0 & 1 & 1 & 1 & 1 & 0 & 1 & 1 & 1 \\ 1 & 0 & 0 & 0 & 1 & 1 & 1 & 0 & 1 & 1 \\ 0 & 1 & 0 & 1 & 0 & 0 & 0 & 0 & 1 & 0 \\ 1 & 0 & 0 & 1 & 1 & 1 & 0 & 1 & 1 & 1 \\ 0 & 1 & 0 & 0 & 0 & 0 & 1 & 1 & 1 & 0 \\ 1 & 1 & 1 & 1 & 0 & 0 & 0 & 0 & 1 & 0 \\ 0 & 0 & 1 & 0 & 1 & 1 & 1 & 0 & 1 & 1 \\ 0 & 1 & 1 & 1 & 1 & 1 & 1 & 0 & 0 & 1 \\ 1 & 0 & 1 & 0 & 0 & 0 & 0 & 0 & 0 & 0 \\ 0 & 0 & 0 & 1 & 0 & 0 & 1 & 1 & 0 & 0 \\ 1 & 1 & 0 & 0 & 1 & 1 & 0 & 1 & 0 & 1 \end{bmatrix}$$

A close inspection of the codewords shows that the minimum distance of the code is $d_{\min} = 2$.

The MATLAB script file for this problem is given next.

M-FILE

```
% MATLAB script for Illustrative Problem 8.8.
% Generate U, denoting all information sequences.
k=4;
for i=1:2^k
  for j=k:-1:1
```

```
    if  rem(i−1,2^(−j+k+1))>=2^(−j+k)
       u(i,j)=1;
    else
       u(i,j)=0;
    end
    echo off ;
  end
end
echo on ;
% Define G, the generator matrix.
g=[1 0 0 1 1 1 0 1 1 1;
   1 1 1 0 0 0 1 1 1 0;
   0 1 1 0 1 1 0 1 0 1;
   1 1 0 1 1 1 1 0 0 1];
% Generate codewords
c=rem(u*g,2);
% Find the minimum distance.
w_min=min(sum((c(2:2^k,:))' ) ) ;
```

A linear block code is in the *systematic form* if its generator matrix is in the following form:

$$G = \begin{bmatrix} 1 & 0 & \cdots & 0 & p_{1,1} & p_{1,2} & \cdots & p_{1,n-k} \\ 0 & 1 & \cdots & 0 & p_{2,1} & p_{2,2} & \cdots & p_{2,n-k} \\ \vdots & \vdots & \ddots & \vdots & \vdots & \vdots & \ddots & \vdots \\ 0 & 0 & \cdots & 1 & p_{k,1} & p_{k,2} & \cdots & p_{k,n-k} \end{bmatrix} \qquad (8.3.11)$$

or

$$G = [I_k \mid P] \qquad (8.3.12)$$

where I_k denotes the $k \times k$ identity matrix and P is a $k \times (n - k)$ matrix. In a systematic code, the first k binary symbols in a codeword are the information bits, and the remaining $n - k$ binary symbols are the *parity-check symbols*.

The *parity-check matrix* of a code is any $(n - k) \times n$ binary matrix H such that for all codewords c, we have

$$cH^t = 0 \qquad (8.3.13)$$

Obviously, we will have

$$GH^t = 0 \qquad (8.3.14)$$

and if G is in systematic form, then

$$H = [P^t \mid I_k] \qquad (8.3.15)$$

Hamming Codes

Hamming codes are $(2^m - 1, 2^m - m - 1)$ linear block codes with minimum distance 3 and a very simple parity-check matrix. The parity-check matrix, which is an $m \times (2^m - 1)$ matrix, has all binary sequences of length m, except the all-0 sequence, as its columns. For instance, for $m = 3$, we have a $(7, 4)$ code whose parity-check matrix in the systematic form is

$$H = \begin{bmatrix} 1 & 0 & 1 & 1 & 1 & 0 & 0 \\ 1 & 1 & 0 & 1 & 0 & 1 & 0 \\ 0 & 1 & 1 & 1 & 0 & 0 & 1 \end{bmatrix} \tag{8.3.16}$$

From this, we have

$$G = \begin{bmatrix} 1 & 0 & 0 & 0 & 1 & 1 & 0 \\ 0 & 1 & 0 & 0 & 0 & 1 & 1 \\ 0 & 0 & 1 & 0 & 1 & 0 & 1 \\ 0 & 0 & 0 & 1 & 1 & 1 & 1 \end{bmatrix} \tag{8.3.17}$$

ILLUSTRATIVE PROBLEM

Illustrative Problem 8.9 [Hamming codes] Find all the codewords of the $(15, 11)$ Hamming code, and verify that its minimum distance is equal to 3.

SOLUTION

Here

$$H = \begin{bmatrix} 1 & 0 & 0 & 1 & 1 & 0 & 1 & 0 & 1 & 1 & 1 & 1 & 0 & 0 & 0 \\ 1 & 1 & 0 & 0 & 0 & 1 & 1 & 1 & 0 & 1 & 1 & 0 & 1 & 0 & 0 \\ 0 & 1 & 1 & 1 & 0 & 0 & 1 & 1 & 1 & 0 & 1 & 0 & 0 & 1 & 0 \\ 0 & 0 & 1 & 0 & 1 & 1 & 0 & 1 & 1 & 1 & 1 & 0 & 0 & 0 & 1 \end{bmatrix} \tag{8.3.18}$$

and, therefore,

$$G = \begin{bmatrix} 1 & 0 & 0 & 0 & 0 & 0 & 0 & 0 & 0 & 0 & 0 & 1 & 1 & 0 & 0 \\ 0 & 1 & 0 & 0 & 0 & 0 & 0 & 0 & 0 & 0 & 0 & 0 & 1 & 1 & 0 \\ 0 & 0 & 1 & 0 & 0 & 0 & 0 & 0 & 0 & 0 & 0 & 0 & 0 & 1 & 1 \\ 0 & 0 & 0 & 1 & 0 & 0 & 0 & 0 & 0 & 0 & 0 & 1 & 0 & 1 & 0 \\ 0 & 0 & 0 & 0 & 1 & 0 & 0 & 0 & 0 & 0 & 0 & 1 & 0 & 0 & 1 \\ 0 & 0 & 0 & 0 & 0 & 1 & 0 & 0 & 0 & 0 & 0 & 0 & 1 & 0 & 1 \\ 0 & 0 & 0 & 0 & 0 & 0 & 1 & 0 & 0 & 0 & 0 & 1 & 1 & 1 & 0 \\ 0 & 0 & 0 & 0 & 0 & 0 & 0 & 1 & 0 & 0 & 0 & 0 & 1 & 1 & 1 \\ 0 & 0 & 0 & 0 & 0 & 0 & 0 & 0 & 1 & 0 & 0 & 1 & 0 & 1 & 1 \\ 0 & 0 & 0 & 0 & 0 & 0 & 0 & 0 & 0 & 1 & 0 & 1 & 1 & 0 & 1 \\ 0 & 0 & 0 & 0 & 0 & 0 & 0 & 0 & 0 & 0 & 1 & 1 & 1 & 1 & 1 \end{bmatrix}$$

There is a total of $2^{11} = 2048$ codewords, each of length 15. The rate of the code is $\frac{11}{15} = 0.733$. In order to verify the minimum distance of the code, we use a MATLAB script similar to the one used in the Illustrative Problem 8.8. The MATLAB script is given next, and it results in $d_{\min} = 3$.

M-FILE

```
% MATLAB script for Illustrative Problem 8.9.
echo on
k=11;
for i=1:2^k
  for j=k:−1:1
    if rem(i−1,2^(−j+k+1))>=2^(−j+k)
       u(i,j)=1;
    else
       u(i,j)=0;
    end
    echo off ;
  end
end
echo on ;

g=[1 0 0 0 0 0 0 0 0 0 0 1 1 0 0;
   0 1 0 0 0 0 0 0 0 0 0 0 1 1 0;
   0 0 1 0 0 0 0 0 0 0 0 0 0 1 1;
   0 0 0 1 0 0 0 0 0 0 0 1 0 1 0;
   0 0 0 0 1 0 0 0 0 0 1 0 0 1;
   0 0 0 0 0 1 0 0 0 0 0 1 0 1;
   0 0 0 0 0 0 1 0 0 0 0 1 1 1 0;
   0 0 0 0 0 0 0 1 0 0 0 0 1 1 1;
   0 0 0 0 0 0 0 0 1 0 0 1 0 1 1;
   0 0 0 0 0 0 0 0 0 1 0 1 1 0 1;
   0 0 0 0 0 0 0 0 0 0 1 1 1 1 1];

c=rem(u*g,2);
w_min=min(sum((c(2:2^k,:))' ) ) ;
```

Performance of Linear Block Codes

Linear block codes can be decoded using either soft-decision decoding or hard-decision decoding. In a hard-decision decoding scheme, first a bit-by-bit decision is made on the components of the codeword, and then, with a minimum Hamming distance criterion, the decoding is performed. The performance of this decoding scheme depends on the distance structure of the code, but a tight upper bound, particularly at high values of the SNR, can be obtained in terms of the minimum distance of the code.

The (message) error probability of a linear block code with minimum distance $d_{\min}$, in hard-decision decoding, is upper-bounded by

$$p_e \leq (M - 1)\left[4p(1 - p)\right]^{d_{\min}/2} \tag{8.3.19}$$

where p denotes the error probability of the binary channel (error probability in demodulation) and M is the number of codewords ($M = 2^k$).

In soft-decision decoding, the received signal is mapped into the codeword whose corresponding signal is at minimum Euclidean distance from the received signal. The message-error probability in this case is upper-bounded by

$$p_e \leq (M - 1)Q\left(\frac{d^{\mathrm{E}}}{\sqrt{2N_0}}\right) \tag{8.3.20}$$

where $M = 2^k$ is the number of codewords, N_0 is the one-sided noise power-spectral density, and d^{E} is the *minimum Euclidean distance of the code* and is given by

$$d^{\mathrm{E}} = \begin{cases} \sqrt{2d_{\min}\mathcal{E}} & \text{for orthogonal signaling} \\ \sqrt{4d_{\min}\mathcal{E}} & \text{for antipodal signaling} \end{cases} \tag{8.3.21}$$

which results in

$$p_e \leq \begin{cases} (M - 1)Q\left(\sqrt{\dfrac{d_{\min}\mathcal{E}}{N_0}}\right) & \text{for orthogonal signaling} \\[3mm] (M - 1)Q\left(\sqrt{\dfrac{2d_{\min}\mathcal{E}}{N_0}}\right) & \text{for antipodal signaling} \end{cases} \tag{8.3.22}$$

In these inequalities, $d_{\min}$ is the minimum Hamming distance of the code, and $\mathcal{E}$ denotes the energy per each component of the codewords. Since each codeword has n components, the energy per codeword is $n\mathcal{E}$, and since each codeword carries k information bits, the energy per bit, $\mathcal{E}_b$, is given by

$$\mathcal{E}_b = \frac{n\mathcal{E}}{k} = \frac{\mathcal{E}}{R_c} \tag{8.3.23}$$

where $R_c = k/n$ denotes the rate of the code. Therefore, the above relations can be written as

$$p_e \leq \begin{cases} (M - 1)Q\left(\sqrt{\dfrac{d_{\min}R_c\mathcal{E}_b}{N_0}}\right) & \text{for orthogonal signaling} \\[3mm] (M - 1)Q\left(\sqrt{\dfrac{2d_{\min}R_c\mathcal{E}_b}{N_0}}\right) & \text{for antipodal signaling} \end{cases} \tag{8.3.24}$$

The bounds obtained are usually useful only for large values of $\gamma_b = \mathcal{E}_b/N_0$. For smaller γ_b values, the bounds become very loose and can even exceed 1.

— **ILLUSTRATIVE PROBLEM** ————————————————

Illustrative Problem 8.10 [Performance of hard-decision decoding] Assuming that the (15, 11) Hamming code is used with antipodal signaling and hard-decision decoding, plot the message-error probability as a function of $\gamma_b = \mathcal{E}_b/N_0$.

— **SOLUTION** ————————————————————

Since antipodal signaling is employed, the error probability of the binary channel is given by

$$p = Q\left(\sqrt{\frac{2\mathcal{E}}{N_0}}\right) \tag{8.3.25}$$

where $\mathcal{E}$ is the energy per component of the code (energy per dimension) and is derived from $\mathcal{E}_b$ by

$$\mathcal{E} = \mathcal{E}_b R_c \tag{8.3.26}$$

Therefore,

$$p = Q\left(\sqrt{\frac{2R_c\mathcal{E}_b}{N_0}}\right) \tag{8.3.27}$$

where $R_c = k/n = \frac{11}{15} = 0.73333$. Since the minimum distance of Hamming codes is 3, we have

$$p_e \leq \left(2^{11} - 1\right)[4p(1-p)]^{d_{\min}/2}$$

$$= 2047\left[4Q\left(\sqrt{\frac{1.466\mathcal{E}_b}{N_0}}\right)\left(1 - Q\left(\sqrt{\frac{1.466\mathcal{E}_b}{N_0}}\right)\right)\right]^{1.5} \tag{8.3.28}$$

The resulting plot is shown in Figure 8.12.

The MATLAB function for computing the bound on message-error probability of a linear block code when hard-decision decoding and antipodal signaling are employed is given next.

— **M-FILE** ————————————————————

```
function [p_err,gamma_db]=p_e_hd_a(gamma_db_l,gamma_db_h,k,n,d_min)
% p_e_hd_a.m     Matlab function for computing error probability in
%                hard decision decoding of a linear block code
%                when antipodal signaling is used.
%                [p_err,gamma_db]=p_e_hd_a(gamma_db_l,gamma_db_h,k,n,d_min)
%                gamma_db_l=lower E_b/N_0
%                gamma_db_h=higher E_b/N_0
%                k=number of information bits in the code
```

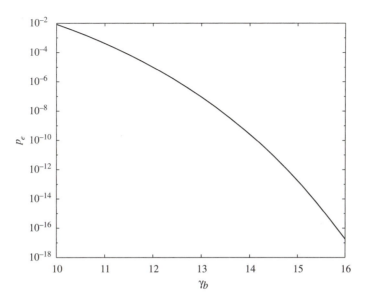

Figure 8.12   Error probability as a function of γ_b for a (15, 11) Hamming code with hard-decision decoding and antipodal signaling

```
%                n=code block length
%                d_min=minimum distance of the code

gamma_db=[gamma_db_l:(gamma_db_h−gamma_db_l)/20:gamma_db_h];
gamma_b=10.^(gamma_db/10);
R_c=k/n;
p_b=q(sqrt(2.*R_c.*gamma_b));
p_err=(2^k−1).*(4*p_b.*(1−p_b)).^(d_min/2);
```

In the MATLAB script given next, the preceding MATLAB function is employed to plot error probability versus γ_b.

━━◖ M-FILE ◗━━━━━━━━━━━━━━━━━━━━━━━━

```
% MATLAB script for Illustrative Problem 8.10.
[p_err_ha,gamma_b]=p_e_hd_a(10,16,11,15,3);
semilogy(gamma_b,p_err_ha)
```

━━◖ILLUSTRATIVE PROBLEM◗━━━━━━━━━━━━━━

Illustrative Problem 8.11 [Hard-decision decoding] If the (15, 11) Hamming code is used with an orthogonal binary modulation scheme instead of the antipodal scheme, plot the message-error probability as a function of $\gamma_b = \mathscr{E}_b/N_0$.

SOLUTION

The problem is similar to Illustrative Problem 8.10, except that the crossover probability of the equivalent binary symmetric channel (after hard-decision decoding) is given by

$$p = Q\left(\sqrt{\frac{\mathcal{E}}{N_0}}\right) \tag{8.3.29}$$

Using the relation

$$\mathcal{E} = \mathcal{E}_b R_c \tag{8.3.30}$$

we obtain

$$p = Q\left(\sqrt{\frac{R_c \mathcal{E}_b}{N_0}}\right) \tag{8.3.31}$$

and, finally,

$$p_e \le \left(2^{11} - 1\right) [4p(1 - p)]^{d_{\min}/2}$$

$$= 2047 \left[4Q\left(\sqrt{\frac{0.733\mathcal{E}_b}{N_0}}\right)\left(1 - Q\left(\sqrt{\frac{0.733\mathcal{E}_b}{N_0}}\right)\right)\right]^{1.5} \tag{8.3.32}$$

The p_e-versus-$\mathcal{E}_b/N_0$ plot is shown in Figure 8.13.
    The MATLAB file for this problem is given next.

M-FILE

```
% MATLAB script for Illustrative Problem 8.11.
echo on
gamma_b_db=[-4:1:14];
gamma_b=10.^(gamma_b_db/10);
qq=q(sqrt(0.733.*gamma_b));
p_err=2047*qq.^2.*(3-2.*qq);
pause % Press a key to see p_err versus gamma_b curve
loglog(gamma_b,p_err)
```

As we observe from Figure 8.13, for lower values of γ_b, the derived bound is too loose. In fact, for these values of γ_b, the bound on the error probability is larger than 1. It is also instructive to plot the two error probability bounds for orthogonal and antipodal signaling on the same figure. This is done in Figure 8.14. The superior performance of the antipodal signaling compared to orthogonal signaling is readily seen from a comparison of these two plots.
    The MATLAB function for computing the message-error probability in the case of hard-decision decoding with orthogonal signaling is given next.

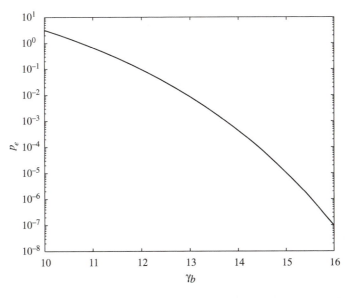

Figure 8.13   Error probability versus γ_b of a (15,11) code with orthogonal signaling and hard-decision decoding

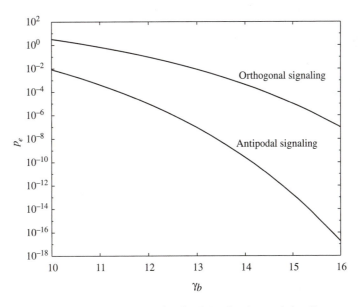

Figure 8.14   Comparison of antipodal and orthogonal signaling

---
M-FILE

```
function [p_err,gamma_db]=p_e_hd_o(gamma_db_l,gamma_db_h,k,n,d_min)
% p_e_hd_o.m    Matlab function for computing error probability in
%               hard decision decoding of a linear block code
%               when orthogonal signaling is used.
%               [p_err,gamma_db]=p_e_hd_o(gamma_db_l,gamma_db_h,k,n,d_min)
%               gamma_db_l=lower E_b/N_0
%               gamma_db_h=higher E_b/N_0
%               k=number of information bits in the code
%               n=code block length
%               d_min=minimum distance of the code

gamma_db=[gamma_db_l:(gamma_db_h-gamma_db_l)/20:gamma_db_h];
gamma_b=10.^(gamma_db/10);
R_c=k/n;
p_b=q(sqrt(R_c.*gamma_b));
p_err=(2^k-1).*(4*p_b.*(1-p_b)).^(d_min/2);
```

ILLUSTRATIVE PROBLEM

Illustrative Problem 8.12 [Soft-decision decoding] Solve Illustrative Problem 8.11 when soft-decision decoding is used instead of hard-decision decoding.

SOLUTION

In this case, we have to use Equation (8.3.24) to find an upper bound for the error probability. In the problem under study, $d_{\min} = 3$, $R_c = \frac{11}{15}$, and $M = 2^{11} - 1 = 2047$. Therefore, we have

$$
p_e \leq
\begin{cases}
(M-1)Q\left(\sqrt{\dfrac{d_{\min} R_c \mathcal{E}_b}{N_0}}\right) & \text{for orthogonal signaling} \\[3ex]
(M-1)Q\left(\sqrt{\dfrac{2 d_{\min} R_c \mathcal{E}_b}{N_0}}\right) & \text{for antipodal signaling}
\end{cases}
$$

$$
\leq
\begin{cases}
2047\, Q\left(\sqrt{\dfrac{11}{5}\dfrac{\mathcal{E}_b}{N_0}}\right) & \text{for orthogonal signaling} \\[3ex]
2047\, Q\left(\sqrt{\dfrac{22}{5}\dfrac{\mathcal{E}_b}{N_0}}\right) & \text{for antipodal signaling}
\end{cases}
$$

The corresponding plots are shown in Figure 8.15. The superior performance of antipodal signaling is obvious from these plots.

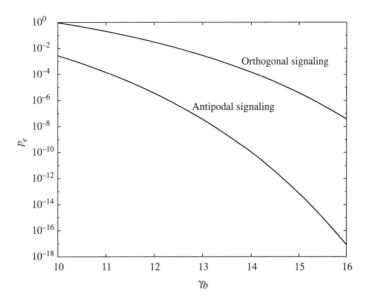

Figure 8.15   Message-error probability versus γ_b for soft-decision decoding

Two MATLAB functions, one for computing the error probability for antipodal signaling and one for computing the error probability for orthogonal signaling when soft-decision decoding is employed, are given next.

M-FILE

```
function [p_err,gamma_db]=p_e_sd_a(gamma_db_l,gamma_db_h,k,n,d_min)
% p_e_sd_a.m      Matlab function for computing error probability in
%                 soft decision decoding of a linear block code
%                 when antipodal signaling is used.
%                 [p_err,gamma_db]=p_e_sd_a(gamma_db_l,gamma_db_h,k,n,d_min)
%                 gamma_db_l=lower E_b/N_0
%                 gamma_db_h=higher E_b/N_0
%                 k=number of information bits in the code
%                 n=code block length
%                 d_min=minimum distance of the code

gamma_db=[gamma_db_l:(gamma_db_h-gamma_db_l)/20:gamma_db_h];
gamma_b=10.^(gamma_db/10);
R_c=k/n;
p_err=(2^k-1).*q(sqrt(2.*d_min.*R_c.*gamma_b));
```

M-FILE

```
function [p_err,gamma_db]=p_e_sd_o(gamma_db_l,gamma_db_h,k,n,d_min)
% p_e_sd_o.m      Matlab function for computing error probability in
%                 soft decision decoding of a linear block code
%                 when orthogonal signaling is used.
%                 [p_err,gamma_db]=p_e_sd_o(gamma_db_l,gamma_db_h,k,n,d_min)
%                 gamma_db_l=lower E_b/N_0
%                 gamma_db_h=higher E_b/N_0
%                 k=number of information bits in the code
%                 n=code block length
%                 d_min=minimum distance of the code

gamma_db=[gamma_db_l:(gamma_db_h−gamma_db_l)/20:gamma_db_h];
gamma_b=10.^(gamma_db/10);
R_c=k/n;
p_err=(2^k−1).*q(sqrt(d_min.*R_c.*gamma_b));
```

In Figure 8.16, four plots corresponding to antipodal and orthogonal signaling with soft- and hard-decision decoding are shown.

The MATLAB script that generates this figure is given next.

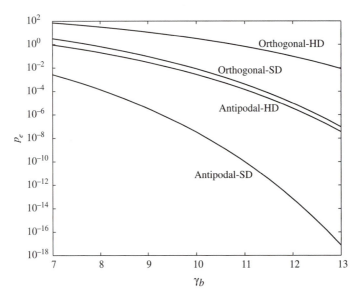

Figure 8.16   Comparison of antipodal/orthogonal signaling and soft/hard-decision decoding

M-FILE

```
% MATLAB script for Illustrative Problem 8.12.
[p_err_ha,gamma_b]=p_e_hd_a(7,13,11,15,3);
[p_err_ho,gamma_b]=p_e_hd_o(7,13,11,15,3);
[p_err_so,gamma_b]=p_e_sd_o(7,13,11,15,3);
[p_err_sa,gamma_b]=p_e_sd_a(7,13,11,15,3);
semilogy(gamma_b,p_err_sa,gamma_b,p_err_so,gamma_b,p_err_ha,gamma_b,p_err_ho)
```

8.3.2   Convolutional Codes

In block codes, each sequence of k information bits is mapped into a sequence of n channel inputs in a fixed way regardless of the previous information bits. In convolutional codes, each sequence of k_0 information bits is mapped into a channel input sequence of length n_0, but the channel input sequence depends not only on the most recent k_0 information bits but also on the last $(L-1)k_0$ inputs of the encoder. Therefore, the encoder has the structure of a finite-state machine, where at each time instance, the output sequence depends not only on the input sequence but also on the state of the encoder, which is determined by the last $(L-1)k_0$ inputs of the encoder. The parameter L is called the *constraint length* of the convolutional code.[1] A binary convolutional code is, therefore, a finite-state machine with $2^{k_0(L-1)}$ states. The schematic diagram for a convolutional code with $k_0 = 2$, $n_0 = 3$, and $L = 4$ is shown in Figure 8.17.

In this convolutional encoder, the information bits are loaded into the shift register 2 bits at a time, and the last 2 information bits in the shift register move out. The 3 encoded bits are then computed as shown in the figure and are transmitted over the channel. The rate of this code is, therefore, $R = \frac{2}{3}$. Note that the 3 encoder outputs transmitted over the channel depend on the 2 information bits loaded into the shift register as well as the contents of the first three stages (6 bits) of the shift register. The contents of the last stage (2 bits) have no effect on the output because they leave the shift register as soon as the 2 information bits are loaded into it.

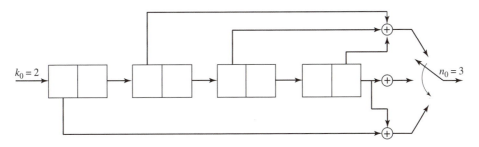

Figure 8.17   A convolutional code with $k_0 = 2$, $n_0 = 3$, and $L = 4$

[1] Some authors define $m = Lk_0$ as the constraint length, and some prefer $(L-1)k_0$ as the constraint length.

A convolutional code is usually defined in terms of the *generator sequences* of the convolutional code, denoted by $g_1, g_2, \ldots, g_n$. The ith component of g_j, $1 \leq i \leq k_0 L$ and $1 \leq j \leq n$, is 1 if the ith element of the shift register is connected to the combiner corresponding to the jth bit in the output and 0 otherwise. For example, in the convolutional code depicted in Figure 8.17, we have

$$\begin{aligned}
g_1 &= [0 \quad 0 \quad 1 \quad 0 \quad 1 \quad 0 \quad 0 \quad 1] \\
g_2 &= [0 \quad 0 \quad 0 \quad 0 \quad 0 \quad 0 \quad 0 \quad 1] \\
g_3 &= [1 \quad 0 \quad 0 \quad 0 \quad 0 \quad 0 \quad 0 \quad 1]
\end{aligned}$$

As soon as $g_1, g_2, \ldots, g_n$ are specified, the convolutional code is uniquely determined.

We also define the generator matrix of the convolutional code as

$$G = \begin{bmatrix} g_1 \\ g_2 \\ \vdots \\ g_n \end{bmatrix}$$

which is, in general, an $n \times k_0 L$ matrix. For the convolutional code shown in Figure 8.17, we have

$$G = \begin{bmatrix} 0 & 0 & 1 & 0 & 1 & 0 & 0 & 1 \\ 0 & 0 & 0 & 0 & 0 & 0 & 0 & 1 \\ 1 & 0 & 0 & 0 & 0 & 0 & 0 & 1 \end{bmatrix}$$

It is helpful to assume the shift register that generates the convolutional code is loaded with 0's before the first information bit enters it (i.e., the encoder is initially in zero state) and that the information bit sequence is padded with $(L - 1)k_0$ 0's to bring back the convolutional encoder to the all-0 state. We also assume that the length of the information-bit sequence (input to the convolutional encoder) is a multiple of k_0. If the length of the input sequence is not a multiple of k_0, we pad it with 0's such that the resulting length is a multiple of k_0. This is done before adding the $(L - 1)k_0$ 0's indicated earlier in this paragraph. If, after the first zero-padding, the length of the input sequence is nk_0, the length of the output sequence will be $(n + L - 1)n_0$, so the rate of the code will be

$$\frac{nk_0}{(n + L - 1)n_0}$$

In practice, n is much larger than L; hence, the above expression is well approximated by

$$R_c = \frac{k_0}{n_0}$$

The MATLAB function cnv_encd.m given next generates the output sequence of the convolutional encoder when G, k_0, and the input sequence are given. Note that zero-padding of the input sequence is done by the MATLAB function. The input sequence (denoted by the parameter "input") starts with the first information bit that enters the encoder. The parameters n_0 and L are derived from the matrix **G**.

◖ M-FILE ◗ ──────────────────────────────

```
function output=cnv_encd(g,k0,input)
%                   cnv_encd(g,k0,input)
%                   determines the output sequence of a binary convolutional encoder
%                   g is the generator matrix of the convolutional code
%                   with n0 rows and l*k0 columns. Its rows are g1,g2,...,gn.
%                   k0 is the number of bits entering the encoder at each clock cycle.
%                   input The binary input seq.

%   check to see if extra zero padding is necessary
if rem(length(input),k0) > 0
    input=[input,zeros(size(1:k0−rem(length(input),k0)))];
end
n=length(input)/k0;
%   check the size of matrix g
if rem(size(g,2),k0) > 0
    error('Error, g is not of the right size.')
end
%   determine l and n0
l=size(g,2)/k0;
n0=size(g,1);
%   add extra zeros
u=[zeros(size(1:(l−1)*k0)),input,zeros(size(1:(l−1)*k0))];
%   generate uu, a matrix whose columns are the contents of
%   conv. encoder at various clock cycles.
u1=u(l*k0:−1:1);
for i=1:n+l−2
    u1=[u1,u((i+l)*k0:−1:i*k0+1)];
end
uu=reshape(u1,l*k0,n+l−1);
%   determine the output
output=reshape(rem(g*uu,2),1,n0*(l+n−1));
```

──────────────────────────────

◖ILLUSTRATIVE PROBLEM◗ ──────────────────────────────

Illustrative Problem 8.13 [Convolutional encoder] Determine the output of the convolutional encoder shown in Figure 8.17 when the information sequence is

$$1\ 0\ 0\ 1\ 1\ 1\ 0\ 0\ 1\ 1\ 0\ 0\ 0\ 0\ 1\ 1\ 1$$

---**SOLUTION**--

Here, the length of the information sequence is 17, which is not a multiple of $k_0 = 2$; therefore, extra zero-padding will be done. In this case, it is sufficient to add one 0, which gives a length of 18. Thus, we have the following information sequence:

$$1\ 0\ 0\ 1\ 1\ 1\ 0\ 0\ 1\ 1\ 0\ 0\ 0\ 0\ 1\ 1\ 1\ 0$$

Now, since we have

$$G = \begin{bmatrix} 0 & 0 & 1 & 0 & 1 & 0 & 0 & 1 \\ 0 & 0 & 0 & 0 & 0 & 0 & 0 & 1 \\ 1 & 0 & 0 & 0 & 0 & 0 & 0 & 1 \end{bmatrix}$$

we obtain $n_0 = 3$ and $L = 4$. (This is also obvious from Figure 8.17.) The length of the output sequence is, therefore,

$$\left(\frac{18}{2} + 4 - 1\right) \times 3 = 36$$

The zero-padding required to make sure that the encoder starts from the all-0 state and returns to the all-0 state adds $(L-1)k_0$ 0's to the beginning and end of the input sequence. Therefore, the sequence under study becomes

$$0\ 0\ 0\ 0\ 0\ 0\ 1\ 0\ 0\ 1\ 1\ 1\ 0\ 0\ 1\ 1\ 0\ 0\ 0\ 0\ 1\ 1\ 1\ 0\ 0\ 0\ 0\ 0\ 0\ 0$$

Using the function cnv_encd.m, we find that the output sequence is

$$0\ 0\ 0\ 0\ 0\ 1\ 1\ 0\ 1\ 1\ 1\ 1\ 1\ 0\ 1\ 0\ 1\ 1\ 1\ 0\ 0\ 1\ 1\ 0\ 1\ 0\ 0\ 1\ 0\ 0\ 1\ 1\ 1\ 1\ 1\ 1$$

The MATLAB script to solve this problem is given next.

---**M-FILE**--

```
k0=2;
g=[0 0 1 0 1 0 0 1;0 0 0 0 0 0 0 1;1 0 0 0 0 0 0 1];
input=[1 0 0 1 1 1 0 0 1 1 0 0 0 0 1 1 1];
output=cnv_encd(g,k0,input);
```

Representation of Convolutional Codes

We have seen that a convolutional code can be represented by either the structure of the encoder or G, the generator matrix. We have also seen that a convolutional encoder can be represented as a finite-state machine and, therefore, can be described by a state transition diagram representing the finite-state machine. A more widely used method for representation of convolutional codes is in terms of their *trellis diagram*. A trellis diagram is a state transition diagram plotted versus time. Therefore, a trellis diagram is a sequence of $2^{(L-1)k_0}$ states, shown as dots, for each clock cycle and branches corresponding to transitions between these states.

Consider the convolutional code with $k_0 = 1$, $n_0 = 2$, and $L = 3$ shown in Figure 8.18. Obviously, this code can be represented by a finite-state machine with four states corresponding to different possible contents of the first two elements of the shift register—namely, 00, 01, 10, and 11. Let us represent these four states by the letters a, b, c, and d, respectively. In order to draw the trellis diagram for this code, we have to draw four dots corresponding to each state for each clock cycle and then connect them according to various transitions that can take place between states. The trellis diagram for this code is shown in Figure 8.19.

As we can see in Figure 8.19, on the time axis, which corresponds to clock cycles, the four states are denoted by black dots, and the transitions between states are indicated by branches connecting these dots. On each branch connecting two states, two binary symbols indicate the encoder output corresponding to that transition. Also note that we always start from the all-0 state (state a), move through the trellis following the branches corresponding to the given input sequence, and return to the all-0 state. Therefore, codewords of a convolutional code correspond to paths through the corresponding trellis, starting at the all-0 state and returning to the all-0 state.

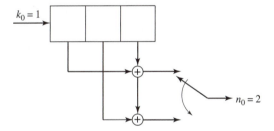

Figure 8.18   A convolutional encoder with $k_0 = 1$, $n_0 = 2$, and $L = 3$

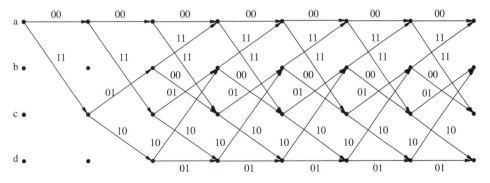

Figure 8.19   The trellis diagram for the convolutional code shown in Figure 8.18

The number of states in the trellis increases exponentially with the constraint length of the convolutional code. For example, for the convolutional encoder shown in Figure 8.17, the number of states is $2^6 = 64$; therefore, the structure of the trellis is much more complex.

The Transfer Function of a Convolutional Code

For each convolutional code, the transfer function gives information about the various paths through the trellis that start from the all-0 state and return to this state for the first time. According to the coding convention described before, any codeword of a convolutional encoder corresponds to a path through the trellis that starts from the all-0 state and returns to the all-0 state. As we will see later, the transfer function of a convolutional code plays a major role in bounding the error probability of the code. To obtain the transfer function of a convolutional code, we split the all-0 state into two states, one denoting the starting state and one denoting the first return to the all-0 state. All the other states are denoted as in-between states. Corresponding to each branch connecting two states, a function of the form $D^\alpha N^\beta J$ is defined, where α denotes the number of 1's in the output bit sequence and β is the number of 1's in the corresponding input sequence for that branch. The *transfer function* of the convolutional code is then the transfer function of the flow graph between the starting all-0 state and the final all-0 state and is denoted by $T(D, N, J)$. Each term of $T(D, N, J)$ corresponds to a path through the trellis starting from the all-0 state and ending at the all-0 state. The exponent of J indicates the number of branches spanned by that path, the exponent of D shows the number of 1's in the codeword corresponding to that path (or equivalently the Hamming distance of the codeword with the all-0 codeword), and the exponent of N indicates the number of 1's in the input information sequence. $T(D, N, J)$ indicates the properties of all paths through the trellis starting from the all-0 path and returning to it *for the first time*, so, in deriving it, any self-loop at the all-0 state is ignored. To obtain the transfer function of the convolutional code, we can use all rules that can be used to obtain the transfer function of a flow graph. For more details on deriving the transfer function of a convolutional code, see [1].

Following the rules for deriving the transfer function, we can easily show that the transfer function of the code shown in Figure 8.18 is given by

$$T(D, N, J) = \frac{D^5 N J^3}{1 - DNJ - DNJ^2}$$

which, when expanded, can be expressed as

$$T(D, N, J) = D^5 N J^3 + D^6 N^2 J^4 + D^6 N^2 J^5 + D^7 N^3 J^5 + \cdots$$

From this expression for $T(D, N, J)$, we can see that there exists one codeword with Hamming weight 5, two codewords with Hamming weight 6, and so on. It also shows, for example, that the codeword with Hamming weight 5 corresponds to an input sequence of Hamming weight 1 and length 3. The smallest power of D in the expansion of $T(D, N, J)$ is called the *free distance* of the convolutional code and is denoted by d_{free}. In this example, $d_{\text{free}} = 5$.

Decoding of Convolutional Codes

There exist many algorithms for decoding of convolutional codes. The Viterbi algorithm is probably the most widely used decoding method of convolutional codes. This algorithm is particularly interesting because it is a maximum-likelihood decoding algorithm, which—upon receiving the channel output—searches through the trellis to find the path that is most likely to have generated the received sequence. If hard-decision decoding is used, this algorithm finds the path that is at the minimum Hamming distance from the received sequence, and if soft-decision decoding is employed, the Viterbi algorithm finds the path that is at the minimum Euclidean distance from the received sequence.

In hard-decision decoding of convolutional codes, we want to choose a path through the trellis whose codeword, denoted by c, is at minimum Hamming distance from the quantized received sequence y. In hard-decision decoding, the channel is binary memoryless. (The fact that the channel is memoryless follows from the fact that the channel noise is assumed to be white.) Since the desired path starts from the all-0 state and returns back to the all-0 state, we assume that this path spans a total of m branches, and since each branch corresponds to n_0 bits of the encoder output, the total number of bits in c and y is mn_0. We denote the sequence of bits corresponding to the ith branch by c_i and y_i, respectively, where $1 \leq i \leq m$, and each c_i and y_i is of length n_0. The Hamming distance between c and y is, therefore,

$$d(c,y) = \sum_{i=1}^{m} d(c_i,y_i) \tag{8.3.33}$$

In soft-decision decoding, we have a similar situation, with three differences:

1. Instead of y, we are dealing directly with the vector r, the vector output of the optimal (matched filter type or correlator type) digital demodulator.

2. Instead of the binary 0, 1 sequence c, we are dealing with the corresponding sequence c' with

$$c'_{ij} = \begin{cases} \sqrt{\mathcal{E}}, & \text{if } c_{ij} = 1 \\ -\sqrt{\mathcal{E}}, & \text{if } c_{ij} = 0 \end{cases}$$

for $1 \leq i \leq m$ and $1 \leq j \leq n$.

3. Instead of Hamming distance, we are using Euclidean distance. This is a consequence of the fact that the channel under study is an additive white Gaussian noise channel.

From the preceding, we have

$$d_E^2(c',r) = \sum_{i=1}^{m} d_E^2(c'_i, r_i) \tag{8.3.34}$$

From Equations (8.3.33) and (8.3.34), we see that the generic form of the problem we have to solve is: Given a vector a, find a path through the trellis starting at the all-0 state

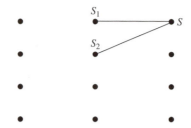

Figure 8.20   Justification of the Viterbi algorithm

and ending at the all-0 state such that some distance measure between a and a sequence b corresponding to that path is minimized. The important fact that makes this problem easy to solve is that the distance between a and b in both cases of interest can be written as the sum of distances corresponding to individual branches of the path. This is easily observed from (8.3.33) and (8.3.34).

Now let us assume that we are dealing with a convolutional code with $k_0 = 1$. This means that there are only two branches entering each state in the trellis. If the optimal path at a certain point passes through state S, there are two paths that connect the previous states S_1 and S_2 to this state. (See Figure 8.20.)

If we want to see which one of these two branches is a good candidate to minimize the overall distance, we have to add the overall (minimum) metrics at states S_1 and S_2 to the metrics of the branches connecting these two states to the state S. Then, obviously, the branch that has the minimum total metric accumulation up to state S is a candidate to be considered for the state after the state S. This branch is called a *survivor* at state S, and the other branch is simply not a suitable candidate and is deleted. Now, after the survivor at state S is determined, we also save the minimum metric up to this state, and we can move to the next state. This procedure is continued until we reach the all-0 state at the end of the trellis. For cases where $k_0 > 1$, the only difference is that at each stage we have to choose one survivor path from among 2^{k_0} branches leading to state S.

The preceding procedure can be summarized in the following algorithm, known as the *Viterbi algorithm*.

1. Parse the received sequence into m subsequences each of length n_0.

2. Draw a trellis of depth m for the code under study. For the last $L - 1$ stages of the trellis, draw only paths corresponding to the all-0 input sequences. [This is done because we know that the input sequence has been padded with $k_0(L - 1)$ 0's.]

3. Set $l = 1$ and set the metric of the initial all-0 state equal to 0.

4. Find the distance of the lth subsequence of the received sequence to all branches connecting lth-stage states to the $(l + 1)$st-stage states of the trellis.

5. Add these distances to the metrics of the lth-stage states to obtain the metric candidates for the $(l + 1)$st-stage states. For each state of the $(l + 1)$st stage, there are 2^{k_0} candidate metrics, each corresponding to one branch ending at that state.

6. For each state at the $(l+1)$st stage, choose the minimum of the candidate metrics and label the branch corresponding to this minimum value as the *survivor*, and assign the minimum of the metric candidates as the metrics of the $(l + 1)$st-stage states.

7. If $l = m$, go to the next step; otherwise, increase l by 1 and go to step 4.

8. Starting with the all-0 state at the $(m+1)$st stage, go back through the trellis along the survivors to reach the initial all-0 state. This path is the optimal path, and the input-bit sequence corresponding to that is the maximum-likelihood decoded information sequence. To obtain the best guess about the input-bit sequence, remove the last $k_0(L - 1)$ 0's from this sequence.

As we can see from the algorithm, the decoding delay and the amount of memory required for decoding a long information sequence are unacceptable. The decoding cannot be started until the whole sequence (which in the case of convolutional codes can be very long) is received, and the total surviving paths have to be stored. In practice, a suboptimal solution that does not cause these problems is desirable. One such approach, which is referred to as *path memory truncation*, is that the decoder at each stage searches only δ stages back in the trellis and not to the start of the trellis. With this approach, at the $(\delta + 1)$st stage, the decoder makes a decision on the input bits corresponding to the first stage of the trellis (the first k_0 bits), and future received bits do not change this decision. This means that the decoding delay will be $k_0\delta$ bits, and it is required to keep the surviving paths corresponding to the last δ stages. Computer simulations have shown that if $\delta \geq 5L$, the degradation in performance due to path memory truncation is negligible.

───(ILLUSTRATIVE PROBLEM)──────────────────────

Illustrative Problem 8.14 [Viterbi decoding] Let us assume that in hard-decision decoding, the quantized received sequence is

$$y = 01101111010001$$

The convolutional code is the one given in Figure 8.18. Find the maximum-likelihood information sequence and the number of errors.

───(SOLUTION)──────────────────────

The code is a (2, 1) code with $L = 3$. The length of the received sequence y is 14. This means that $m = 7$, and we have to draw a trellis of depth 7. Also note that since the input information sequence is padded with $k_0(L - 1) = 2$ 0's, for the final two stages of the trellis we will draw only the branches corresponding to all-0 inputs. This also means that the actual length of the input sequence is 5, which, after padding with two 0's, has increased to 7. The trellis diagram for this case is shown in Figure 8.21.

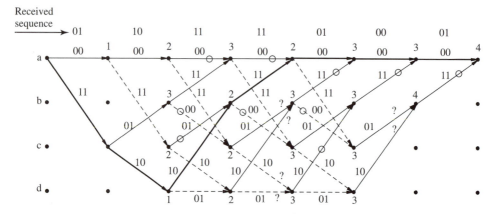

Figure 8.21   The trellis diagram for Viterbi decoding of the sequence
(01101111010001)

The parsed received sequence y is also shown in this figure. Note that in drawing the trellis in the last two stages, we have considered only the zero inputs to the encoder. (Notice that in the final two stages, there exist no dashed lines corresponding to 1 inputs.) Now the metric of the initial all-0 state is set to 0, and the metrics of the next stage are computed. In this step, there is only one branch entering each state; therefore, there is no comparison, and the metrics (which are the Hamming distances between that part of the received sequence and the branches of the trellis) are added to the metric of the previous state. In the next stage, there exists no comparison either. In the fourth stage, for the first time we have two branches entering each state. This means that a comparison has to be made here, and survivors are to be chosen. From the two branches that enter each state, one that corresponds to the least total accumulated metric remains as a survivor, and the other branches are deleted (marked by a small circle on the trellis). If, at any stage, two paths result in the same metric, each one of them can be a survivor. Such cases have been marked by a question mark in the trellis diagram. The procedure is continued to the final all-0 state of the trellis; then, starting from that state, we move along the surviving paths to the initial all-0 state. This path, which is denoted by a heavy path through the trellis, is the optimal path. The input-bit sequence corresponding to this path is 1100000, where the last two 0's are not information bits but were added to return the encoder to the all-0 state. Therefore, the information sequence is 11000. The corresponding codeword for the selected path is 11101011000000, which is at Hamming distance 4 from the received sequence. All other paths through the trellis correspond to codewords that are at greater Hamming distance from the received sequence.

For soft-decision decoding, a similar procedure is followed, with squared Euclidean distances substituted for Hamming distances.

The MATLAB function viterbi.m given next employs the Viterbi algorithm to decode a channel output. This algorithm can be used for both soft-decision and hard-decision decoding of convolutional codes. The separate file metric.m defines the metric used in the decoding process. For hard-decision decoding, this metric is the Hamming distance; for soft-decision decoding, it is the Euclidean distance. For cases where the channel output is quantized, the metric is usually the negative of the log-likelihood, $-\log p$(channel output | channel input). A number of short m-files called by viterbi.m are also given next.

M-FILE

```
function [decoder_output,survivor_state,cumulated_metric]=viterbi(G,k,channel_output)
%VITERBI        The Viterbi decoder for convolutional codes
%               [decoder_output,survivor_state,cumulated_metric]=viterbi(G,k,channel_output)
%               G is a n x Lk matrix each row of which
%               determines the connections from the shift register to the
%               n-th output of the code, k/n is the rate of the code.
%               survivor_state is a matrix showing the optimal path through
%               the trellis. The metric is given in a separate function metric(x,y)
%               and can be specified to accommodate hard and soft decision.
%               This algorithm minimizes the metric rather than maximizing
%               the likelihood.

n=size(G,1);
%   check the sizes
if rem(size(G,2),k) ~=0
    error('Size of G and k do not agree')
end
if rem(size(channel_output,2),n) ~=0
    error('channel output not of the right size')
end
L=size(G,2)/k;
number_of_states=2^((L-1)*k);
%   generate state transition matrix, output matrix, and input matrix
for j=0:number_of_states-1
    for l=0:2^k-1
        [next_state,memory_contents]=nxt_stat(j,l,L,k);
        input(j+1,next_state+1)=l;
        branch_output=rem(memory_contents*G',2);
        nextstate(j+1,l+1)=next_state;
        output(j+1,l+1)=bin2deci(branch_output);
    end
end
state_metric=zeros(number_of_states,2);
depth_of_trellis=length(channel_output)/n;
channel_output_matrix=reshape(channel_output,n,depth_of_trellis);
survivor_state=zeros(number_of_states,depth_of_trellis+1);
%   start decoding of non-tail channel outputs
for i=1:depth_of_trellis-L+1
    flag=zeros(1,number_of_states);
    if i <= L
        step=2^((L-i)*k);
    else
        step=1;
    end
    for j=0:step:number_of_states-1
```

```
    for l=0:2^k−1
      branch_metric=0;
      binary_output=deci2bin(output(j+1,l+1),n);
      for ll=1:n
        branch_metric=branch_metric+metric(channel_output_matrix(ll,i),binary_output(ll));
      end
      if((state_metric(nextstate(j+1,l+1)+1,2) > state_metric(j+1,1)...
        +branch_metric) | flag(nextstate(j+1,l+1)+1)==0)
        state_metric(nextstate(j+1,l+1)+1,2) = state_metric(j+1,1)+branch_metric;
        survivor_state(nextstate(j+1,l+1)+1,i+1)=j;
        flag(nextstate(j+1,l+1)+1)=1;
      end
    end
  end
  state_metric=state_metric(:,2:−1:1);
end
%  start decoding of the tail channel-outputs
for i=depth_of_trellis−L+2:depth_of_trellis
  flag=zeros(1,number_of_states);
  last_stop=number_of_states/(2^((i−depth_of_trellis+L−2)*k));
  for j=0:last_stop−1
    branch_metric=0;
    binary_output=deci2bin(output(j+1,1),n);
    for ll=1:n
      branch_metric=branch_metric+metric(channel_output_matrix(ll,i),binary_output(ll));
    end
    if((state_metric(nextstate(j+1,1)+1,2) > state_metric(j+1,1)...
      +branch_metric) | flag(nextstate(j+1,1)+1)==0)
      state_metric(nextstate(j+1,1)+1,2) = state_metric(j+1,1)+branch_metric;
      survivor_state(nextstate(j+1,1)+1,i+1)=j;
      flag(nextstate(j+1,1)+1)=1;
    end
  end
  state_metric=state_metric(:,2:−1:1);
end
%  generate the decoder output from the optimal path
state_sequence=zeros(1,depth_of_trellis+1);
state_sequence(1,depth_of_trellis)=survivor_state(1,depth_of_trellis+1);
for i=1:depth_of_trellis
  state_sequence(1,depth_of_trellis−i+1)=survivor_state((state_sequence(1,depth_of_trellis+2−i)...
  +1),depth_of_trellis−i+2);
end
decoder_output_matrix=zeros(k,depth_of_trellis−L+1);
for i=1:depth_of_trellis−L+1
  dec_output_deci=input(state_sequence(1,i)+1,state_sequence(1,i+1)+1);
  dec_output_bin=deci2bin(dec_output_deci,k);
  decoder_output_matrix(:,i)=dec_output_bin(k:−1:1)' ;
end
decoder_output=reshape(decoder_output_matrix,1,k*(depth_of_trellis−L+1));
cumulated_metric=state_metric(1,1);
```

◖ M-FILE ◗

```
function  distance=metric(x,y)
if  x==y
   distance=0;
else
   distance=1;
end
```

◖ M-FILE ◗

```
function  [next_state,memory_contents]=nxt_stat(current_state,input,L,k)
binary_state=deci2bin(current_state,k*(L−1));
binary_input=deci2bin(input,k);
next_state_binary=[binary_input,binary_state(1:(L−2)*k)];
next_state=bin2deci(next_state_binary);
memory_contents=[binary_input,binary_state];
```

◖ M-FILE ◗

```
function  y=bin2deci(x)
l=length(x);
y=(l−1:−1:0);
y=2.^y;
y=x*y' ;
```

◖ M-FILE ◗

```
function  y=deci2bin(x,l)
y = zeros(1,l);
i = 1;
while  x>=0  &  i<=l
        y(i)=rem(x,2);
        x=(x−y(i))/2;
        i=i+1;
end
y=y(l:−1:1);
```

◖ ILLUSTRATIVE PROBLEM ◗

Illustrative Problem 8.15 Repeat Illustrative Problem 8.14 using the MATLAB function viterbi.m.

SOLUTION

It is enough to use the m-file viterbi.m with the following inputs:

$$G = \begin{bmatrix} 1 & 0 & 1 \\ 1 & 1 & 1 \end{bmatrix}$$

$$k = 1$$

$$\text{channel\_output} = [0 \quad 1 \quad 1 \quad 0 \quad 1 \quad 1 \quad 1 \quad 1 \quad 0 \quad 1 \quad 0 \quad 0 \quad 0 \quad 1]$$

which results in decoder_output = $[1 \quad 1 \quad 0 \quad 0 \quad 0]$ and an accumulated metric of 4.

Error Probability Bounds for Convolutional Codes

Finding bounds on the error performance of convolutional codes is different from the method used to find error bounds for block codes because here we are dealing with se-quences of very large length; since the free distance of these codes is usually small, some errors will eventually occur. The number of errors is a random variable that de-pends on both the channel characteristics (signal-to-noise ratio in soft-decision decoding and crossover probability in hard-decision decoding) and the length of the input sequence. The longer the input sequence, the higher the probability of making errors. Therefore, it makes sense to normalize the number of bit errors to the length of the input sequence. A measure that is usually adopted for comparing the performance of convolutional codes is the expected number of bits received in error per input bit. To find a bound on the average number of bits in error for each input bit, we first derive a bound on the average number of bits in error for each input sequence of length k. To determine this, let us assume that the all-0 sequence is transmitted[2] and, up to stage l in the decoding, there has been no error. Now k information bits enter the encoder and result in moving to the next stage in the trellis. We are interested in finding a bound on the expected number of errors that can occur due to this input block of length k. Since we are assuming that up to stage l there has been no error, up to this stage the all-0 path through the trellis has the minimum metric. Now, as we move to the next stage (stage $(l + 1)$st), it is possible that another path through the trellis will have a metric less than the all-0 path and will therefore cause errors. If this happens, we must have a path through the trellis that merges with the all-0 path for the first time at the $(l + 1)$st stage and has a metric less than the all-0 path. Such an event is called the *first error event*, and the corresponding probability is called the *first error event probability*. This situation is depicted in Figure 8.22.

Our first step is bounding the first error event probability. Let $P_2(d)$ denote the prob-ability that a path through the trellis that is at Hamming distance d from the all-0 path is the survivor at the $(l + 1)$st stage. Since d is larger than d_{free}, we can bound the first error event probability by

$$P_e \leq \sum_{d=d_{\text{free}}}^{\infty} a_d P_2(d)$$

---

[2]Because of the linearity of convolutional codes, we can, without loss of generality, make this assumption.

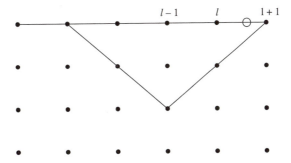

Figure 8.22   The path corresponding to the first error event

where on the right-hand side, we have included all paths through the trellis that merge with the all-0 path at the $(l + 1)$st stage. $P_2(d)$ denotes the error probability for a path at Hamming distance d from the all-0 path, and a_d denotes the number of paths at Hamming distance d from the all-0 path. The value of $P_2(d)$ depends on whether soft- or hard-decision decoding is employed.

For soft-decision decoding, if antipodal signaling (binary PSK) is used, we have

$$P_2(d) = Q\left(\frac{d^E}{\sqrt{2N_0}}\right)$$

$$= Q\left(\sqrt{\frac{2\mathcal{E}d}{N_0}}\right)$$

$$= Q\left(\sqrt{2R_c d \frac{\mathcal{E}_b}{N_0}}\right)$$

and, therefore,

$$P_e \leq \sum_{d=d_{\text{free}}}^{\infty} a_d Q\left(\sqrt{2R_c d \frac{\mathcal{E}_b}{N_0}}\right)$$

Using the well-known upper bound on the Q function

$$Q(x) \leq \frac{1}{2} e^{-x^2/2}$$

we obtain

$$Q\left(\sqrt{2R_c d \frac{\mathcal{E}_b}{N_0}}\right) \leq \frac{1}{2} e^{-R_c d \mathcal{E}_b/N_0}$$

Now, noting that

$$e^{-R_c d \mathcal{E}_b/N_0} = D^d \Big|_{D=e^{-R_c \mathcal{E}_b/N_0}}$$

we finally obtain

$$P_e \le \frac{1}{2} \sum_{d=d_{\text{free}}}^{\infty} a_d D^d \Bigg|_{D=e^{-R_c \mathcal{E}_b/N_0}} = \frac{1}{2} T_1(D) \Bigg|_{D=e^{-R_c \mathcal{E}_b/N_0}}$$

where

$$T_1(D) = T(D, N, J)|_{N=J=1}$$

This is a bound on the first error event probability. To find a bound on the average number of bits in error for k input bits, $\bar{P}_b(k)$, we note that each path through the trellis causes a certain number of input bits to be decoded erroneously. For a general $D^d N^{f(d)} J^{g(d)}$ in the expansion of $T(D, N, J)$,[3] there is a total of $f(d)$ nonzero input bits. This means that the average number of input bits in error can be obtained by multiplying the probability of choosing each path by the total number of input errors that would result if that path were chosen. Hence, the average number of bits in error, in the soft-decision case, can be bounded by

$$\bar{P}_b(k) \le \sum_{d=d_{\text{free}}}^{\infty} a_d f(d) P_2(d)$$

$$= \sum_{d=d_{\text{free}}}^{\infty} a_d f(d) Q\left(\sqrt{2 R_c d \frac{\mathcal{E}_b}{N_0}}\right)$$

$$\le \frac{1}{2} \sum_{d=d_{\text{free}}}^{\infty} a_d f(d) e^{-R_c d \mathcal{E}_b/N_0} \qquad (8.3.35)$$

If we define

$$T_2(D, N) = T(D, N, J)|_{J=1}$$

$$= \sum_{d=d_{\text{free}}}^{\infty} a_d D^d N^{f(d)}$$

we have

$$\frac{\partial T_2(D, N)}{\partial N}\Bigg|_{N=1} = \sum_{d=d_{\text{free}}}^{\infty} a_d f(d) D^d \qquad (8.3.36)$$

Therefore, using (8.3.35) and (8.3.36), we obtain

$$\bar{P}_b(k) \le \frac{1}{2} \frac{\partial T_2(D, N)}{\partial N}\Bigg|_{N=1, D=e^{-R_c \mathcal{E}_b/N_0}}$$

---

[3]Here we are somewhat sloppy in notation. The power of N is not strictly a function of d, but we are denoting it by $f(d)$. This, however, does not have any effect on the final result.

To obtain the average number of bits in error for each input bit, we have to divide this bound by k. Thus, the final result is

$$\bar{P}_b = \frac{1}{2k} \left. \frac{\partial T_2(D, N)}{\partial N} \right|_{N=1, D=e^{-R_c \mathcal{E}_b/N_0}}$$

For hard-decision decoding, the basic procedure follows the above derivation. The only difference is the bound on $P_2(d)$. It can be shown (see [1]) that $P_2(d)$ can be bounded by

$$P_2(d) \leq [4p(1 - p)]^{d/2}$$

From this result, it is straightforward to show that in hard-decision decoding, the probability of error is upper-bounded as

$$\bar{P}_b \leq \frac{1}{k} \left. \frac{\partial T_2(D, N)}{\partial N} \right|_{N=1, D=\sqrt{4p(1-p)}}$$

A comparison of hard-decision decoding and soft-decision decoding for convolutional codes shows that here, as in the case for linear block codes, soft-decision decoding outperforms hard-decision decoding by a margin of roughly 2 dB in additive white Gaussian noise channels.

Problems

8.1 Write a MATLAB script to plot the capacity of a binary symmetric channel with crossover probability p as a function of p for $0 \leq p \leq 1$. For what value of p is the capacity minimized, and what is the minimum value?

8.2 A binary nonsymmetric channel is characterized by the conditional probabilities $p(0 \mid 1) = 0.2$ and $p(1 \mid 0) = 0.4$. Plot the mutual information $I(X; Y)$ between the input and the output of this channel as a function of $p = P(X = 1)$. For what value of p is the mutual information maximized, and what is the value of this maximum?

8.3 A Z-channel is a binary input, binary output channel, with input and output alphabets $\mathcal{X} = \mathcal{Y} = \{0, 1\}$ and characterized by $p(0 \mid 1) = \epsilon$ and $p(1 \mid 0) = 0$. Plot $I(X; Y)$ as a function of $p = P(X = 1)$ for $\epsilon = 0, 0.1, 0.2, 0.3, 0.4, 0.5, 0.7, 0.9, 1$. Determine the capacity of the channel in each case.

8.4 A binary input, ternary output channel is characterized by the input and output alphabets $\mathcal{X} = \{0, 1\}$ and $\mathcal{Y} = \{0, 1, 2\}$ and transition probabilities $p(0 \mid 0) = 0.05$, $p(1 \mid 0) = 0.2$, $p(0 \mid 1) = 0.1$, and $p(1 \mid 1) = 0.1$. Plot $I(X; Y)$ as a function of $p = P(X = 1)$, and determine the channel capacity.

8.5 A ternary input, binary output channel is characterized by the input and output alphabets $\mathcal{X} = \{0, 1, 2\}$ and $\mathcal{Y} = \{0, 1\}$ and transition probabilities $p(0 \mid 0) = 0.05$, $p(1 \mid 1) = 0.2$, and $p(0 \mid 2) = 0.1$. Plot $I(X; Y)$ as a function of $p_1 = P(X = 1)$ and $p_2 = P(X = 2)$, and determine the channel capacity.

8.6 Plot the capacity of a binary symmetric channel that employs binary orthogonal signaling as a function of $\mathcal{E}_b/N_0$.

8.7 Repeat Illustrative Problem 8.3, but assume that the two transmitted signals are equal-energy and *orthogonal*. How do your results differ from those obtained in Illustrative Problem 8.3?

8.8 Compare the plots of the capacity for hard decision and soft decision when orthogonal signals are employed. Compare these results with those obtained for antipodal signals.

8.9 Plot the capacity of a binary symmetric channel that uses orthogonal signals as a function of $\mathcal{E}_b/N_0$. Do this once with the assumption of coherent detection and once with the assumption of noncoherent detection. Show the two plots on the same figure and compare the results.

8.10 Write a MATLAB script that generates the generator matrix of a Hamming code in the systematic form for any given m.

8.11 Repeat Illustrative Problem 8.10 using orthogonal signaling with coherent and non-coherent detection. Plot the results on the same figure.

8.12 Use Monte Carlo simulation to plot the error probability versus γ_b in Illustrative Problem 8.10.

8.13 Repeat Illustrative Problem 8.12, but instead of orthogonal and antipodal signaling, compare the performance of coherent and noncoherent demodulation of orthogonal signals under soft-decision decoding.

8.14 Use Monte Carlo simulation to plot the error probability versus γ_b in Illustrative Problem 8.12.

8.15 Use MATLAB to find the output of the convolutional encoder shown in Figure 8.18 when the input sequence is

$$1100101010100101111010111111010$$

8.16 A convolutional code is described by

$$G = \begin{bmatrix} 1 & 0 & 1 & 1 \\ 0 & 1 & 1 & 0 \\ 1 & 1 & 0 & 1 \\ 1 & 1 & 1 & 1 \end{bmatrix}$$

a. If $k = 1$, determine the output of the encoder when the input sequence is

$$1100101010100101111010111111010$$

b. Repeat part (a) with $k = 2$.

8.17 In Problem 8.15, after obtaining the output of the encoder, change the first 6 bits of the received sequence, and decode the result using Viterbi decoding. Compare the decoder output with the transmitted sequence. How many errors have occurred? Repeat the problem once by changing the last 6 bits in the received sequence and once by changing the first 3 bits and the last 3 bits, and compare the results. In all cases, the Hamming metric is used.

8.18 Generate an equiprobable binary sequence of length 1000. Encode the sequence using the convolutional code shown in Figure 8.18. Generate four random binary error sequences, each of length 2000, with probability of 1 equal to 0.01, 0.05, 0.1, and 0.2, respectively. Add (modulo 2) each of these error sequences to the encoded sequence, and use the Viterbi algorithm to decode the result. In each case, compare the decoded sequence with the encoder input, and determine the bit-error rate.

8.19 Use Monte Carlo simulation to plot the bit-error rate versus γ_b in a convolutional encoder using the code shown in Figure 8.18. Assume that the modulation scheme is binary antipodal, once with hard-decision and once with soft-decision decoding. Let γ_b be in the interval from 3 dB to 9 dB, and choose the length of your information sequence appropriately. Compare your results with the theoretical bounds.

8.20 The encoder of Figure 8.18 is used to transmit information over a channel with two inputs and three outputs. The outputs are denoted by 0, 1, and 2. This is the case where the output of a Gaussian channel is quantized to three levels. The conditional probabilities of the channel are given by $p(0|0) = p(1|1) = 0.9$, $p(2|0) = p(2|1) = 0.09$. Use the Viterbi algorithm to decode the received sequence

020201102021120022201101010200111112

Chapter 9

Spread Spectrum Communication Systems

9.1 Preview

Spread spectrum signals for digital communications were originally developed and used for military communications either (1) to provide resistance to hostile jamming; (2) to hide the signal by transmitting it at low power, thus, making it difficult for an unintended listener to detect its presence in noise; or (3) to make it possible for multiple users to communicate through the same channel. Today, however, spread spectrum signals are being used to provide reliable communications in a variety of commercial applications, including mobile vehicular communications and interoffice wireless communications.

The basic elements of a spread spectrum digital communication system are illustrated in Figure 9.1. We observe that the channel encoder and decoder and the modulator and demodulator are the basic elements of a conventional digital communication system. In addition to these elements, a spread spectrum system employs two identical pseudorandom sequence generators, one of which interfaces with the modulator at the transmitting end and the second of which interfaces with the demodulator at the receiving end. These two generators produce a pseudorandom or pseudonoise (PN) binary-valued sequence that is

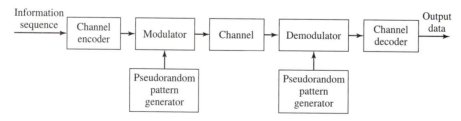

Figure 9.1   Model of spread spectrum digital communication system

used to spread the transmitted signal in frequency at the modulator and to despread the received signal at the demodulator.

Time synchronization of the PN sequence generated at the receiver with the PN sequence contained in the received signal is required to properly despread the received spread spectrum signal. In a practical system, synchronization is established prior to the transmission of information by transmitting a fixed PN bit pattern that is designed so that the receiver will detect it with high probability in the presence of interference. After time synchronization of the PN sequence generators is established, the transmission of information commences. In the data mode, the communication system usually tracks the timing of the incoming received signal and keeps the PN sequence generator in synchronism.

In this chapter, we consider two basic types of spread spectrum signals for digital communications—namely, direct-sequence (DS) spread spectrum and frequency-hopped (FH) spread spectrum.

Two types of digital modulation are considered in conjunction with spread spectrum—namely, PSK and FSK. PSK modulation is generally used with DS spread spectrum and is appropriate for applications where phase coherence between the transmitted signal and the received signal can be maintained over a time interval that spans several symbol (or bit) intervals. On the other hand, FSK modulation is commonly used with FH spread spectrum and is appropriate in applications where phase coherence of the carrier cannot be maintained due to time variations in the transmission characteristics of the communications channel.

9.2   Direct-Sequence Spread Spectrum Systems

Let us consider the transmission of a binary information sequence by means of binary PSK. The information rate is R bits per second, and the bit interval is $T_b = 1/R$ seconds. The available channel bandwidth is B_c hertz, where $B_c \gg R$. At the modulator, the bandwidth of the information signal is expanded to $W = B_c$ hertz by shifting the phase of the carrier pseudorandomly at a rate of W times per second according to the pattern of the PN generator. The resulting modulated signal is called a *direct-sequence* (DS) *spread spectrum signal*.

The information-bearing baseband signal is denoted as $v(t)$ and is expressed as

$$v(t) = \sum_{n=-\infty}^{\infty} a_n g_T(t - nT_b) \qquad (9.2.1)$$

where $\{a_n = \pm 1, -\infty < n < \infty\}$ and $g_T(t)$ is a rectangular pulse of duration T_b. This signal is multiplied by the signal from the PN sequence generator, which may be expressed as

$$c(t) = \sum_{n=-\infty}^{\infty} c_n p(t - nT_c) \qquad (9.2.2)$$

where $\{c_n\}$ represents the binary PN code sequence of ± 1's and $p(t)$ is a rectangular pulse of duration T_c, as illustrated in Figure 9.2. This multiplication operation serves

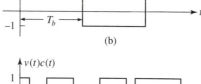

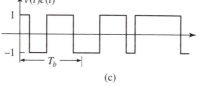

Figure 9.2 Generation of a DS spread spectrum signal: (a) PN signal; (b) data signal; (c) product signal

to spread the bandwidth of the information-bearing signal (whose bandwidth is approximately R hertz) into the wider bandwidth occupied by PN generator signal $c(t)$ (whose bandwidth is approximately $1/T_c$). The spectrum spreading is illustrated in Figure 9.3, which shows, in simple terms using rectangular spectra, the convolution of the two spectra, the narrow spectrum corresponding to the information-bearing signal and the wide spectrum corresponding to the signal from the PN generator.

The product signal $v(t)c(t)$, also illustrated in Figure 9.2, is used to amplitude-modulate the carrier $A_c \cos 2\pi f_c t$ and, thus, to generate the DSB-SC signal

$$u(t) = A_c v(t)c(t) \cos 2\pi f_c t \qquad (9.2.3)$$

Since $v(t)c(t) = \pm 1$ for any t, it follows that the carrier-modulated transmitted signal may also be expressed as

$$u(t) = A_c \cos[2\pi f_c t + \theta(t)] \qquad (9.2.4)$$

where $\theta(t) = 0$ when $v(t)c(T) = 1$ and $\theta(t) = \pi$ when $v(t)c(t) = -1$. Therefore, the transmitted signal is a binary PSK signal whose phase varies at the rate $1/T_c$.

The rectangular pulse $p(t)$ is usually called a *chip*, and its time duration, T_c, is called the *chip interval*. The reciprocal $1/T_c$ is called the *chip rate* and corresponds (approximately) to the bandwidth W of the transmitted signal. The ratio of the bit interval T_b to the chip interval T_c is usually selected to be an integer in practical spread spectrum systems. We denote this ratio as

$$L_c = \frac{T_b}{T_c} \qquad (9.2.5)$$

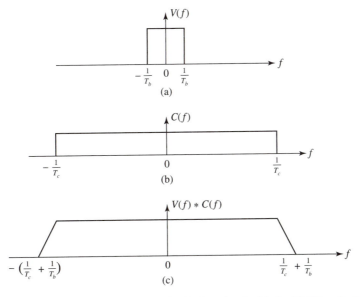

Figure 9.3   Convolution of spectra of the (a) data signal with the (b) PN code signal

Hence, L_c is the number of chips of the PN code sequence per information bit. Another interpretation is that L_c represents the number of possible $180°$ phase transitions in the transmitted signal during the bit interval T_b.

9.2.1   Signal Demodulation

The demodulation of the signal is performed as illustrated in Figure 9.4. The received signal is first multiplied by a replica of the waveform $c(t)$ generated by the PN code sequence generator at the receiver, which is synchronized to the PN code in the received signal. This operation is called (spectrum) *despreading*, since the effect of multiplication by $c(t)$ at the receiver is to undo the spreading operation at the transmitter. Thus, we have

$$A_c v(t) c^2(t) \cos 2\pi f_c t = A_c v(t) \cos 2\pi f_c t \qquad (9.2.6)$$

since $c^2(t) = 1$ for all t. The resulting signal $A_c v(t) \cos 2\pi f_c t$ occupies a bandwidth (approximately) of R hertz, which is the bandwidth of the information-bearing signal. Therefore, the demodulator for the despread signal is simply the conventional cross-correlator or matched filter that was described in Chapters 5 and 7. Since the demodulator has a bandwidth that is identical to the bandwidth of the despread signal, the only additive noise that corrupts the signal at the demodulator is the noise that falls within the information bandwidth of the received signal.

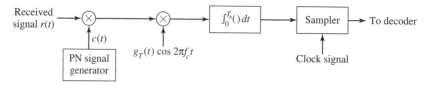

Figure 9.4   Demodulation of DS spread spectrum signal

Effects of Despreading on a Narrowband Interference

It is interesting to investigate the effect of an interfering signal on the demodulation of the desired information-bearing signal. Suppose that the received signal is

$$r(t) = A_c v(t) c(t) \cos 2\pi f_c t + i(t) \tag{9.2.7}$$

where $i(t)$ denotes the interference. The despreading operation at the receiver yields

$$r(t)c(t) = A_c v(t) \cos 2\pi f_c t + i(t)c(t) \tag{9.2.8}$$

The effect of multiplying the interference $i(t)$ with $c(t)$ is to spread the bandwidth of $i(t)$ to W hertz.

As an example, let us consider a sinusoidal interfering signal of the form

$$i(t) = A_J \cos 2\pi f_J t \tag{9.2.9}$$

where f_J is a frequency within the bandwidth of the transmitted signal. Its multiplication with $c(t)$ results in a wideband interference with power-spectral density $J_0 = P_J/W$, where $P_J = A_J^2/2$ is the average power of the interference. Since the desired signal is demodulated by a matched filter (or correlator) that has a bandwidth R, the total power in the interference at the output of the demodulator is

$$J_0 R = \frac{P_J R}{W} = \frac{P_J}{W/R} = \frac{P_J}{T_b/T_c} = \frac{P_J}{L_c} \tag{9.2.10}$$

Therefore, the power in the interfering signal is reduced by an amount equal to the bandwidth expansion factor W/R. The factor $W/R = T_b/T_c = L_c$ is called the *processing gain* of the spread spectrum system. The reduction in interference power is the basic reason for using spread spectrum signals to transmit digital information over channels with interference.

In summary, the PN code sequence is used at the transmitter to spread the information-bearing signal into a wide bandwidth for transmission over the channel. When the received signal is multiplied by a synchronized replica of the PN code signal, the desired signal is despread back to a narrow bandwidth, whereas any interference signals are spread over a wide bandwidth. The net effect is a reduction in the interference power by the factor W/R, which is the processing gain of the spread spectrum system.

The PN code sequence $\{c_n\}$ is assumed to be known only to the intended receiver. Any other receiver that does not have knowledge of the PN code sequence cannot demodulate the signal. Consequently, the use of a PN code sequence provides a degree of privacy (or security) that is not possible to achieve with conventional modulation. The primary cost for this security and performance gain against interference is an increase in channel bandwidth utilization and in the complexity of the communication system.

9.2.2   Probability of Error

In an AWGN channel, the probability of error for a DS spread spectrum system employing binary PSK is identical to the probability of error for conventional (unspread) binary PSK—that is,

$$P_b = Q\left(\sqrt{\frac{2\mathcal{E}_b}{N_0}}\right) \tag{9.2.11}$$

On the other hand, if the interference is the sinusoidal signal given by (9.2.9) with power P_J, the probability of error is (approximately)

$$P_b = Q\left(\sqrt{\frac{2\mathcal{E}_b}{P_J/W}}\right) = Q\left(\sqrt{\frac{2\mathcal{E}_b}{J_0}}\right) \tag{9.2.12}$$

Thus, the interference power is reduced by the factor of the spread spectrum signal bandwidth W. In this case, we have ignored the AWGN, which is assumed to be negligible—that is, $N_0 \ll P_J/W$. If we account for the AWGN in the channel, the error probability is expressed as

$$P_b = Q\left(\sqrt{\frac{2\mathcal{E}_b}{N_0 + P_J/W}}\right)$$
$$= Q\left(\sqrt{\frac{2\mathcal{E}_b}{N_0 + J_0}}\right) \tag{9.2.13}$$

The Jamming Margin

When the interference signal is a jamming signal, we may express $\mathcal{E}_b/J_0$ as

$$\frac{\mathcal{E}_b}{J_0} = \frac{P_S T_b}{P_J/W} = \frac{P_S/R}{P_J/W} = \frac{W/R}{P_J/P_S} \tag{9.2.14}$$

Now, suppose we specify a required $\mathcal{E}_b/J_0$ to achieve a desired performance. Then, using a logarithmic scale, we may express (9.2.14) as

$$10\log\frac{P_J}{P_S} = 10\log\frac{W}{R} - 10\log\left(\frac{\mathcal{E}_b}{J_0}\right)$$
$$\left(\frac{P_J}{P_S}\right)_{\text{dB}} = \left(\frac{W}{R}\right)_{\text{dB}} - \left(\frac{\mathcal{E}_b}{J_0}\right)_{\text{dB}} \tag{9.2.15}$$

The ratio $(P_J/P_S)_{\text{dB}}$ is called the *jamming margin*. This is the relative power advantage that a jammer may have without disrupting the communication system.

ILLUSTRATIVE PROBLEM

Illustrative Problem 9.1 [Processing gain and jamming margin] Suppose that we require $\mathcal{E}_b/J_0 = 10$ dB to achieve reliable communication with binary PSK. Determine the processing gain that is necessary to provide a jamming margin of 20 dB.

SOLUTION

By using (9.2.15), we find that the processing gain $(W/R)_{\text{dB}} = 30$ dB—that is, $W/R = L_c = 1000$. This means that with $W/R = 1000$, the average jamming power at the receiver may be 100 times the power P_S of the desired signal, and we would still maintain reliable communication.

Performance of Coded Spread Spectrum Signals

As shown in Chapter 8, when the transmitted information is coded by a binary linear (block or convolutional) code, the SNR at the output of a soft-decision decoder is increased by the coding gain, defined as

$$\text{Coding gain} = R_c d_{\text{min}}^H \qquad (9.2.16)$$

where R_c is the code rate and d_{min}^H is the minimum Hamming distance of the code. Therefore, the effect of the coding is to increase the jamming margin by the coding gain. Thus, (9.2.15) may be modified as

$$\left(\frac{P_J}{P_S}\right)_{\text{dB}} = \left(\frac{W}{R}\right)_{\text{dB}} + (CG)_{\text{dB}} - \left(\frac{\mathcal{E}_b}{J_0}\right)_{\text{dB}} \qquad (9.2.17)$$

9.2.3   Two Applications of DS Spread Spectrum Signals

In this subsection, we briefly describe the use of DS spread spectrum signals in two applications. First, we consider an application in which the signal is transmitted at very low power, so that a listener would encounter great difficulty in trying to detect the presence of the signal. A second application is multiple-access radio communications.

Low-Detectability Signal Transmission

In this application, the information-bearing signal is transmitted at a very low power level relative to the background channel noise and thermal noise that are generated in the front end of a receiver. If the DS spread spectrum signal occupies a bandwidth W and the power-spectral density of the additive noise is N_0 watts/hertz, the average noise power in the bandwidth W is $P_N = W N_0$.

The average received signal power at the intended receiver is P_R. If we wish to hide the presence of the signal from receivers that are in the vicinity of the intended receiver, the signal is transmitted at a power level such that $P_R/P_N \ll 1$. The intended receiver can recover the weak information-bearing signal from the background noise with the aid of the processing gain and the coding gain. However, any other receiver that has no knowledge of the PN code sequence is unable to take advantage of the processing gain and the coding gain. Consequently, the presence of the information-bearing signal is difficult to detect. We say that the transmitted signal has a *low probability of being intercepted* (LPI), and it is called an *LPI signal*.

The probability of error given in Section 9.2.2 applies as well to the demodulation and decoding of LPI signals at the intended receiver.

ILLUSTRATIVE PROBLEM

Illustrative Problem 9.2 [DS spread spectrum system design] A DS spread spectrum signal is to be designed such that the power ratio at the intended receiver is $P_R/P_N = 0.01$ for an AWGN channel. The desired value of $\mathscr{E}_b/N_0$ is for acceptable performance. Determine the minimum value of the processing gain required to achieve an $\mathscr{E}_b/N_0$ of 10.

SOLUTION

We may write $\mathscr{E}_b/N_0$ as

$$\frac{\mathscr{E}_b}{N_0} = \frac{P_R T_b}{N_0} = \frac{P_R L_c T_c}{N_0} = \left(\frac{P_R}{W N_0} \right) L_c = \left(\frac{P_R}{P_N} \right) L_c \tag{9.2.18}$$

Since $\mathscr{E}_b/N_0 = 10$ and $P_R/P_N = 10^{-2}$, it follows that the necessary processing gain is $L_c = 1000$.

Code Division Multiple Access

The enhancement in performance obtained from a DS spread spectrum signal through the processing gain and the coding gain can be used to enable many DS spread spectrum signals to occupy the same channel bandwidth, provided that each signal has its own pseudo-random (signature) sequence. Thus, it is possible to have several users transmit messages simultaneously over the same channel bandwidth. This type of digital communication, in which each transmitter/receiver user pair has its own distinct signature code for transmitting over a common channel bandwidth, is called *code division multiple access* (CDMA).

In digital cellular communications, a base station transmits signals to N_u mobile receivers using N_u orthogonal PN sequences, one for each intended receiver. These N_u

signals are perfectly synchronized at transmission so that they arrive at each mobile receiver in synchronism. Consequently, due to the orthogonality of the N_u PN sequences, each intended receiver can demodulate its own signal without interference from the other transmitted signals that share the same bandwidth. However, this type of synchronism cannot be maintained in the signals transmitted from the mobile transmitters to the base station (the uplink, or reverse link). In the demodulation of each DS spread spectrum signal at the base station, the signals from the other simultaneous users of the channel appear as additive interference. Let us determine the number of simultaneous signals that can be accommodated in a CDMA system. We assume that all signals have identical average powers at the base station. In many practical systems, the received signal's power level from each user is monitored at the base station, and power control is exercised over all simultaneous users by use of a control channel that instructs the users on whether to increase or decrease their power levels. With such power control, if there are N_u simultaneous users, the desired signal-to-noise interference power ratio at a given receiver is

$$\frac{P_S}{P_N} = \frac{P_S}{(N_u - 1)P_S} = \frac{1}{N_u - 1} \tag{9.2.19}$$

From this relation, we can determine the number of users that can be accommodated simultaneously.

In determining the maximum number of simultaneous users of the channel, we implicitly assumed that the pseudorandom code sequences used by the various users are orthogonal and that the interference from other users adds on a power basis only. However, orthogonality of the pseudorandom sequences among the N_u users generally is difficult to achieve, especially if N_u is large. In fact, the design of a large set of pseudorandom sequences with good correlation properties is an important problem that has received considerable attention in the technical literature. We briefly treat this problem in Section 9.3.

ILLUSTRATIVE PROBLEM

Illustrative Problem 9.3 [Maximum number of users in CDMA] Suppose that the desired level of performance for a user in a CDMA system is achieved when $\mathcal{E}_b/J_0 = 10$. Determine the maximum number of simultaneous users that can be accommodated in the CDMA system if the bandwidth-to-bit-rate ratio is 100 and the coding gain is 6 dB.

SOLUTION

From the basic relationship given in (9.2.17), we have

$$\left(\frac{P_N}{P_S}\right)_{\text{dB}} = \left(\frac{W}{R}\right)_{\text{dB}} + (CG)_{\text{dB}} - \left(\frac{\mathcal{E}_b}{J_0}\right)_{\text{dB}}$$
$$= 20 + 6 - 10 = 16 \text{ dB}$$

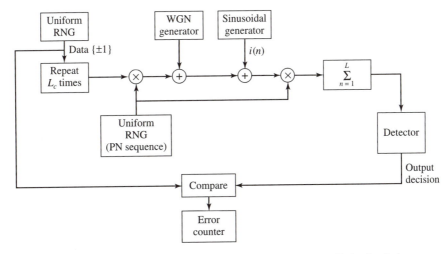

Figure 9.5   Model of DS spread spectrum system for Monte Carlo simulation

Consequently,

$$\frac{1}{N_u - 1} = \frac{P_S}{P_N} = \frac{1}{40}$$

and hence,

$$N_u = 41 \text{ users}$$

─────◖ILLUSTRATIVE PROBLEM◗──────────────────

Illustrative Problem 9.4 [DS spread spectrum simulation] The objective of this problem is to demonstrate the effectiveness of a DS spread spectrum signal in suppressing sinusoidal interference via Monte Carlo simulation. The block diagram of the system to be simulated is illustrated in Figure 9.5.

─────◖SOLUTION◗──────────────────

A uniform random number generator (RNG) is used to generate a sequence of binary information symbols (± 1). Each information bit is repeated L_c times, where L_c corresponds to the number of PN chips per information bit. The resulting sequence, which contains L_c repetitions per bit, is multiplied by a PN sequence $c(n)$ generated by another uniform RNG. To this product sequence, we add white Gaussian noise with variance $\sigma^2 = N_0/2$ and sinusoidal interference of the form

$$i(n) = A \sin \omega_0 n$$

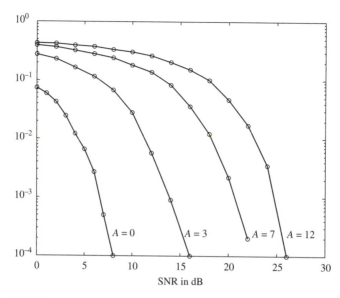

Figure 9.6   Error-rate performance of the system from the Monte Carlo simulation

where $0 < \omega_0 < \pi$ and the amplitude of the sinusoid is selected to satisfy $A < L_c$. The demodulator performs the cross-correlation with the PN sequence and sums (integrates) the blocks of L_c signal samples that constitute each information bit. The output of the summer is fed to the detector, which compares this signal with the threshold of zero and decides on whether the transmitted bit is $+1$ or -1. The error counter counts the number of errors made by the detector. The results of the Monte Carlo simulation are shown in Figure 9.6 for three different values of the amplitude of the sinusoidal interference with $L_c = 20$. Also shown in Figure 9.6 is the measured error rate when the sinusoidal interference is removed. The variance of the additive noise was kept fixed in these simulations, and the level of the desired signal was scaled to achieve the desired SNR for each simulation run.

The MATLAB scripts for the simulation program are given next.

```
M-FILE
```

```
% MATLAB script for Illustrative Problem 9.4.
echo on
Lc=20;                      % number of chips per bit
A1=3;                       % amplitude of the first sinusoidal interference
A2=7;                       % amplitude of the second sinusoidal interference
A3=12;                      % amplitude of the third sinusoidal interference
A4=0;                       % fourth case: no interference
w0=1;                       % frequency of the sinusoidal interference in radians
SNRindB=0:2:30;
for i=1:length(SNRindB),
```

```
  % measured error rates
  smld_err_prb1(i)=ss_Pe94(SNRindB(i),Lc,A1,w0);
  smld_err_prb2(i)=ss_Pe94(SNRindB(i),Lc,A2,w0);
  smld_err_prb3(i)=ss_Pe94(SNRindB(i),Lc,A3,w0);
  echo off ;
end;
echo on ;
SNRindB4=0:1:8;
for  i=1:length(SNRindB4),
  % measured error rate when there is no interference
  smld_err_prb4(i)=ss_Pe94(SNRindB4(i),Lc,A4,w0);
  echo off ;
end;
echo on ;
% Plotting commands follow
```

M-FILE

```
function [p]=ss_Pe94(snr_in_dB, Lc, A, w0)
% [p]=ss_Pe94(snr_in_dB, Lc, A, w0)
%                SS_PE94  finds the measured error rate. The function
%                that returns the measured probability of error for the given value of
%                the snr_in_dB, Lc, A and w0.
snr=10^(snr_in_dB/10);
sgma=1;                          % noise standard deviation is fixed
Eb=2*sgma^2*snr;                 % signal level required to achieve the given
                                 % signal to noise ratio
E_chip=Eb/Lc;                    % energy per chip
N=10000;                         % number of bits transmitted
% The generation of the data, noise, interference, decoding process and error
% counting is performed all together in order to decrease the run time of the
% program. This is accomplished by avoiding very large sized vectors.
num_of_err=0;
for i=1:N,
  % generate the next data bit
  temp=rand;
  if (temp<0.5),
    data=-1;
  else
    data=1;
  end;
  % repeat it Lc times, i.e. divide it into chips
  for j=1:Lc,
    repeated_data(j)=data;
  end;
  % pn sequence for the duration of the bit is generated next
  for j=1:Lc,
    temp=rand;
    if (temp<0.5),
      pn_seq(j)=-1;
```

```
  else
      pn_seq(j)=1;
    end;
  end;
  % the transmitted signal is
  trans_sig=sqrt(E_chip)*repeated_data.*pn_seq;
  % AWGN with variance sgma^2
  noise=sgma*randn(1,Lc);
  % interference
  n=(i−1)*Lc+1:i*Lc;
  interference=A*sin(w0*n);
  % received signal
  rec_sig=trans_sig+noise+interference;
  % determine the decision variable from the received signal
  temp=rec_sig.*pn_seq;
  decision_variable=sum(temp);
  % making decision
  if (decision_variable<0),
    decision=−1;
  else
    decision=1;
  end;
  % if it is an error, increment the error counter
  if (decision~=data),
    num_of_err=num_of_err+1;
  end;
end;
% then the measured error probability is
p=num_of_err/N;
```

9.3   Generation of PN Sequences

A pseudorandom, or PN, sequence is a code sequence of 1's and 0's whose autocorrelation has properties similar to those of white noise. In this section, we briefly describe the construction of some PN sequences and their autocorrelation and cross-correlation properties.

By far, the most widely known binary PN code sequences are the maximum-length shift-register sequences. A maximum-length shift-register sequence, or m-sequence for short, has a length $L = 2^m - 1$ bits and is generated by an m-stage shift register with linear feedback, as illustrated in Figure 9.7. The sequence is periodic with period L. Each period has a sequence of 2^{m-1} ones and $2^{m-1} - 1$ zeros. Table 9.1 lists the shift register connections for generating maximum-length sequences.

In DS spread spectrum applications, the binary sequence with elements $\{0, 1\}$ is mapped into a corresponding binary sequence with elements $\{-1, 1\}$. We shall call the equivalent sequence $\{c_n\}$ with elements $\{-1, 1\}$ a *bipolar sequence*.

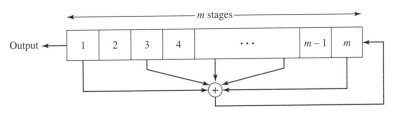

Figure 9.7   General *m*-stage shift register with linear feedback

Table 9.1   Shift-register connections for generating maximum-length sequences

m	Stages Connected to Modulo-2 Adder	*m*	Stages Connected to Modulo-2 Adder	*m*	Stages Connected to Modulo-2 Adder
2	1, 2	13	1, 10, 11, 13	24	1, 18, 23, 24
3	1, 3	14	1, 5, 9, 14	25	1, 23
4	1, 4	15	1, 15	26	1, 21, 25, 26
5	1, 4	16	1, 5, 14, 16	27	1, 23, 26, 27
6	1, 6	17	1, 15	28	1, 26
7	1, 7	18	1, 12	29	1, 28
8	1, 5, 6, 7	19	1, 15, 18, 19	30	1, 8, 29, 30
9	1, 6	20	1, 18	31	1, 29
10	1, 8	21	1, 20	32	1, 11,31,32
11	1, 10	22	1, 22	33	1, 21
12	1, 7, 9, 12	23	1, 19	34	1, 8, 33, 34

An important characteristic of a periodic PN sequence is its autocorrelation function, which is usually defined in terms of the bipolar sequences $\{c_n\}$ as

$$R_c(m) = \sum_{n=1}^{L} c_n c_{n+m}, \qquad 0 \le m \le L - 1 \tag{9.3.1}$$

where L is the period of the sequence. Since the sequence $\{c_n\}$ is periodic with period L, the autocorrelation sequence $\{R_c(m)\}$ is also periodic with period L.

Ideally, a PN sequence should have an autocorrelation function that has correlation properties similar to those of white noise. That is, the ideal autocorrelation sequence for $\{c_n\}$ is $R_c(0) = L$ and $R_c(m) = 0$ for $1 \le m \le L - 1$. In the case of *m*-sequences, the autocorrelation sequence is

$$R_c(m) = \begin{cases} L, & m = 0 \\ -1, & 1 \le m \le L - 1 \end{cases} \tag{9.3.2}$$

For long *m*-sequences, the size of the off-peak values of $R_c(m)$ relative to the peak value $R_c(0)$—that is, the ratio $R_c(m)/R_c(0) = -1/L$—is small and, from a practical view-

Table 9.2   Peak cross-correlations of m-sequences and Gold sequences

m	$L = 2^m - 1$	m-sequences			Gold sequences	
		Number	R_{max}	$R_{max}/R(0)$	R_{max}	$R_{max}/R(0)$
3	7	2	5	0.71	5	0.71
4	15	2	9	0.60	9	0.60
5	31	6	11	0.35	9	0.29
6	63	6	23	0.36	17	0.27
7	127	18	41	0.32	17	0.13
8	255	16	95	0.37	33	0.13
9	511	48	113	0.22	33	0.06
10	1023	60	383	0.37	65	0.06
11	2047	176	287	0.14	65	0.03
12	4095	144	1407	0.34	129	0.03

point, inconsequential. Therefore, m-sequences are very close to ideal PN sequences when viewed in terms of their autocorrelation function.

In some applications, the cross-correlation properties of PN sequences are as important as the autocorrelation properties. For example, in CDMA each user is assigned a particular PN sequence. Ideally, the PN sequences among users should be mutually orthogonal so that the level of interference experienced by one user from transmissions of other users is zero. However, the PN sequences used in practice by different users exhibit some correlation.

To be specific, let us consider the class of m-sequences. It is known that the periodic cross-correlation function between a pair of m-sequences of the same period can have relatively large peaks. Table 9.2 lists the peak magnitude R_{max} for the periodic cross-correlation between pairs of m-sequences for $3 \leq m \leq 12$. Also listed in Table 9.2 is the number of m-sequences of length $L = 2^m - 1$ for $3 \leq m \leq 12$. We observe that the number of m-sequences of the length L increases rapidly with m. We also observe that, for most sequences, the peak magnitude R_{max} of the cross-correlation function is a large percentage of the peak value of the autocorrelation function. Consequently, m-sequences are not suitable for CDMA communication systems. Although it is possible to select a small subset of m-sequences that have relatively smaller cross-correlation peak values than R_{max}, the number of sequences in the set is usually too small for CDMA applications.

Methods for generating PN sequences with better periodic cross-correlation properties than m-sequences have been developed by Gold [5], [6] and by Kasami [7]. Gold sequences are constructed by taking a pair of specially selected m-sequences, called *preferred m-sequences*, and forming the modulo-2 sum of the two sequences for each of L cyclically shifted versions of one sequence relative to the other sequence. Thus, L Gold sequences are generated as illustrated in Figure 9.8. For large L and m odd, the maximum value of the cross-correlation function between any pair of Gold sequences is $R_{max} = \sqrt{2L}$. For m even, $R_{max} = \sqrt{L}$.

Kasami [7] described a method for constructing PN sequences by decimating an m-sequence. In Kasami's method of construction, every $(2^{m/2} + 1)$st bit of an m-sequence is

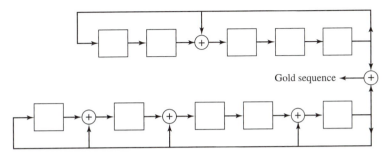

Figure 9.8   Generation of Gold sequences of length 31

selected. This method of construction yields a smaller set of PN sequences compared with Gold sequences, but their maximum cross-correlation value is $R_{max} = \sqrt{L}$.

It is interesting to compare the peak value of the cross-correlation function for Gold sequences and for Kasami sequences with a known lower bound for the maximum cross-correlation between any pair of binary sequences of length L. Given a set of N sequences of period L, a lower bound on their maximum cross-correlation is

$$R_{max} \geq L\sqrt{\frac{N-1}{NL-1}} \qquad (9.3.3)$$

which, for large values of L and N, is well approximated as $R_{max} \geq \sqrt{L}$. Hence, we observe that Kasami sequences satisfy the lower bound, and hence, they are optimal. On the other hand, Gold sequences with m odd have $R_{max} = \sqrt{2L}$. Hence, they are slightly suboptimal.

Besides the well-known Gold sequences and Kasami sequences, there are other binary sequences that are appropriate for CDMA applications. The interested reader is referred to papers by Scholtz [8] and by Sarwate and Pursley [9].

─(**ILLUSTRATIVE PROBLEM**)────────────────────

Illustrative Problem 9.5 [Gold sequence generation] Generate the $L = 31$ Gold sequences that result from taking the modulo-2 sum of the two shift-register outputs shown in Figure 9.8.

─(**SOLUTION**)────────────────────

The MATLAB scripts for performing this computation are given next. The 31 sequences generated are shown in Table 9.3. The maximum value of the cross-correlation for these sequences is $R_{max} = 9$.

Table 9.3   Table of Gold sequences in Illustrative Problem 9.5

```
0 1 1 1 0 0 0 0 1 0 0 0 0 1 1 0 0 1 0 0 1 0 1 1 1 1 0 0 0 0 0
0 1 0 1 1 0 1 1 1 0 0 0 1 0 0 0 0 0 1 0 0 1 0 0 0 1 1 0 0 0 1
0 0 0 0 1 1 0 1 1 0 0 1 0 1 0 0 1 1 1 1 1 0 1 1 0 0 1 0 0 1 1
1 0 1 0 0 0 0 1 1 0 1 0 1 1 0 1 0 1 0 0 0 1 0 1 1 0 1 0 1 1 1
1 1 1 1 1 0 0 1 1 1 0 1 1 1 1 0 0 0 1 1 1 0 0 0 1 0 1 1 1 1 0
0 1 0 0 1 0 0 1 0 0 1 1 1 0 0 0 1 1 0 0 0 0 1 0 1 0 0 1 1 0 0
0 0 1 0 1 0 0 0 1 1 1 1 0 1 0 1 0 0 1 1 0 1 1 0 1 1 0 1 0 0 1
1 1 1 0 1 0 1 1 0 1 1 0 1 1 1 0 1 1 0 1 1 1 1 0 0 1 0 0 0 1 1
0 1 1 0 1 1 0 0 0 1 0 1 1 0 0 1 0 0 0 0 1 1 1 1 0 1 1 0 1 1 0
0 1 1 0 0 0 1 0 0 0 1 1 0 1 1 0 1 0 1 0 1 1 0 1 0 0 1 1 1 0 1
0 1 1 1 1 1 1 0 1 1 1 0 1 0 0 1 1 1 1 0 1 0 0 1 1 0 0 1 0 1 1
0 1 0 0 0 1 1 1 0 1 0 1 0 1 1 1 0 1 1 0 0 0 0 0 1 1 0 0 1 1 1
0 0 1 1 0 1 0 0 0 0 1 0 1 0 1 0 0 1 1 1 0 0 1 0 0 1 1 1 1 1 1
1 1 0 1 0 0 1 0 1 1 0 1 0 0 0 0 0 1 0 1 0 1 1 1 0 0 0 1 1 1 1
0 0 0 1 1 1 1 1 0 0 1 0 0 1 0 0 0 0 0 1 1 1 0 1 1 1 0 1 1 1 0
1 0 0 0 0 1 0 0 1 1 0 0 1 1 0 0 1 0 0 0 1 0 0 0 0 1 0 1 1 0 1
1 0 1 1 0 0 1 1 0 0 0 1 1 1 0 1 1 0 1 0 0 0 1 1 0 1 0 1 0 1 0
1 1 0 1 1 1 0 0 1 0 1 1 1 1 1 1 1 1 0 1 0 1 0 1 0 0 1 0 0
0 0 0 0 0 0 1 1 1 1 1 1 0 1 1 0 1 0 1 1 0 0 1 0 1 1 1 0 0 0
1 0 1 1 1 1 0 1 0 1 1 1 0 0 1 0 0 0 0 0 0 0 0 1 0 0 0 0 0 0 1
1 1 0 0 0 0 0 0 0 1 1 0 0 0 0 0 1 0 1 1 0 0 0 1 1 1 1 0 0 1 0
0 0 1 1 1 0 1 0 0 1 0 0 0 1 0 1 1 1 0 1 0 0 0 0 0 0 1 0 1 0 0
1 1 0 0 1 1 1 0 0 0 0 0 1 1 1 1 0 0 0 1 0 0 1 1 1 0 1 1 0 0 1
0 0 1 0 0 1 1 0 1 0 0 1 1 0 1 0 1 0 0 1 0 1 0 0 1 0 0 0 0 1 0
1 1 1 1 0 1 1 1 1 0 1 1 0 0 0 1 1 0 0 1 1 0 1 0 1 1 1 0 1 0 1
0 1 0 1 0 1 0 1 1 1 1 0 0 1 1 1 1 0 0 0 0 1 1 0 0 0 1 1 0 1 0
0 0 0 1 0 0 0 1 0 1 0 0 1 0 1 1 1 0 1 1 1 1 1 1 0 0 0 1 0 1
1 0 0 1 1 0 0 0 0 0 0 1 0 0 1 1 1 1 0 0 1 1 0 0 1 1 1 1 0 1 1
1 0 0 0 1 0 1 0 1 0 1 0 0 0 1 1 0 0 1 0 1 0 1 0 0 0 0 0 1 1 0
1 0 1 0 1 1 1 1 1 0 0 0 0 1 0 1 1 1 0 0 1 1 1 1 1 1 1 1 0 0
1 1 1 0 0 1 0 1 0 0 0 0 0 0 0 1 0 1 1 1 1 1 0 0 0 0 0 1 0 0 0
```

―――――――――――――― **M-FILE** ――――――――――――――――――――――――――

% *MATLAB script for Illustrative Problem 9.5.*
echo on
% *first determine the maximal length shift register sequences*
% *We'll take the initial shift register content as "00001".*
connections1=[1 0 1 0 0];
connections2=[1 1 1 0 1];

```
sequence1=ss_mlsrs(connections1);
sequence2=ss_mlsrs(connections2);
% cyclically shift the second sequence and add it to the first one
L=2^length(connections1)−1;;
for shift_amount=0:L−1,
   temp=[sequence2(shift_amount+1:L) sequence2(1:shift_amount)];
   gold_seq(shift_amount+1,:)=(sequence1+temp) − floor((sequence1+temp)./2).*2;
   echo off ;
end;
echo on ;
% find the max value of the cross correlation for these sequences
max_cross_corr=0;
for i=1:L−1,
   for j=i+1:L,
      % equivalent sequences
      c1=2*gold_seq(i,:)−1;
      c2=2*gold_seq(j,:)−1;
      for m=0:L−1,
         shifted_c2=[c2(m+1:L) c2(1:m)];
         corr=abs(sum(c1.*shifted_c2));
         if (corr>max_cross_corr),
            max_cross_corr=corr;
         end;
         echo off ;
      end;
   end;
end;
% note that max_cross_corr turns out to be 9 in this example.
```

M-FILE

```
function [seq]=ss_mlsrs(connections);
% [seq]=ss_mlsrs(connections)
%              SS_MLSRS  generates the maximal length shift register sequence when the
%              shift register connections are given as input to the function. A "zero"
%              means not connected, whereas a "one" represents a connection.
m=length(connections);
L=2^m−1;                           % length of the shift register sequence requested
registers=[zeros(1,m−1) 1];        % initial register contents
seq(1)=registers(m);               % first element of the sequence
for i=2:L,
   new_reg_cont(1)=connections(1)*seq(i−1);
   for j=2:m,
      new_reg_cont(j)=registers(j−1)+connections(j)*seq(i−1);
   end;
   registers=new_reg_cont;         % current register contents
   seq(i)=registers(m);            % the next element of the sequence
end;
```

9.4   Frequency-Hopped Spread Spectrum

In frequency-hopped (FH) spread spectrum, the available channel bandwidth W is subdivided into a large number of nonoverlapping frequency slots. In any signaling interval, the transmitted signal occupies one or more of the available frequency slots. The selection of the frequency slot (or slots) in each signal interval is made pseudorandomly according to the output from a PN generator.

A block diagram of the transmitter and receiver for a FH spread spectrum system is shown in Figure 9.9. The modulation is either binary or M-ary FSK (MFSK). For example, if binary FSK is employed, the modulator selects one of two frequencies, say, f_0 or f_1, corresponding to the transmission of a 0 or a 1. The resulting binary FSK signal is translated in frequency by an amount that is determined by the output sequence from a PN generator, which is used to select a frequency f_c that is synthesized by the frequency synthesizer. This frequency-translated signal is transmitted over the channel. For example, by taking m bits from the PN generator, we may specify $2^m - 1$ possible carrier frequencies. Figure 9.10 illustrates a FH signal pattern.

At the receiver, there is an identical PN sequence generator, synchronized with the received signal, which is used to control the output of the frequency synthesizer. Thus, the pseudorandom frequency translation introduced at the transmitter is removed at the demodulator by mixing the synthesizer output with the received signal. The resultant signal is then demodulated by means of an FSK demodulator. A signal for maintaining synchronism of the PN sequence generator with the FH received signal is usually extracted from the received signal.

Although binary PSK modulation generally yields better performance than binary FSK, it is difficult to maintain phase coherence in the synthesis of the frequencies used in the hopping pattern and, also, in the propagation of the signal over the channel as the signal is hopped from one frequency to another over a wide bandwidth. Consequently, FSK modulation with noncoherent demodulation is usually employed in FH spread spectrum systems.

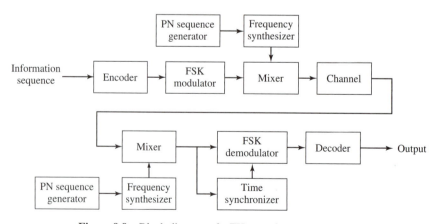

Figure 9.9   Block diagram of a FH spread spectrum system

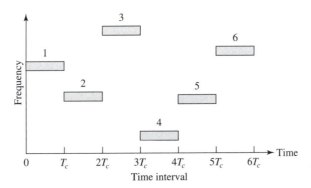

Figure 9.10   An example of a FH pattern

The frequency-hopping rate, denoted as R_h, may be selected to be either equal to the symbol rate, lower than the symbol rate, or higher than the symbol rate. If R_h is equal to or lower than the symbol rate, the FH system is called a *slow-hopping* system. If R_h is higher than the symbol rate—that is, there are multiple hops per symbol—the FH system is called a *fast-hopping* system. We shall consider only a hopping rate equal to the symbol rate.

9.4.1   Probability of Error for FH Signals

Let us consider a FH system in which binary FSK is used to transmit the digital information. The hop rate is 1 hop per bit. The demodulation and detection are noncoherent. In an AWGN channel, the probability of error of such a system is

$$P_b = \frac{1}{2}e^{-\mathcal{E}_b/2N_0} \tag{9.4.1}$$

The same result applies if the interference is a broadband signal or jammer with flat spectrum that covers the entire FH band of width W. In such a case, N_0 is replaced by $N_0 + J_0$, where J_0 is the spectral density of the interference.

As in the case of a DS spread spectrum system, we observe that $\mathcal{E}_b$, the energy per bit, can be expressed as $\mathcal{E}_b = P_S T_b = P_S/R$, where P_S is the average signal power and R is the bit rate. Similarly, $J_0 = P_J/W$, where P_J is the average power of the broadband interference and W is the available channel bandwidth. Therefore, assuming that $J_0 \gg N_0$, we can express the SNR as

$$\frac{\mathcal{E}_b}{J_0} = \frac{W/R}{P_J/P_S} \tag{9.4.2}$$

where W/R is the processing gain and P_J/P_S is the jamming margin for the FH spread spectrum signal.

Slow FH spread spectrum systems are particularly vulnerable to a partial-band interference that may result either from intentional jamming or in FH CDMA systems. To be specific, suppose that the partial-band interference is modeled as a zero-mean Gaussian random process with a flat power-spectral density over a fraction of the total bandwidth W and zero in the remainder of the frequency band. In the region or regions where the power-spectral density is nonzero, its value is $S_J(f) = J_0/\alpha$, where $0 < \alpha \leq 1$. In other words, the interference average power P_J is assumed to be constant, and α is the fraction of the frequency band occupied by the interference.

Suppose that the partial-band interference comes from a jammer that selects α to optimize the effect on the communications system. In an uncoded slow-hopping system with binary FSK modulation and noncoherent detection, the transmitted frequencies are selected with uniform probability in the frequency band W. Consequently, the received signal will be jammed with probability α, and it will not be jammed with probability $1 - \alpha$. When it is jammed, the probability of error is $1/2 \exp(-\alpha \rho_b/2)$, and when it is not jammed, the detection of the signal is assumed to be error-free, where $\rho_b \equiv \mathcal{E}_b/J_0$. Therefore, the average probability of error is

$$P_2(\alpha) = \frac{\alpha}{2} \exp\left(-\frac{\alpha \rho_b}{2}\right)$$
$$= \frac{\alpha}{2} \exp\left(-\frac{\alpha W/R}{2P_J/P_S}\right) \tag{9.4.3}$$

Figure 9.11 illustrates the error rate as a function of the SNR ρ_b for several values of α. The jammer is assumed to optimize its strategy by selecting α to maximize the probability of error. By differentiating $P_2(\alpha)$ and solving for the value of α that maximizes $P_2(\alpha)$, we find that the jammer's best choice of α is

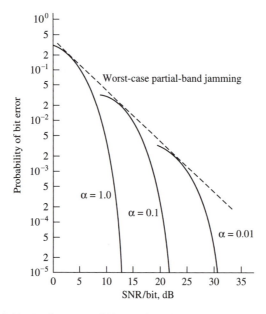

Figure 9.11   Performance of binary FSK with partial-band interference

$$\alpha^* = \begin{cases} 2/\rho_b, & \rho_b \geq 2 \\ 1, & \rho_b < 2 \end{cases} \tag{9.4.4}$$

The corresponding error probability for the worst-case partial-band jammer is

$$P_2 = \begin{cases} e^{-1}/\rho_b, & \rho_b \geq 2 \\ \frac{1}{2}e^{-\rho_b/2}, & \rho_b < 2 \end{cases} \tag{9.4.5}$$

which is also shown in Figure 9.11. Whereas the error probability decreases exponentially for full-band jamming, as given by (9.4.3), the error probability for worst-case partial-band jamming decreases only inversely with $\mathcal{E}_b/J_0$.

◖ILLUSTRATIVE PROBLEM◗

Illustrative Problem 9.6 [FH system simulation] Via Monte Carlo simulation, demonstrate the performance of an FH digital communication system that employs binary FSK and is corrupted by worst-case partial-band interference. The block diagram of the system to be simulated is shown in Figure 9.12.

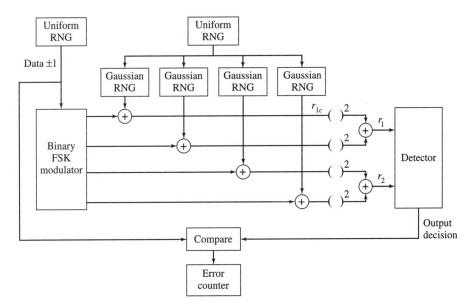

Figure 9.12   Model of a binary FSK system with partial-band interference for the Monte Carlo simulation

SOLUTION

A uniform random number generator (RNG) is used to generate a binary information sequence, which is the input to the FSK modulator. The output of the FSK modulator is corrupted by additive Gaussian noise with probability α, where $0 < \alpha \leq 1$. A second uniform RNG is used to determine when the additive Gaussian noise corrupts the signal and when it does not. In the presence of noise, the input to the detector, assuming that a 0 is transmitted, is

$$r_1 = \left(\sqrt{\mathcal{E}_b}\cos\phi + n_c\right)^2 + \left(\sqrt{\mathcal{E}_b}\sin\phi + n_s\right)^2$$
$$r_2 = n_{2c}^2 + n_{2s}^2$$

where ϕ represents the channel-phase shift and n_{1c}, n_{1s}, n_{2c}, n_{2s} represent the additive noise components. In the absence of noise, we have

$$r_1 = \mathcal{E}_b, \qquad r_2 = 0$$

and hence, no errors occur at the detector. The variance of each of the noise components is $\sigma^2 = \alpha J_0/2$, where α is given by (9.4.4). For simplicity, we may set $\phi = 0$ and normalize J_0 by setting it equal to unity. Then $\rho_b = \mathcal{E}_b/J_0 = \mathcal{E}_b$. Since $\sigma^2 = J_0/2\alpha$ and $\alpha = 2/\rho_b$, it follows that, in the presence of partial-band interference, $\sigma^2 = \mathcal{E}_b/4$ and $\alpha = 2/\mathcal{E}_b$, where $\mathcal{E}_b$ is constrained to $\mathcal{E}_b \geq 2$. Figure 9.13 illustrates the error rate that results from the Monte Carlo simulation. Also shown in the figure is the theoretical value of the probability of error given by (9.4.5).

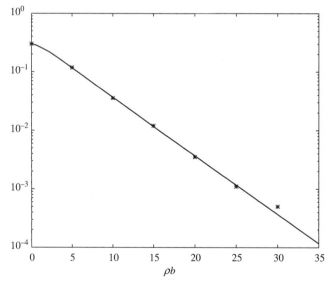

Figure 9.13   Error-rate performance of FH binary FSK system with partial-band interference—Monte Carlo simulation

The MATLAB scripts for the simulation program are given next.

M-FILE

```
% MATLAB script for Illustrative Problem 9.6.
echo on
rho_b1=0:5:35;                          % rho in dB for the simulated error rate
rho_b2=0:0.1:35;                        % rho in dB for theoretical error rate computation
for i=1:length(rho_b1),
   smld_err_prb(i)=ss_pe96(rho_b1(i));   % simulated error rate
   echo off ;
end;
echo on ;
for i=1:length(rho_b2),
   temp=10^(rho_b2(i)/10);
   if (temp>2)
      theo_err_rate(i)=1/(exp(1)*temp);  % theoretical error rate if rho>2
   else
      theo_err_rate(i)=(1/2)*exp(-temp/2);% theoretical error rate if rho<2
   end;
   echo off ;
end;
echo on ;
% Plotting commands follow
```

M-FILE

```
function [p]=ss_Pe96(rho_in_dB)
% [p]=ss_Pe96(rho_in_dB)
%               SS_PE96  finds the measured error rate. The value of
%               signal per interference ratio in dB is given as an
%               input to the function.
rho=10^(rho_in_dB/10);
Eb=rho;                                 % energy per bit
if (rho>2),
   alpha=2/rho;                          % optimal alpha if rho>2
else
   alpha=1;                              % optimal alpha if rho<2
end;
sgma=sqrt(1/(2*alpha));                 % noise standard deviation
N=10000;                                % number of bits transmitted
% generation of the data sequence
for i=1:N,
   temp=rand;
   if (temp<0.5)
      data(i)=1;
   else
      data(i)=0;
   end;
end;
```

```
% find the received signals
for i=1:N,
  % the transmitted signal
  if (data(i)==0),
    r1c(i)=sqrt(Eb);
    r1s(i)=0;
    r2c(i)=0;
    r2s(i)=0;
  else
    r1c(i)=0;
    r1s(i)=0;
    r2c(i)=sqrt(Eb);
    r2s(i)=0;
  end;
  % the received signal is found by adding noise with probability alpha
  if (rand<alpha),
    r1c(i)=r1c(i)+gngauss(sgma);
    r1s(i)=r1s(i)+gngauss(sgma);
    r2c(i)=r2c(i)+gngauss(sgma);
    r2s(i)=r2s(i)+gngauss(sgma);
  end;
end;
% make the decisions and count the number of errors made.
num_of_err=0;
for i=1:N,
  r1=r1c(i)^2+r1s(i)^2;          % first decision variable
  r2=r2c(i)^2+r2s(i)^2;                % second decision variable
  % decision is made next
  if (r1>r2),
    decis=0;
  else
    decis=1;
  end;
  % increment the counter if this is an error
  if (decis~=data(i)),
    num_of_err=num_of_err+1;
  end;
end;
% measured bit error rate is then
p=num_of_err/N;
```

9.4.2   Use of Signal Diversity to Overcome Partial-Band Interference

The performance of the FH system corrupted by partial-band interference as described in the previous section is very poor. For example, for the system to achieve an error probability of 10^{-6}, the SNR required at the detector is almost 60 dB in the presence of worst-case interference. By comparison, in the absence of partial-band interference, the SNR required in AWGN is about 10 dB. Consequently, the loss in SNR due to the presence of the partial-band interference is approximately 50 dB, which is excessively high.

The way to reduce the effect of partial-band interference on the FH spread spectrum system is through signal diversity; that is, the same information bit is transmitted on multiple frequency hops, and the signals from the multiple transmissions are weighed and combined together at the input to the detector. To be specific, suppose that each information bit is transmitted on two successive frequency hops. The resulting system is called a *dual diversity system*. In this case, assuming that a 0 is transmitted, either the two inputs to the combiner are both corrupted by interference, or one of the two transmitted signals is corrupted by interference, or neither of the two transmitted signals is corrupted by interference.

The combiner is assumed to know the level of the interference and thus forms the combined decision variables

$$x = w_1 r_{11} + w_2 r_{12}$$
$$y = w_1 r_{21} + w_2 r_{22} \tag{9.4.6}$$

where r_{11}, r_{21} are the two outputs of the square-law device for the first transmitted signal and r_{12}, r_{22} are the outputs of the square-law device from the second transmitted signal. The weights w_1 and w_2 are set to $1/\sigma^2$, where σ^2 is the variance of the additive noise plus interference. Hence, when σ^2 is large, as would be the case when interference is present, the weight placed on the received signal is small. On the other hand, when σ^2 is small, as would be the case when there is no interference, the weight placed on the received signal is large. Thus, the combiner de-emphasizes the received signal components that are corrupted by interference.

The two components x and y from the combiner are fed to the detector, which decides in favor of the larger signal component.

The performance of the FH signal with dual diversity is now dominated by the case in which both signal transmissions are corrupted by interference. However, the probability of this event is proportional to α^2, which is significantly smaller than α. As a consequence, the probability of error for the worst-case partial-band interference has the form

$$P_2(2) = \frac{K_2}{\rho_b^2}, \qquad \rho_b > 2 \tag{9.4.7}$$

where K_2 is a constant and $\rho_b = \mathcal{E}_b/J_0$. In this case, the probability of error for dual diversity decreases inversely as the square of the SNR. In other words, an increase in SNR by a factor of 10 (10 dB) results in a decrease of the error probability by a factor of 100. Consequently, an error probability of 10^{-6} can be achieved with an SNR of about 30 dB with dual diversity, compared to 60 dB (1000 times larger) for no diversity.

More generally, if each information bit is transmitted on D frequency hops, where D is the order of diversity, the probability of error has the form

$$P_2(D) = \frac{K_D}{\rho_b^D}, \qquad \rho_b > 2 \tag{9.4.8}$$

where K_D is a constant.

Since signal diversity as described above is a trivial form of coding (repetition coding), it is not surprising to observe that instead of repeating the transmission of each information bit D times, we may use a code with a minimum Hamming distance equal to D and soft-decision decoding of the outputs from the square-law devices.

ILLUSTRATIVE PROBLEM

Illustrative Problem 9.7 [Diversity in FH systems] Repeat the Monte Carlo simulation for the FH system considered in Illustrative Problem 9.6, but now employ dual diversity.

SOLUTION

In the absence of interference, the weight used in the combiner is set to $w = 10$, which corresponds to $\sigma^2 = 0.1$, a value that may be typical of the level of the additive Gaussian noise. On the other hand, when interference is present, the weight is set to $w = 1/\sigma^2 = 2/\mathcal{E}$, where $\mathcal{E}$ is constrained to be $\mathcal{E} \geq 4$. The SNR per hop is $\mathcal{E}$, and the total energy per bit in the two hops is $\mathcal{E}_b = 2\mathcal{E}$. Therefore, the error probability is plotted as a function of $\mathcal{E}_b/J_0$. The results of the Monte Carlo simulation are illustrated in Figure 9.14.

The MATLAB scripts for the simulation program are given next.

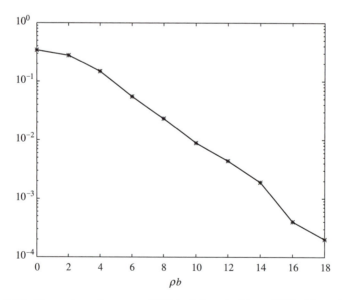

Figure 9.14   Error-rate performance of FH dual diversity binary FSK with partial-band
interference—Monte Carlo simulation

M-FILE

```
% MATLAB script for Illustrative Problem 9.7.
echo on
rho_b=0:2:24;                              % rho in dB
for i=1:length(rho_b),
   smld_err_prb(i)=ss_Pe97(rho_b(i));      % simulated error rate
   echo off ;
end;
echo on ;
% Plotting commands follow
```

M-FILE

```
function [p]=ss_Pe97(rho_in_dB)
% [p]=ss_Pe97(rho_in_dB)
%                  SS_PE97  finds the measured error rate. The value of
%                  signal per interference ratio in dB is given as an input
%                  to the function.
rho=10^(rho_in_dB/10);
Eb=rho;                                    % energy per information bit
E=Eb/2;                                    % energy per symbol transmitted
% the optimal value of alpha
if (rho>2),
   alpha=2/rho;
else
   alpha=1;
end;
% the variance of the additive noise
if (E>1),
   sgma=sqrt(E/2);
else
   sgma=sqrt(1/2);
end;
N=10000;                                   % number of bits transmitted
% generation of the data sequence
for i=1:N,
   temp=rand;
   if (temp<0.5)
     data(i)=1;
   else
     data(i)=0;
   end;
end;
% find the transmitted signals
for i=1:N,
   if (data(i)==0),
     tr11c(i)=sqrt(E);   tr12c(i)=sqrt(E);
     tr11s(i)=0;         tr12s(i)=0;
     tr21c(i)=0;         tr22c(i)=0;
     tr21s(i)=0;         tr22s(i)=0;
```

```
      else
         tr11c(i)=0;       tr12c(i)=0;
         tr11s(i)=0;       tr12s(i)=0;
         tr21c(i)=sqrt(E);          tr22c(i)=sqrt(E);
         tr21s(i)=0;       tr22s(i)=0;
      end;
   end;

% find the received signals, make the decisions and count the number of errors made.
num_of_err=0;
for i=1:N,
   % determine if there is jamming
   if (rand<alpha),
      jamming1=1;                            % jamming present on the second transmission
   else
      jamming1=0;                            % jamming not present on the first transmission
   end;
   if (rand<alpha),
      jamming2=1;                            % jamming present on the second transmission
   else
      jamming2=0;                            % jamming not present on the second transmission
   end;
   % the received signals are
   if (jamming1==1)
      r11c=tr11c(i)+gngauss(sgma);       r11s=tr11s(i)+gngauss(sgma);
      r21c=tr21c(i)+gngauss(sgma);       r21s=tr21s(i)+gngauss(sgma);
   else
      r11c=tr11c(i);       r11s=tr11s(i);
      r21c=tr21c(i);       r21s=tr21s(i);
   end;
   if (jamming2==1)
      r12c=tr12c(i)+gngauss(sgma);       r12s=tr12s(i)+gngauss(sgma);
      r22c=tr22c(i)+gngauss(sgma);       r22s=tr22s(i)+gngauss(sgma);
   else
      r12c=tr12c(i);       r12s=tr12s(i);
      r22c=tr22c(i);       r22s=tr22s(i);
   end;
   % compute the decision variables, first the weights
   if (jamming1==1),
      w1=1/sgma^2;
   else
      w1=10;
   end;
   if (jamming2==1),
      w2=1/sgma^2;
   else
      w2=10;
   end;
   % the intermediate decision variables are computed as follows
   r11=r11c^2+r11s^2;
   r12=r12c^2+r12s^2;
   r21=r21c^2+r21s^2;
   r22=r22c^2+r22s^2;
   % finally, the resulting decision variables x and y are computed
```

```
   x=w1*r11+w2*r12;
   y=w1*r21+w2*r22;
   % make the decision
   if (x>y),
      decis=0;
   else
      decis=1;
   end;
   % increment the counter if this is an error
   if (decis~=data(i)),
      num_of_err=num_of_err+1;
   end;
end;
% the measured bit error rate is then
p=num_of_err/N;
```

Problems

9.1 Write a MATLAB program to perform a Monte Carlo simulation of a DS spread spectrum system that transmits the information via binary PSK through an AWGN channel. Assume that the processing gain is 10. Plot a graph of the measured error rate versus the SNR, and thus, demonstrate that there is no performance gain from the spread spectrum signal.

9.2 Write a MATLAB program to perform a Monte Carlo simulation of a DS spread spectrum system that operates in an LPI mode. The processing gain is 20 (13 dB), and the desired power signal-to-noise ratio P_S/P_N at the receiver prior to despreading the signal is -5 dB or smaller. Plot the measured error rate as a function of the SNR.

9.3 Repeat the Monte Carlo simulation described in Illustrative Problem 9.4 for a processing gain of 10 and plot the measured error rate.

9.4 Write a MATLAB program that implements an $m = 12$-stage maximum-length shift register, and generate three periods of the sequence. Compute and graph the periodic autocorrelation function of the equivalent bipolar sequence given by (9.3.1).

9.5 Write a MATLAB program that implements an $m = 3$-stage and an $m = 4$-stage maximum-length shift register, and form the modulo-2 sum of their output sequences. Is the resulting sequence periodic? If so, what is the period of the sequence? Compute and sketch the autocorrelation sequence of the resulting (bipolar) sequence using (9.3.1).

9.6 Write a MATLAB program to compute the autocorrelation sequence of the $L = 31$ Gold sequences that were generated in Illustrative Problem 9.5.

9.7 An FH binary orthogonal FSK system employs an $m = 7$-state shift register to generate a periodic maximum-length sequence of length $L = 127$. Each stage of the shift register selects one of $N = 127$ nonoverlapping frequency bands in the hopping pattern. Write a

MATLAB program that simulates the selection of the center frequency and the generation of the two frequencies in each of the $N = 127$ frequency bands. Show the frequency selection pattern for the first 10 bit intervals.

9.8 Write a Monte Carlo program to simulate a FH digital communication system that employs binary FSK with noncoherent (square-law) detection. The system is corrupted by partial-band interference with spectral density J_0/α, where $\alpha = 0.1$. The interference is spectrally flat over the frequency band $0 < \alpha \le 0.1$. Plot the measured error rate for this system versus the SNR $\mathcal{E}_b/J_0$.

9.9 Repeat the Monte Carlo simulation in Illustrative Problem 9.7 when the weight used at the combiner in the absence of interference is set to $w = 100$ and with interference the weight is $w = 1/\sigma^2 = 2/\mathcal{E}$, where the signal energy is $\mathcal{E} \ge 4$. Plot the measured error rate from the Monte Carlo simulation for this dual diversity system, and compare this performance with that obtained in Illustrative Problem 9.7.

References

1. Proakis, J. G., and Salehi, M. *Communication Systems Engineering*. Upper Saddle River, NJ: Prentice Hall, 1994.

2. Max, J. "Quantization for Minimum Distortion," *IRE Transactions on Information Theory*, vol. IT-6 (March 1960): 7–12.

3. Proakis, J. G. *Digital Communications* (3rd ed.). New York: McGraw-Hill, 1995.

4. Forney, G. D., Jr. "Maximum-Likelihood Sequence Estimation of Digital Sequences in the Presence of Intersymbol Interference," *IEEE Trans. Inform. Theory*, vol. IT-18 (May 1972): 363–378.

5. Gold, R. "Optimal Binary Sequences for Spread Spectrum Multiplexing," *IEEE Trans. Inform. Theory*, vol. IT-13 (October 1967): 619– 621.

6. Gold, R. "Maximal Recursive Sequences with 3-Valued Recursive Cross Correlation Functions," *IEEE Trans. Inform. Theory*, vol. IT-14 (January 1968): 154–156.

7. Kasami, T. "Weight Distribution Formula for Some Class of Cyclic Codes," Coordinated Science Laboratory, University of Illinois, Urbana, Ill., Technical Report No. R-285, April 1966.

8. Scholtz, R. A. "Optimal CDMA Codes," *1979 National Telecommunication Conference Records*, Washington, D.C. (November 1979): 54.2.1–54.2.4.

9. Sarwate, D. V., and Pursley, M. B. "Crosscorrelation Properties of Pseudorandom and Related Sequences," *Proceedings of the IEEE*, vol. 68 (May 1980): 593–619.

Index